Selected Titles in This Se

Volume

4 **W. J. Kaczor and M. T. Nowak**
  Problems in mathematical analysis I: Real numbers, sequences and series
  2000

3 **Roger Knobel**
  An introduction to the mathematical theory of waves
  2000

2 **Gregory F. Lawler and Lester N. Coyle**
  Lectures on contemporary probability
  1999

1 **Charles Radin**
  Miles of tiles
  1999

Problems in Mathematical Analysis I

Real Numbers,
Sequences and Series

STUDENT MATHEMATICAL LIBRARY
Volume 4

Problems in Mathematical Analysis I

Real Numbers, Sequences and Series

W. J. Kaczor
M. T. Nowak

American Mathematical Society

Editorial Board

David Bressoud   Carl Pomerance
Robert Devaney, Chair   Hung-Hsi Wu

Originally published in Polish, as
Zadania z Analizy Matematycznej. Część Pierwsza.
Liczby Rzeczywiste, Ciągi i Szeregi Liczbowe
© 1996, Wydawnictwo Uniwersytetu Marii Curie–Skłodowskiej, Lublin.

Translated, revised and augmented by the authors.

2000 *Mathematics Subject Classification*. Primary 00A07, 40–01.

Library of Congress Cataloging-in-Publication Data
Kaczor, W. J. (Wiesława J.), 1949–
 [Zadania z analizy matematycznej. English]
 Problems in mathematical analysis. I. Real numbers, sequences and series / W. J. Kaczor, M. T. Nowak.
   p. cm. — (Student mathematical library, ISSN 1520-9121 ; v. 4)
 Includes bibliographical references.
 ISBN 0-8218-2050-8 (softcover : alk. paper)
 1. Mathematical analysis. I. Nowak, M. T. (Maria T.), 1951– II. Title.
III. Series.
QA300 K32513   2000
515′.076–dc21   99-087039

Copying and reprinting. Individual readers of this publication, and nonprofit libraries acting for them, are permitted to make fair use of the material, such as to copy a chapter for use in teaching or research. Permission is granted to quote brief passages from this publication in reviews, provided the customary acknowledgment of the source is given.

Republication, systematic copying, or multiple reproduction of any material in this publication is permitted only under license from the American Mathematical Society. Requests for such permission should be addressed to the Assistant to the Publisher, American Mathematical Society, P. O. Box 6248, Providence, Rhode Island 02940-6248. Requests can also be made by e-mail to reprint-permission@ams.org.

© 2000 by the American Mathematical Society. All rights reserved.
The American Mathematical Society retains all rights
except those granted to the United States Government.
Printed in the United States of America.

∞ The paper used in this book is acid-free and falls within the guidelines
established to ensure permanence and durability.
Visit the AMS home page at URL: http://www.ams.org/
10 9 8 7 6 5 4 3 2 1   05 04 03 02 01 00

Contents

Preface   xi

Notation and Terminology   xiii

Problems

Chapter 1. Real Numbers
    1.1. Supremum and Infimum of Sets of Real Numbers. Continued Fractions   3
    1.2. Some Elementary Inequalities   8

Chapter 2. Sequences of Real Numbers
    2.1. Monotonic Sequences   19
    2.2. Limits. Properties of Convergent Sequences   26
    2.3. The Toeplitz Transformation, the Stolz Theorem and their Applications   35
    2.4. Limit Points. Limit Superior and Limit Inferior   40
    2.5. Miscellaneous Problems   47

Chapter 3. Series of Real Numbers
    3.1. Summation of Series   63

3.2.	Series of Nonnegative Terms	72
3.3.	The Integral Test	88
3.4.	Series of Positive and Negative Terms - Convergence, Absolute Convergence. Theorem of Leibniz	92
3.5.	The Dirichlet and Abel Tests	99
3.6.	Cauchy Product of Infinite Series	102
3.7.	Rearrangement of Series. Double Series	105
3.8.	Infinite Products	112

Solutions

Chapter 1. Real Numbers

1.1.	Supremum and Infimum of Sets of Real Numbers. Continued Fractions	125
1.2.	Some Elementary Inequalities	136

Chapter 2. Sequences of Real Numbers

2.1.	Monotonic Sequences	151
2.2.	Limits. Properties of Convergent Sequences	162
2.3.	The Toeplitz Transformation, the Stolz Theorem and their Applications	181
2.4.	Limit Points. Limit Superior and Limit Inferior	189
2.5.	Miscellaneous Problems	208

Chapter 3. Series of Real Numbers

3.1.	Summation of Series	245
3.2.	Series of Nonnegative Terms	269
3.3.	The Integral Test	302
3.4.	Series of Positive and Negative Terms - Convergence, Absolute Convergence. Theorem of Leibniz	309
3.5.	The Dirichlet and Abel Tests	324
3.6.	Cauchy Product of Infinite Series	333

Contents

3.7. Rearrangement of Series. Double Series	342
3.8. Infinite Products	360
Bibliography - Books	379

Preface

This book is an enlarged and revised English edition of a Polish version published in 1996 by the Publishing House of Maria Curie-Skłodowska University in Lublin, Poland. It is the first volume of a planned series of books of problems in mathematical analysis. The second volume, already published in Polish, is under translation into English. The series is mainly intended for students who take courses in basic principles of analysis. The choice and arrangement of the material make it suitable for self-study, and instructors may find it useful as an aid in organizing tutorials and seminars.

This volume covers three topics: real numbers, sequences, and series. It does not contain problems concerning metric and topological spaces, which we intend to present in subsequent volumes.

The book is divided into two parts. The first part is a collection of exercises and problems, and the second contains their solutions. Complete solutions are given in most cases. Where no difficulties could be expected or when an analogous problem has already been solved, only a hint or simply an answer is given. Very often various solutions of a given problem are possible; we present here only one, hoping students themselves will find others.

With the student in mind, we have tried to keep things at an elementary level whenever possible. For example, we present an elementary proof of the Toeplitz theorem about the so-called regular transformation of sequences, which in many texts is proved by methods of functional analysis. The proof presented is taken from Toeplitz's original paper, published in 1911 in Prace Matematyczno-Fizyczne, Vol. 22. We hope that our presentation of this part of real analysis will be more accessible to readers and will ensure wider understanding.

All the notations and definitions used in this volume are standard and commonly used. The reader can find them, for example, in the textbooks [12] and [23], in which all necessary theoretical background can be found. However, to make the book consistent and to avoid ambiguity, a list of notations and definitions is included.

We have borrowed freely from many textbooks, problem books and problem sections of journals like the American Mathematical Monthly, Mathematics Today (Russian) and Delta (Polish). A complete list is given in the bibliography. It was beyond the authors' scope to trace all original sources, and we may have overlooked some contributions. If this has happened, we offer our sincere apologies.

We are deeply indebted to all our friends and colleagues from the Department of Mathematics of Maria Curie-Skłodowska University who offered stimulating suggestions. We have had many fruitful conversations with M. Koter-Mórgowska, T.Kuczumow, W. Rzymowski, S. Stachura and W. Zygmunt. Our sincere thanks are also due to Professor Jan Krzyż for his help in preparing the first version of the English manuscript. We are pleased to express our gratitude to Professor Kazimierz Goebel for his encouragement and active interest in the project. It is our pleasure to thank Professor Richard J. Libera, University of Delaware, for his invaluable and most generous help with the English translation and for all his suggestions and corrections which greatly improved the final version of the book.

W. J. Kaczor, M. T. Nowak

Notation and Terminology

- $\mathbb{R}$ - the set of all real numbers
- $\mathbb{R}_+$ - the set of all positive real numbers
- $\mathbb{Z}$ - the set of all integers
- $\mathbb{N}$ - the set of all positive integers
- $\mathbb{Q}$ - the set of all rationals
- (a,b) - open interval with the endpoints a and b
- $[a,b]$ - closed interval with the endpoints a and b
- $[x]$ - the integral part of a real number x
- For $x \in \mathbb{R}$,

$$\operatorname{sgn} x = \begin{cases} 1 & \text{for } x > 0, \\ -1 & \text{for } x < 0, \\ 0 & \text{for } x = 0. \end{cases}$$

- For $n \in \mathbb{N}$,
  $n! = 1 \cdot 2 \cdot 3 \cdot ... \cdot n$,
  $(2n)!! = 2 \cdot 4 \cdot 6 \cdot ... \cdot (2n-2)(2n)$ and
  $(2n-1)!! = 1 \cdot 3 \cdot 5 \cdot ... \cdot (2n-3)(2n-1)$.
- If $\mathbf{A} \subset \mathbb{R}$ is nonempty and bounded from above, then $\sup \mathbf{A}$ denotes the least upper bound of $\mathbf{A}$. If a nonempty set $\mathbf{A}$ is not bounded above, then we assume that $\sup \mathbf{A} = +\infty$.

- If $\mathbf{A} \subset \mathbb{R}$ is nonempty and bounded from below, then $\inf \mathbf{A}$ denotes the greatest lower bound of $\mathbf{A}$. If a nonempty set $\mathbf{A}$ is not bounded below, then we assume that $\inf \mathbf{A} = -\infty$.
- A sequence $\{a_n\}$ of real numbers is said to be monotonically increasing (monotonically decreasing) if $a_{n+1} \geq a_n$ for all $n \in \mathbb{N}$ ($a_{n+1} \leq a_n$ for all $n \in \mathbb{N}$). The class of monotonic sequences consists of the increasing and the decreasing sequences.
- A number c is a limit point of the sequence $\{a_n\}$ if there is a subsequence $\{a_{n_k}\}$ of $\{a_n\}$ converging to c.
- Let $\mathbf{S}$ be the set of all the limit points of $\{a_n\}$. The limit inferior, $\varliminf_{n \to \infty} a_n$, and the limit superior, $\varlimsup_{n \to \infty} a_n$, of the sequence $\{a_n\}$ are defined as follows:

$$\varlimsup_{n \to \infty} a_n = \begin{cases} +\infty & \text{if } \{a_n\} \text{ is not bounded above,} \\ -\infty & \text{if } \{a_n\} \text{ is bounded above and } \mathbf{S} = \emptyset, \\ \sup \mathbf{S} & \text{if } \{a_n\} \text{ is bounded above and } \mathbf{S} \neq \emptyset, \end{cases}$$

$$\varliminf_{n \to \infty} a_n = \begin{cases} -\infty & \text{if } \{a_n\} \text{ is not bounded below,} \\ +\infty & \text{if } \{a_n\} \text{ is bounded below and } \mathbf{S} = \emptyset, \\ \inf \mathbf{S} & \text{if } \{a_n\} \text{ is bounded below and } \mathbf{S} \neq \emptyset. \end{cases}$$

- An infinite product $\prod_{n=1}^{\infty} a_n$ is said to be convergent if there exists $n_0 \in \mathbb{N}$ such that $a_n \neq 0$ for $n \geq n_0$ and the sequence $\{a_{n_0} a_{n_0+1} \cdot \ldots \cdot a_{n_0+n}\}$ converges, as $n \to \infty$, to a limit P_0 other than zero. The number $P = a_1 a_2 \cdot \ldots \cdot a_{n_0-1} \cdot P_0$ is called the value of the infinite product.

Problems

Chapter 1

Real Numbers

1.1. Supremum and Infimum of Sets of Real Numbers. Continued Fractions

1.1.1. Show that
$$\sup\{x \in \mathbb{Q} : x > 0,\ x^2 < 2\} = \sqrt{2}.$$

1.1.2. Let $\mathbf{A} \subset \mathbb{R}$ be a nonempty set. Define $-\mathbf{A} = \{x : -x \in \mathbf{A}\}$. Show that
$$\sup(-\mathbf{A}) = -\inf \mathbf{A},$$
$$\inf(-\mathbf{A}) = -\sup \mathbf{A}.$$

1.1.3. Let $\mathbf{A}, \mathbf{B} \subset \mathbb{R}$ be nonempty. Define
$$\mathbf{A} + \mathbf{B} = \{z = x + y : x \in \mathbf{A},\ y \in \mathbf{B}\},$$
$$\mathbf{A} - \mathbf{B} = \{z = x - y : x \in \mathbf{A},\ y \in \mathbf{B}\}.$$
Show that
$$\sup(\mathbf{A} + \mathbf{B}) = \sup \mathbf{A} + \sup \mathbf{B},$$
$$\sup(\mathbf{A} - \mathbf{B}) = \sup \mathbf{A} - \inf \mathbf{B}.$$
Establish analogous formulas for $\inf(\mathbf{A} + \mathbf{B})$ and $\inf(\mathbf{A} - \mathbf{B})$.

1.1.4. Given nonempty subsets $\mathbf{A}$ and $\mathbf{B}$ of positive real numbers, define
$$\mathbf{A} \cdot \mathbf{B} = \{z = x \cdot y : x \in \mathbf{A},\ y \in \mathbf{B}\}$$
$$\frac{1}{\mathbf{A}} = \left\{z = \frac{1}{x} : x \in \mathbf{A}\right\}.$$
Show that
$$\sup(\mathbf{A} \cdot \mathbf{B}) = \sup \mathbf{A} \cdot \sup \mathbf{B}.$$
Show also that, if $\inf \mathbf{A} > 0$, then
$$\sup\left(\frac{1}{\mathbf{A}}\right) = \frac{1}{\inf \mathbf{A}}$$
and, if $\inf \mathbf{A} = 0$, then $\sup\left(\frac{1}{\mathbf{A}}\right) = +\infty$. Additionally, show that if $\mathbf{A}$ and $\mathbf{B}$ are bounded sets of real numbers, then

$\sup(\mathbf{A} \cdot \mathbf{B})$
$= \max\{\sup \mathbf{A} \cdot \sup \mathbf{B},\ \sup \mathbf{A} \cdot \inf \mathbf{B},\ \inf \mathbf{A} \cdot \sup \mathbf{B},\ \inf \mathbf{A} \cdot \inf \mathbf{B}\}.$

1.1.5. Let $\mathbf{A}$ and $\mathbf{B}$ be nonempty subsets of real numbers. Show that
$$\sup(\mathbf{A} \cup \mathbf{B}) = \max\{\sup \mathbf{A}, \sup \mathbf{B}\}$$
and
$$\inf(\mathbf{A} \cup \mathbf{B}) = \min\{\inf \mathbf{A}, \inf \mathbf{B}\}.$$

1.1.6. Find the least upper bound and the greatest lower bound of $\mathbf{A}_1$, $\mathbf{A}_2$ defined by setting
$$\mathbf{A}_1 = \left\{2(-1)^{n+1} + (-1)^{\frac{n(n+1)}{2}}\left(2 + \frac{3}{n}\right) : n \in \mathbb{N}\right\},$$
$$\mathbf{A}_2 = \left\{\frac{n-1}{n+1} \cos \frac{2n\pi}{3} : n \in \mathbb{N}\right\}.$$

1.1.7. Find the supremum and the infimum of the sets $\mathbf{A}$ and $\mathbf{B}$, where $\mathbf{A} = \{0.2, 0.22, 0.222, \dots\}$ and $\mathbf{B}$ is the set of decimal fractions between 0 and 1 whose only digits are zeros and ones.

1.1.8. Find the greatest lower and the least upper bounds of the set of numbers $\frac{(n+1)^2}{2^n}$, where $n \in \mathbb{N}$.

1.1. Supremum and Infimum. Continued Fractions

1.1.9. Find the supremum and the infimum of the set of numbers $\frac{(n+m)^2}{2^{nm}}$, where $n, m \in \mathbb{N}$.

1.1.10. Determine the least upper and the greatest lower bounds of the following sets:

(a) $\quad \mathbf{A} = \left\{ \dfrac{m}{n} : m, n \in \mathbb{N},\ m < 2n \right\}$,

(b) $\quad \mathbf{B} = \left\{ \sqrt{n} - [\sqrt{n}] : n \in \mathbb{N} \right\}$.

1.1.11. Find

(a) $\sup\{x \in \mathbb{R} : x^2 + x + 1 > 0\}$,

(b) $\inf\{z = x + x^{-1} : x > 0\}$,

(c) $\inf\{z = 2^x + 2^{\frac{1}{x}} : x > 0\}$.

1.1.12. Find the supremum and the infimum of the following sets:

(a) $\quad \mathbf{A} = \left\{ \dfrac{m}{n} + \dfrac{4n}{m} : m, n \in \mathbb{N} \right\}$,

(b) $\quad \mathbf{B} = \left\{ \dfrac{mn}{4m^2 + n^2} : m \in \mathbb{Z},\ n \in \mathbb{N} \right\}$,

(c) $\quad \mathbf{C} = \left\{ \dfrac{m}{m+n} : m, n \in \mathbb{N} \right\}$,

(d) $\quad \mathbf{D} = \left\{ \dfrac{m}{|m|+n} : m \in \mathbb{Z},\ n \in \mathbb{N} \right\}$,

(e) $\quad \mathbf{E} = \left\{ \dfrac{mn}{1+m+n} : m, n \in \mathbb{N} \right\}$.

1.1.13. Let $n \geq 3$ be an arbitrarily fixed integer. Take all the possible finite sequences $(a_1, \ldots, a_n)$ of positive numbers. Find the least upper and the greatest lower bounds of the set of numbers

$$\sum_{k=1}^{n} \frac{a_k}{a_k + a_{k+1} + a_{k+2}},$$

where we put $a_{n+1} = a_1$ and $a_{n+2} = a_2$.

1.1.14. Show that for any irrational number α and for any positive integer n there exist a positive integer q_n and an integer p_n such that
$$\left|\alpha - \frac{p_n}{q_n}\right| < \frac{1}{nq_n}.$$
Show also that $\{p_n\}$ and $\{q_n\}$ can be chosen in such a way that we have
$$\left|\alpha - \frac{p_n}{q_n}\right| < \frac{1}{q_n^2}.$$

1.1.15. Let α be irrational. Show that $\mathbf{A} = \{m + n\alpha : m, n \in \mathbb{Z}\}$ is dense in $\mathbb{R}$, i.e. in any open interval there is at least one element of $\mathbf{A}$.

1.1.16. Show that $\{\cos n : n \in \mathbb{N}\}$ is dense in $[-1, 1]$.

1.1.17. Let $x \in \mathbb{R} \setminus \mathbb{Z}$. Define the sequence $\{x_n\}$ by setting
$$x = [x] + \frac{1}{x_1}, \quad x_1 = [x_1] + \frac{1}{x_2}, \ldots, \quad x_{n-1} = [x_{n-1}] + \frac{1}{x_n}.$$
Then
$$x = [x] + \cfrac{1}{[x_1] + \cfrac{1}{[x_2] + \cfrac{1}{\ddots + \cfrac{1}{[x_{n-1}] + \cfrac{1}{x_n}}}}}.$$
Show that x is rational if and only if there exists $n \in \mathbb{N}$ for which x_n is an integer.

Remark. The above representation of x is said to be *a finite continued fraction*. The expression
$$a_0 + \cfrac{1}{a_1 + \cfrac{1}{a_2 + \cfrac{1}{\ddots + \cfrac{1}{a_{n-1} + \cfrac{1}{a_n}}}}}$$

1.1. Supremum and Infimum. Continued Fractions

will also be written in the more convenient form
$$a_0 + \frac{1|}{|a_1} + \frac{1|}{|a_2} + ... + \frac{1|}{|a_n}.$$

1.1.18. For positive real numbers $a_1, a_2, ..., a_n$ set

$p_0 = a_0,$ $\qquad q_0 = 1,$
$p_1 = a_0 a_1 + 1,$ $\qquad q_1 = a_1,$
$p_k = p_{k-1} a_k + p_{k-2},$ $q_k = q_{k-1} a_k + q_{k-2},$ with $k = 2, 3, ..., n,$

and define
$$R_0 = a_0, \quad R_k = a_0 + \frac{1|}{|a_1} + \frac{1|}{|a_2} + ... + \frac{1|}{|a_k}, \quad k = 1, 2, ..., n.$$

$\left(R_k \text{ is called } \textit{the kth convergent to } a_0 + \frac{1|}{|a_1} + \frac{1|}{|a_2} + ... + \frac{1|}{|a_n}\right).$
Show that
$$R_k = \frac{p_k}{q_k} \quad \text{for} \quad k = 0, 1, ..., n.$$

1.1.19. Show that if p_k, q_k are defined as in the foregoing problem and the $a_0, a_1, ..., a_n$ are integers, then
$$p_{k-1} q_k - q_{k-1} p_k = (-1)^k \quad \text{for} \quad k = 1, 2, ..., n.$$
Use this equality to conclude that p_k and q_k are co-prime.

1.1.20. For an irrational number x we define a sequence $\{x_n\}$ by
$$x_1 = \frac{1}{x - [x]}, \quad x_2 = \frac{1}{x_1 - [x_1]}, ..., \quad x_n = \frac{1}{x_{n-1} - [x_{n-1}]}, ...$$
Moreover, we put $a_0 = [x]$, $a_n = [x_n]$, $n = 1, 2, ...,$ and
$$R_n = a_0 + \frac{1|}{|a_1} + \frac{1|}{|a_2} + ... + \frac{1|}{|a_n}.$$
Show that the difference between the number x and its nth convergent is given by
$$x - R_n = \frac{(-1)^n}{(q_n x_{n+1} + q_{n-1}) q_n},$$
where p_n, q_n are defined in 1.1.18. Conclude that x is between its two consecutive convergents.

Problems. 1: Real Numbers

1.1.21. Show that the set $\{\sin n : n \in \mathbb{N}\}$ is dense in $[-1, 1]$.

1.1.22. Apply the result in 1.1.20 to prove that for every irrational number x there exists a sequence $\{\frac{p_n}{q_n}\}$ of rational numbers, with odd q_n, such that
$$\left| x - \frac{p_n}{q_n} \right| < \frac{1}{q_n^2}.$$
(Compare with 1.1.14.)

1.1.23. Derive the following formula for the difference of two consecutive convergents:
$$R_{n+1} - R_n = \frac{(-1)^n}{q_n q_{n+1}}.$$

1.1.24. Let x be irrational. Show that its convergents R_n defined in 1.1.20 are successively closer to x, that is,
$$|x - R_{n+1}| < |x - R_n|, \quad n = 0, 1, 2, \ldots.$$

1.1.25. Prove that the convergent $R_n = p_n/q_n$ is the best approximation of x of all rational fractions with denominator q_n or less. That is: if r/s is a rational number with a positive denominator such that $|x - r/s| < |x - R_n|$, then $s > q_n$.

1.1.26. Expand each of the following as infinite continued fractions: $\sqrt{2}, \frac{\sqrt{5}-1}{2}$.

1.1.27. For a positive integer k, find the representation of $\sqrt{k^2 + k}$ by an infinite continued fraction.

1.1.28. Find all the numbers $x \in (0, 1)$ whose infinite continued representations have a_1 (see Problem 1.1.20) equal to a given positive integer n.

1.2. Some Elementary Inequalities

1.2.1. Show that if $a_k > -1$, $k = 1, \ldots, n$, are all positive or negative, then
$$(1 + a_1) \cdot (1 + a_2) \cdot \ldots \cdot (1 + a_n) \geq 1 + a_1 + a_2 + \ldots + a_n.$$

1.2. Some Elementary Inequalities

Remark. Note that if $a_1 = a_2 = ... = a_n = a$ then we get the well known *Bernoulli inequality:* $(1+a)^n \geq 1 + na$, $a > -1$.

1.2.2. Using induction, prove the following result: If $a_1, a_2, ..., a_n$ are positive real numbers such that $a_1 \cdot a_2 \cdot ... \cdot a_n = 1$, then $a_1 + a_2 + ... + a_n \geq n$.

1.2.3. Let A_n, G_n and H_n denote the arithmetic, geometric and harmonic means of n positive real numbers $a_1, a_2, ..., a_n$; that is,

$$A_n = \frac{a_1 + a_2 + ... + a_n}{n},$$

$$G_n = \sqrt[n]{a_1 \cdot ... \cdot a_n},$$

$$H_n = \frac{n}{\frac{1}{a_1} + \frac{1}{a_2} + ... + \frac{1}{a_n}}.$$

Show that $A_n \geq G_n \geq H_n$.

1.2.4. Using the result $(G_n \leq A_n)$ in the foregoing problem, establish the Bernoulli inequality

$$(1+x)^n \geq 1 + nx \quad \text{for} \quad x > 0.$$

1.2.5. For $n \in \mathbb{N}$, verify the following claims:

(a) $\dfrac{1}{n} + \dfrac{1}{n+1} + \dfrac{1}{n+2} + ... + \dfrac{1}{2n} > \dfrac{2}{3}$,

(b) $\dfrac{1}{n+1} + \dfrac{1}{n+2} + \dfrac{1}{n+3} + ... + \dfrac{1}{3n+1} > 1$,

(c) $\dfrac{1}{2} < \dfrac{1}{3n+1} + \dfrac{1}{3n+2} + ... + \dfrac{1}{5n} + \dfrac{1}{5n+1} < \dfrac{2}{3}$,

(d) $n(\sqrt[n]{n+1} - 1) < 1 + \dfrac{1}{2} + ... + \dfrac{1}{n}$

$$< n\left(1 - \dfrac{1}{\sqrt[n]{n+1}} + \dfrac{1}{n+1}\right), \quad n > 1.$$

1.2.6. Show that for any $x > 0$ and $n \in \mathbb{N}$ we have
$$\frac{x^n}{1 + x + x^2 + x^3 + \ldots + x^{2n}} \leq \frac{1}{2n+1}.$$

1.2.7. Let $\{a_n\}$ be an arithmetic progression with positive terms. Show that
$$\sqrt{a_1 a_n} \leq \sqrt[n]{a_1 a_2 \ldots a_n} \leq \frac{a_1 + a_n}{2}.$$

1.2.8. Show that
$$\sqrt{n} \leq \sqrt[n]{n!} \leq \frac{n+1}{2}, \quad n \in \mathbb{N}.$$

1.2.9. Let a_k, $k = 1, 2, \ldots, n$, be positive numbers and let $\sum_{k=1}^{n} a_k \leq 1$. Show that
$$\sum_{k=1}^{n} \frac{1}{a_k} \geq n^2.$$

1.2.10. Let $a_k > 0$, $k = 1, 2, \ldots n$, where $n > 1$, and set $s = \sum_{k=1}^{n} a_k$. Verify the following claims:

(a) $\quad n\left(\sum_{k=1}^{n} \frac{a_k}{s - a_k}\right)^{-1} \leq n - 1 \leq \frac{1}{n}\sum_{k=1}^{n} \frac{s - a_k}{a_k},$

(b) $\quad \sum_{k=1}^{n} \frac{s}{s - a_k} \geq \frac{n^2}{n-1},$

(c) $\quad n\left(\sum_{k=1}^{n} \frac{a_k}{s + a_k}\right)^{-1} \geq n + 1.$

1.2.11. Show that if $a_k > 0$, $k = 1, \ldots, n$, and $a_1 \cdot a_2 \cdot \ldots \cdot a_n = 1$, then
$$(1 + a_1)(1 + a_2) \cdot \ldots \cdot (1 + a_n) \geq 2^n.$$

1.2. Some Elementary Inequalities

1.2.12. Prove the following *Cauchy inequality*:

$$\left(\sum_{k=1}^{n} a_k b_k\right)^2 \leq \sum_{k=1}^{n} a_k^2 \sum_{k=1}^{n} b_k^2.$$

1.2.13. Show that

$$\left(\left(\sum_{k=1}^{n} a_k\right)^2 + \left(\sum_{k=1}^{n} b_k\right)^2\right)^{\frac{1}{2}} \leq \sum_{k=1}^{n} (a_k^2 + b_k^2)^{\frac{1}{2}}.$$

1.2.14. Show that if $\sum_{k=1}^{n} a_k^2 = \sum_{k=1}^{n} b_k^2 = 1$, then

$$\left|\sum_{k=1}^{n} a_k b_k\right| \leq 1.$$

1.2.15. For $a_k > 0$, $k = 1, 2, ..., n$, verify the following claims:

(a) $\quad \sum_{k=1}^{n} a_k \sum_{k=1}^{n} \dfrac{1}{a_k} \geq n^2,$

(b) $\quad \sum_{k=1}^{n} a_k \sum_{k=1}^{n} \dfrac{1 - a_k}{a_k} \geq n \sum_{k=1}^{n} (1 - a_k),$

(c) $\quad (\log_a a_1)^2 + (\log_a a_2)^2 + ... + (\log_a a_n)^2 \geq \dfrac{1}{n},$

provided $\quad a_1 \cdot a_2 \cdot ... \cdot a_n = a \neq 1.$

1.2.16. For $\alpha > 0$, show that

$$\left|\sum_{k=1}^{n} a_k b_k\right| \leq \dfrac{1}{\alpha} \sum_{k=1}^{n} a_k^2 + \dfrac{\alpha}{4} \sum_{k=1}^{n} b_k^2.$$

1.2.17. Establish the following inequalities:

$$\sum_{k=1}^{n} |a_k| \leq \sqrt{n} \left(\sum_{k=1}^{n} a_k^2\right)^{\frac{1}{2}} \leq \sqrt{n} \sum_{k=1}^{n} |a_k|.$$

1.2.18. Show that

(a) $$\left(\sum_{k=1}^{n} a_k b_k\right)^2 \leq \sum_{k=1}^{n} k a_k^2 \sum_{k=1}^{n} \frac{b_k^2}{k},$$

(b) $$\left(\sum_{k=1}^{n} \frac{a_k}{k}\right)^2 \leq \sum_{k=1}^{n} k^3 a_k^2 \sum_{k=1}^{n} \frac{1}{k^5}.$$

1.2.19. Show that
$$\left(\sum_{k=1}^{n} a_k^p\right)^2 \leq \sum_{k=1}^{n} a_k^{p+q} \sum_{k=1}^{n} a_k^{p-q},$$
for any real p, q and any positive $a_1, a_2, ..., a_n$.

1.2.20. Find the minimum of the sum $\sum_{k=1}^{n} a_k^2$ subject to the constraint $\sum_{k=1}^{n} a_k = 1$.

1.2.21. Let $p_1, p_2, ..., p_n$ be given positive numbers. Find the minimum of $\sum_{k=1}^{n} p_k a_k^2$ subject to the constraint $\sum_{k=1}^{n} a_k = 1$.

1.2.22. Show that
$$\left(\sum_{k=1}^{n} a_k\right)^2 \leq (n-1)\left(\sum_{k=1}^{n} a_k^2 + 2a_1 a_2\right).$$

1.2.23. Verify the following claims:

(a) $$\left(\sum_{k=1}^{n}(a_k + b_k)^2\right)^{\frac{1}{2}} \leq \left(\sum_{k=1}^{n} a_k^2\right)^{\frac{1}{2}} + \left(\sum_{k=1}^{n} b_k^2\right)^{\frac{1}{2}},$$

(b) $$\left|\left(\sum_{k=1}^{n} a_k^2\right)^{\frac{1}{2}} - \left(\sum_{k=1}^{n} b_k^2\right)^{\frac{1}{2}}\right| \leq \sum_{k=1}^{n} |a_k - b_k|.$$

1.2. Some Elementary Inequalities

1.2.24. Let $p_1, p_2, ..., p_n$ be given positive numbers. Find the minimum of

$$\sum_{k=1}^{n} a_k^2 + \left(\sum_{k=1}^{n} a_k\right)^2 \quad \text{subject to the constraint} \quad \sum_{k=1}^{n} p_k a_k = 1.$$

1.2.25.. Prove the following *Chebyshev inequality*: If $a_1 \geq a_2 \geq ... \geq a_n$ and $b_1 \geq b_2 \geq ... \geq b_n$ (or $a_1 \leq a_2 \leq ... \leq a_n$ and $b_1 \leq b_2 \leq ... \leq b_n$), then

$$\sum_{k=1}^{n} a_k \sum_{k=1}^{n} b_k \leq n \sum_{k=1}^{n} a_k b_k.$$

1.2.26. Assuming $a_k \geq 0$, $k = 1, 2, ..., n$, and $p \in \mathbb{N}$, show that

$$\left(\frac{1}{n}\sum_{k=1}^{n} a_k\right)^p \leq \frac{1}{n}\sum_{k=1}^{n} a_k^p.$$

1.2.27. Establish the inequality

$$(a+b)^2 \leq (1+c)a^2 + \left(1+\frac{1}{c}\right)b^2$$

for positive c and any real a and b.

1.2.28. Show that $\left|\sqrt{a^2+b^2} - \sqrt{a^2+c^2}\right| \leq |b-c|$.

1.2.29. For positive a, b, c, verify the following claims:

(a) $\dfrac{bc}{a} + \dfrac{ac}{b} + \dfrac{ab}{c} \geq (a+b+c)$,

(b) $\dfrac{1}{a} + \dfrac{1}{b} + \dfrac{1}{c} \geq \dfrac{1}{\sqrt{bc}} + \dfrac{1}{\sqrt{ca}} + \dfrac{1}{\sqrt{ab}}$,

(c) $\dfrac{2}{b+c} + \dfrac{2}{c+a} + \dfrac{2}{a+b} \geq \dfrac{9}{a+b+c}$,

(d) $\dfrac{b^2-a^2}{c+a} + \dfrac{c^2-b^2}{a+b} + \dfrac{a^2-c^2}{b+c} \geq 0$,

(e) $\dfrac{1}{8}\dfrac{(a-b)^2}{a} \leq \dfrac{a+b}{2} - \sqrt{ab} \leq \dfrac{1}{8}\dfrac{(a-b)^2}{b}$ provided $b \leq a$.

1.2.30. For $a_k \in \mathbb{R}$, $b_k > 0$, $k = 1, 2, ..., n$, set
$$m = \min\left\{\frac{a_k}{b_k} : k = 1, 2, ..., n\right\} \text{ and } M = \max\left\{\frac{a_k}{b_k} : k = 1, 2, ..., n\right\}.$$
Show that
$$m \leq \frac{a_1 + a_2 + ... + a_n}{b_1 + b_2 + ... + b_n} \leq M.$$

1.2.31. Show that, if $0 < \alpha_1 < \alpha_2 < ... < \alpha_n < \frac{\pi}{2}$, $n > 1$, then
$$\tan \alpha_1 < \frac{\sin \alpha_1 + \sin \alpha_2 + ... + \sin \alpha_n}{\cos \alpha_1 + \cos \alpha_2 + ... + \cos \alpha_n} < \tan \alpha_n.$$

1.2.32. For positive $c_1, c_2, ..., c_n$, and $k_1, k_2, ..., k_n \in \mathbb{N}$, set
$$S = \max\{\sqrt[k_1]{c_1}, \sqrt[k_2]{c_2}, ..., \sqrt[k_n]{c_n}\},$$
$$s = \min\{\sqrt[k_1]{c_1}, \sqrt[k_2]{c_2}, ..., \sqrt[k_n]{c_n}\}.$$
Show that
$$s \leq (a_1 \cdot a_2 \cdot ... \cdot a_n)^{\frac{1}{k_1 + k_2 + ... + k_n}} \leq S.$$

1.2.33. For $a_k > 0$, $b_k > 0$, $k = 1, 2, ..., n$, set
$$M = \max\left\{\frac{a_k}{b_k} : k = 1, 2, ..., n\right\}.$$
Show that
$$\frac{a_1 + a_2^2 + ... + a_n^n}{b_1 + M b_2^2 + ... + M^{n-1} b_n^n} \leq M.$$

1.2.34. Show that, if x is greater than any of the numbers $a_1, a_2, ..., a_n$, then
$$\frac{1}{x - a_1} + \frac{1}{x - a_2} + ... + \frac{1}{x - a_n} \geq \frac{n}{x - \frac{a_1 + a_2 + ... + a_n}{n}}.$$

1.2.35. Let $c_k = \binom{n}{k}$, $k = 0, 1, 2, ..., n$, be the binomial coefficients. Establish the inequality
$$\sqrt{c_1} + \sqrt{c_2} + ... + \sqrt{c_n} \leq \sqrt{n(2^n - 1)}.$$

1.2. Some Elementary Inequalities

1.2.36. For $n \geq 2$, show that
$$\prod_{k=0}^{n}\binom{n}{k} \leq \left(\frac{2^n - 2}{n - 1}\right)^{n-1}.$$

1.2.37. Let $a_k > 0$, $k = 1, 2, ..., n$, and let A_n be their arithmetic mean. Show that for any integer $p > 1$,
$$\sum_{k=1}^{n} A_k^p \leq \frac{p}{p-1} \sum_{k=1}^{n} A_k^{p-1} a_k.$$

1.2.38. For positive a_k, $k = 1, 2, ..., n$, we set $a = a_1 + a_2 + ... + a_n$. Show that
$$\sum_{k=1}^{n-1} a_k a_{k+1} \leq \frac{a^2}{4}.$$

1.2.39. Show that for any rearrangement $b_1, b_2, ..., b_n$ of the positive numbers $a_1, a_2, ..., a_n$,
$$\frac{a_1}{b_1} + \frac{a_2}{b_2} + ... + \frac{a_n}{b_n} \geq n.$$

1.2.40. Prove the *Weierstrass inequalities*: If $0 < a_k < 1$, $k = 1, 2, ..., n$, and $a_1 + a_2 + ... + a_n < 1$, then

(a) $$1 + \sum_{k=1}^{n} a_k < \prod_{k=1}^{n}(1 + a_k) < \frac{1}{1 - \sum_{k=1}^{n} a_k},$$

(b) $$1 - \sum_{k=1}^{n} a_k < \prod_{k=1}^{n}(1 - a_k) < \frac{1}{1 + \sum_{k=1}^{n} a_k}.$$

1.2.41. Assume that $0 < a_k < 1$, $k = 1, 2, ..., n$, and set $a_1 + a_2 + ... + a_n = a$. Show that
$$\sum_{k=1}^{n} \frac{a_k}{1 - a_k} \geq \frac{na}{n - a}.$$

1.2.42. Let $0 < a_k \leq 1$, $k = 1, 2, ..., n$, and $n \geq 2$. Verify the inequality

$$\sum_{k=1}^{n} \frac{1}{1+a_k} \leq \frac{n \sum_{k=1}^{n} a_k}{\sum_{k=1}^{n} a_k + n \prod_{k=1}^{n} a_k}.$$

1.2.43. For nonnegative a_k, $k = 1, 2, ..., n$, such that $a_1 + a_2 + ... + a_n = 1$, show that

(a) $$\prod_{k=1}^{n}(1+a_k) \geq (n+1)^n \prod_{k=1}^{n} a_k,$$

(b) $$\prod_{k=1}^{n}(1-a_k) \geq (n-1)^n \prod_{k=1}^{n} a_k.$$

1.2.44. Show that if $a_k > 0$, $k = 1, 2, ..., n$, and $\sum_{k=1}^{n} \frac{1}{1+a_k} = n-1$, then

$$\prod_{k=1}^{n} \frac{1}{a_k} \geq (n-1)^n.$$

1.2.45. Prove that under the assumptions of 1.2.43, we have

$$\frac{\prod_{k=1}^{n}(1+a_k)}{(n+1)^n} \geq \frac{\prod_{k=1}^{n}(1-a_k)}{(n-1)^n}, \quad n > 1.$$

1.2.46. Show that for positive $a_1, a_2, ..., a_n$,

$$\frac{a_1}{a_2+a_3} + \frac{a_2}{a_3+a_4} + ... + \frac{a_{n-2}}{a_{n-1}+a_n} + \frac{a_{n-1}}{a_n+a_1} + \frac{a_n}{a_1+a_2} \geq \frac{n}{4}.$$

1.2.47. Let t and $a_1, a_2, ..., a_n$ be any real numbers. Establish the inequality

$$\sum_{k=1}^{n} \frac{\sqrt{|a_k - t|}}{2^k} \geq \sum_{k=2}^{n} \frac{\sqrt{|a_k - a_1|}}{2^k}.$$

1.2. Some Elementary Inequalities 17

1.2.48. Show that for positive $a_1, a_2, ..., a_n$, and $b_1, b_2, ..., b_n$, we have
$$\sqrt[n]{(a_1+b_1)(a_2+b_2)...(a_n+b_n)} \geq \sqrt[n]{a_1 a_2 ... a_n} + \sqrt[n]{b_1 b_2 ... b_n}.$$

1.2.49. Assume that $0 < a_1 < a_2 < ... < a_n$ and $p_1, p_2, ..., p_n$ are nonnegative such that $\sum_{k=1}^{n} p_k = 1$. Establish the inequality
$$\left(\sum_{k=1}^{n} p_k a_k\right)\left(\sum_{k=1}^{n} p_k \frac{1}{a_k}\right) \leq \frac{A^2}{G^2},$$
where $A = \frac{1}{2}(a_1 + a_n)$ and $G = \sqrt{a_1 a_n}$.

1.2.50. For a positive integer n, let $\sigma(n)$ and $\tau(n)$ denote the sum of all the positive divisors of n and the number of these divisors, respectively. Show that $\frac{\sigma(n)}{\tau(n)} \geq \sqrt{n}$.

Chapter 2

Sequences of Real Numbers

2.1. Monotonic Sequences

2.1.1. Show that

(a) if a sequence $\{a_n\}$ is monotonically increasing, then $\lim\limits_{n\to\infty} a_n = \sup\{a_n : n \in \mathbb{N}\}$;

(b) if a sequence $\{a_n\}$ is monotonically decreasing, then $\lim\limits_{n\to\infty} a_n = \inf\{a_n : n \in \mathbb{N}\}$.

2.1.2. Let $a_1, a_2, ..., a_p$ be fixed positive numbers. Consider the sequences

$$s_n = \frac{a_1^n + a_2^n + ... + a_p^n}{p} \quad \text{and} \quad x_n = \sqrt[n]{s_n}, \quad n \in \mathbb{N}.$$

Show that the sequence $\{x_n\}$ is monotonically increasing.
Hint. First establish monotonicity of the sequence $\left\{\frac{s_n}{s_{n-1}}\right\}$, $n \geq 2$.

2.1.3. Show that the sequence $\{a_n\}$, where $a_n = \frac{n}{2^n}$, $n > 1$, strictly decreases and find its limit.

2.1.4. Let $\{a_n\}$ be a bounded sequence which satisfies the condition $a_{n+1} \geq a_n - \frac{1}{2^n}$, $n \in \mathbb{N}$. Show that the sequence $\{a_n\}$ is convergent.
Hint. Consider the sequence $\left\{a_n - \frac{1}{2^{n-1}}\right\}$.

2.1.5. Prove the convergence of the sequences:

(a) $\quad a_n = -2\sqrt{n} + \left(\dfrac{1}{\sqrt{1}} + \dfrac{1}{\sqrt{2}} + ... + \dfrac{1}{\sqrt{n}}\right);$

(b) $\quad b_n = -2\sqrt{n+1} + \left(\dfrac{1}{\sqrt{1}} + \dfrac{1}{\sqrt{2}} + ... + \dfrac{1}{\sqrt{n}}\right).$

Hint. First establish the inequalities

$$2(\sqrt{n+1} - 1) < \dfrac{1}{\sqrt{1}} + \dfrac{1}{\sqrt{2}} + ... + \dfrac{1}{\sqrt{n}} < 2\sqrt{n}, \quad n \in \mathbb{N}.$$

2.1.6. Show that the sequence $\{a_n\}$ defined recursively by

$$a_1 = \dfrac{3}{2}, \quad a_n = \sqrt{3a_{n-1} - 2} \quad \text{for} \quad n \geq 2,$$

converges and find its limit.

2.1.7. For $c > 2$, define the sequence $\{a_n\}$ recursively as follows:

$$a_1 = c^2, \quad a_{n+1} = (a_n - c)^2, \quad n \geq 1.$$

Show that the sequence $\{a_n\}$ strictly increases.

2.1.8. Suppose that the sequence $\{a_n\}$ satisfies the conditions

$$0 < a_n < 1, \quad a_n(1 - a_{n+1}) > \dfrac{1}{4} \quad \text{for} \quad n \in \mathbb{N}.$$

Establish the convergence of the sequence and find its limit.

2.1.9. Establish the convergence and find the limit of the sequence defined by

$$a_1 = 0, \quad a_{n+1} = \sqrt{6 + a_n} \quad \text{for} \quad n \geq 1.$$

2.1.10. Show that the sequence defined by

$$a_1 = 0, \quad a_2 = \dfrac{1}{2}, \quad a_{n+1} = \dfrac{1}{3}(1 + a_n + a_{n-1}^3) \quad \text{for} \quad n > 1$$

converges and determine its limit.

2.1. Monotonic Sequences

2.1.11. Study the monotonicity of the sequence
$$a_n = \frac{n!}{(2n+1)!!}, \quad n \geq 1,$$
and determine its limit.

2.1.12. Determine the convergence or divergence of the sequence
$$a_n = \frac{(2n)!!}{(2n+1)!!}, \quad n \geq 1.$$

2.1.13. Prove the convergence of the sequences

(a) $\quad a_n = 1 + \dfrac{1}{2^2} + \dfrac{1}{3^2} + \ldots + \dfrac{1}{n^2}, \quad n \in \mathbb{N};$

(b) $\quad a_n = 1 + \dfrac{1}{2^2} + \dfrac{1}{3^3} + \ldots + \dfrac{1}{n^n}, \quad n \in \mathbb{N}.$

2.1.14. Show the convergence of the sequence $\{a_n\}$, where
$$a_n = \frac{1}{\sqrt{n(n+1)}} + \frac{1}{\sqrt{(n+1)(n+2)}} + \ldots + \frac{1}{\sqrt{(2n-1)2n}}, \quad n \in \mathbb{N}.$$

2.1.15. For $p \in \mathbb{N}$, $a > 0$ and $a_1 > 0$, define the sequence $\{a_n\}$ by setting
$$a_{n+1} = \frac{1}{p}\left((p-1)a_n + \frac{a}{a_n^{p-1}}\right), \quad n \in \mathbb{N}.$$
Determine $\lim\limits_{n\to\infty} a_n$.

2.1.16. Define $\{a_n\}$ recursively by
$$a_1 = \sqrt{2}, \quad a_{n+1} = \sqrt{2 + \sqrt{a_n}} \quad \text{for } n \geq 1.$$
Prove the convergence of the sequence $\{a_n\}$ and find its limit.

2.1.17. Define the recursive sequence $\{a_n\}$ as follows:
$$a_1 = 1, \quad a_{n+1} = \frac{2(2a_n + 1)}{a_n + 3} \quad \text{for } n \in \mathbb{N}.$$
Establish the convergence of the sequence $\{a_n\}$ and find its limit.

2.1.18. Determine all $c > 0$ such that the recursive sequence $\{a_n\}$ defined by setting
$$a_1 = \frac{c}{2}, \quad a_{n+1} = \frac{1}{2}(c + a_n^2) \quad \text{for } n \in \mathbb{N}$$
converges. In case of convergence find $\lim_{n \to \infty} a_n$.

2.1.19. Let $a > 0$ be fixed and define the sequence $\{a_n\}$ by setting
$$a_1 > 0 \quad \text{and} \quad a_{n+1} = a_n \frac{a_n^2 + 3a}{3a_n^2 + a} \quad \text{for } n \in \mathbb{N}.$$
Determine all a_1 for which the sequence converges and in such a case find its limit.

2.1.20. Let $\{a_n\}$ be defined recursively by
$$a_{n+1} = \frac{1}{4 - 3a_n} \quad \text{for } n \geq 1.$$
Determine for which a_1 the sequence converges and in case of convergence find its limit.

2.1.21. Let a be arbitrarily fixed and let $\{a_n\}$ be defined as follows:
$$a_1 \in \mathbb{R} \quad \text{and} \quad a_{n+1} = a_n^2 + (1 - 2a)a_n + a^2 \quad \text{for } n \in \mathbb{N}.$$
Determine all a_1 such that the sequence converges and in such a case find its limit.

2.1.22. For $c > 0$ and $b > a > 0$, define the recursive sequence $\{a_n\}$ by setting
$$a_1 = c, \quad a_{n+1} = \frac{a_n^2 + ab}{a + b} \quad \text{for } n \in \mathbb{N}.$$
Determine for which values a, b and c the sequence converges and find its limit.

2.1.23. Prove the convergence and find the limit of the sequence $\{a_n\}$ defined inductively by
$$a_1 > 0, \quad a_{n+1} = 6\frac{1 + a_n}{7 + a_n}, \quad n \in \mathbb{N}.$$

2.1. Monotonic Sequences

2.1.24. For $c \geq 0$, define the sequence $\{a_n\}$ as follows:
$$a_1 = 0, \quad a_{n+1} = \sqrt{c + a_n}, \quad n \in \mathbb{N}.$$
Show the convergence of the sequence and determine its limit.

2.1.25. Investigate the convergence of the sequence defined by
$$a_1 = \sqrt{2}, \quad a_{n+1} = \sqrt{2a_n}, \quad n \in \mathbb{N}.$$

2.1.26. Let $k \in \mathbb{N}$ be fixed. Study the convergence of the sequence $\{a_n\}$ defined by setting
$$a_1 = \sqrt[k]{5}, \quad a_{n+1} = \sqrt[k]{5a_n}, \quad n \in \mathbb{N}.$$

2.1.27. Investigate the convergence of the sequence $\{a_n\}$ given by
$$1 \leq a_1 \leq 2, \quad a_{n+1}^2 = 3a_n - 2, \quad n \in \mathbb{N}.$$

2.1.28. For $c > 1$, define the sequences $\{a_n\}$ and $\{b_n\}$ as follows:
(a) $\quad a_1 = \sqrt{c(c-1)}, \quad a_{n+1} = \sqrt{c(c-1) + a_n}, \quad n \geq 1;$
(b) $\quad b_1 = \sqrt{c}, \quad b_{n+1} = \sqrt{cb_n}, \quad n \geq 1.$
Prove that both sequences tend to c.

2.1.29. Given $a > 0$ and $b > 0$, define the sequence $\{a_n\}$ by setting
$$0 < a_1 < b, \quad a_{n+1} = \sqrt{\frac{ab^2 + a_n^2}{a+1}} \quad \text{for} \quad n \geq 1.$$
Find $\lim\limits_{n \to \infty} a_n$.

2.1.30. Prove the convergence of $\{a_n\}$ defined inductively by
$$a_1 = 2, \quad a_{n+1} = 2 + \frac{1}{3 + \frac{1}{a_n}} \quad \text{for} \quad n \geq 1$$
and find its limit.

2.1.31. The recursive sequence $\{a_n\}$ is given by setting

$$a_1 = 1, \quad a_2 = 2, \quad a_{n+1} = \sqrt{a_{n-1}} + \sqrt{a_n} \quad \text{for} \quad n \geq 2.$$

Show that the sequence is bounded and strictly increasing. Find its limit.

2.1.32. The recursive sequence $\{a_n\}$ is given by setting

$$a_1 = 9, \quad a_2 = 6, \quad a_{n+1} = \sqrt{a_{n-1}} + \sqrt{a_n} \quad \text{for} \quad n \geq 2.$$

Show that the sequence is bounded and strictly decreasing. Find its limit.

2.1.33. Define the sequences $\{a_n\}$ and $\{b_n\}$ as follows:

$$0 < b_1 < a_1, \quad a_{n+1} = \frac{a_n + b_n}{2} \quad \text{and} \quad b_{n+1} = \sqrt{a_n b_n} \quad \text{for} \quad n \in \mathbb{N}.$$

Show that $\{a_n\}$ and $\{b_n\}$ both tend to the same limit. (This limit is called the *arithmetic-geometric mean of a_1 and b_1*.)

2.1.34. Show that the sequences $\{a_n\}$ and $\{b_n\}$ given by

$$0 < b_1 < a_1, \quad a_{n+1} = \frac{a_n^2 + b_n^2}{a_n + b_n} \quad \text{and} \quad b_{n+1} = \frac{a_n + b_n}{2} \quad \text{for} \quad n \in \mathbb{N}$$

are both monotonic and have the same limit.

2.1.35. Let the recursive sequences $\{a_n\}$ and $\{b_n\}$ be given by setting

$$0 < b_1 < a_1, \quad a_{n+1} = \frac{a_n + b_n}{2} \quad \text{and} \quad b_{n+1} = \frac{2 a_n b_n}{a_n + b_n} \quad \text{for} \quad n \in \mathbb{N}.$$

Prove the monotonicity of these sequences and show that both of them tend to the arithmetic-geometric mean of a_1 and b_1. (See Problem 2.1.33.)

2.1.36. Show the convergence and find the limit of $\{a_n\}$, where

$$a_n = \frac{n+1}{2^{n+1}} \left(\frac{2}{1} + \frac{2^2}{2} + \ldots + \frac{2^n}{n} \right) \quad \text{for} \quad n \in \mathbb{N}.$$

2.1. Monotonic Sequences

2.1.37. Suppose that a bounded sequence $\{a_n\}$ is such that

$$a_{n+2} \leq \frac{1}{3}a_{n+1} + \frac{2}{3}a_n \quad \text{for} \quad n \geq 1.$$

Prove the convergence of the sequence $\{a_n\}$.

2.1.38. Let $\{a_n\}$ and $\{b_n\}$ be defined as follows:

$$a_n = \left(1 + \frac{1}{n}\right)^n, \quad b_n = \left(1 + \frac{1}{n}\right)^{n+1} \quad \text{for} \quad n \in \mathbb{N}.$$

Using the arithmetic-geometric-harmonic mean inequalities, show that

(a) $a_n < b_n$ for $n \in \mathbb{N}$.

(b) the sequence $\{a_n\}$ is strictly increasing,

(c) the sequence $\{b_n\}$ is strictly decreasing,

Show also that $\{a_n\}$ and $\{b_n\}$ both have the same limit, defined to be Euler's number e.

2.1.39. Let

$$a_n = \left(1 + \frac{x}{n}\right)^n \quad \text{for} \quad n \in \mathbb{N}.$$

(a) Show that if $x > 0$, then the sequence $\{a_n\}$ is bounded and strictly increasing.

(b) Let x be any real number. Show that the sequence $\{a_n\}$ is bounded and strictly increasing for $n > -x$.

The number e^x is defined to be the limit of this sequence.

2.1.40. Suppose that $x > 0$, $l \in \mathbb{N}$ and $l > x$. Show that the sequence $\{b_n\}$, where

$$b_n = \left(1 + \frac{x}{n}\right)^{l+n} \quad \text{for} \quad n \in \mathbb{N},$$

is strictly decreasing.

2.1.41. Establish the monotonicity of the sequences $\{a_n\}$ and $\{b_n\}$, where

$$a_n = 1 + \frac{1}{2} + \ldots + \frac{1}{n-1} - \ln n \quad \text{for} \quad n \in \mathbb{N},$$

$$b_n = 1 + \frac{1}{2} + \ldots + \frac{1}{n-1} + \frac{1}{n} - \ln n \quad \text{for} \quad n \in \mathbb{N}.$$

Show that both of them tend to the same limit γ, which is known as *Euler's constant*.

Hint. Apply the inequality $\left(1 + \frac{1}{n}\right)^n < e < \left(1 + \frac{1}{n}\right)^{n+1}$, which follows from 2.1.38.

2.1.42. Given $x > 0$, set $a_n = \sqrt[2^n]{x}$, $n \in \mathbb{N}$. Show that the sequence $\{a_n\}$ is bounded. Show also that it is strictly increasing if $x < 1$ and strictly decreasing if $x > 1$. Compute $\lim_{n\to\infty} a_n$.

Moreover, put

$$c_n = 2^n(a_n - 1) \quad \text{and} \quad d_n = 2^n\left(1 - \frac{1}{a_n}\right) \quad \text{for} \quad n \in \mathbb{N}.$$

Show that $\{c_n\}$ is decreasing and $\{d_n\}$ is increasing and both sequences have the same limit.

2.2. Limits. Properties of Convergent Sequences

2.2.1. Calculate:

(a) $\quad \lim_{n\to\infty} \sqrt[n]{1^2 + 2^2 + \ldots + n^2},$

(b) $\quad \lim_{n\to\infty} \dfrac{n + \sin n^2}{n + \cos n},$

(c) $\quad \lim_{n\to\infty} \dfrac{1 - 2 + 3 - 4 + \ldots + (-2n)}{\sqrt{n^2 + 1}},$

(d) $\quad \lim_{n\to\infty} (\sqrt{2} - \sqrt[3]{2})(\sqrt{2} - \sqrt[5]{2}) \cdot \ldots \cdot (\sqrt{2} - \sqrt[2n+1]{2}),$

(e) $\quad \lim_{n\to\infty} \dfrac{n}{2\sqrt{n}},$

2.2. Limits. Properties of Convergent Sequences

(f) $\lim\limits_{n\to\infty} \dfrac{n!}{2^{n^2}}$,

(g) $\lim\limits_{n\to\infty} \dfrac{1}{\sqrt{n}}\left(\dfrac{1}{\sqrt{1}+\sqrt{3}} + \dfrac{1}{\sqrt{3}+\sqrt{5}} + \ldots + \dfrac{1}{\sqrt{2n-1}+\sqrt{2n+1}}\right)$,

(h) $\lim\limits_{n\to\infty}\left(\dfrac{1}{n^2+1} + \dfrac{2}{n^2+2} + \ldots + \dfrac{n}{n^2+n}\right)$,

(i) $\lim\limits_{n\to\infty}\left(\dfrac{n}{n^3+1} + \dfrac{2n}{n^3+2} + \ldots + \dfrac{nn}{n^3+n}\right)$.

2.2.2. Let $s > 0$ and $p > 0$. Show that
$$\lim_{n\to\infty} \dfrac{n^s}{(1+p)^n} = 0.$$

2.2.3. For $\alpha \in (0,1)$, calculate $\lim\limits_{n\to\infty}((n+1)^\alpha - n^\alpha)$.

2.2.4. For $\alpha \in \mathbb{Q}$, calculate $\lim\limits_{n\to\infty} \sin(n!\alpha\pi)$.

2.2.5. Show that the limit $\lim\limits_{n\to\infty} \sin n$ does not exist.

2.2.6. Show that for any irrational α the limit $\lim\limits_{n\to\infty} \sin n\alpha\pi$ does not exist.

2.2.7. For $a \in \mathbb{R}$, calculate
$$\lim_{n\to\infty} \dfrac{1}{n}\left(\left(a+\dfrac{1}{n}\right)^2 + \left(a+\dfrac{2}{n}\right)^2 + \ldots + \left(a+\dfrac{n-1}{n}\right)^2\right).$$

2.2.8. Suppose $a_n \neq 1$ for all n and $\lim\limits_{n\to\infty} a_n = 1$. Given a positive integer k, compute
$$\lim_{n\to\infty} \dfrac{a_n + a_n^2 + \ldots + a_n^k - k}{a_n - 1}.$$

2.2.9. Find
$$\lim_{n\to\infty}\left(\dfrac{1}{1\cdot 2\cdot 3} + \dfrac{1}{2\cdot 3\cdot 4} + \ldots + \dfrac{1}{n\cdot(n+1)\cdot(n+2)}\right).$$

2.2.10. Calculate
$$\lim_{n\to\infty} \prod_{k=2}^{n} \frac{k^3-1}{k^3+1}.$$

2.2.11. Determine
$$\lim_{n\to\infty} \sum_{i=1}^{n} \sum_{j=1}^{i} \frac{j}{n^3}.$$

2.2.12. Compute
$$\lim_{n\to\infty} \left(1 - \frac{2}{2\cdot 3}\right)\left(1 - \frac{2}{3\cdot 4}\right) \cdot \ldots \cdot \left(1 - \frac{2}{(n+1)\cdot(n+2)}\right).$$

2.2.13. Calculate
$$\lim_{n\to\infty} \sum_{k=1}^{n} \frac{k^3 + 6k^2 + 11k + 5}{(k+3)!}.$$

2.2.14. For $x \neq -1$ and $x \neq 1$, find
$$\lim_{n\to\infty} \sum_{k=1}^{n} \frac{x^{2^{k-1}}}{1 - x^{2^k}}.$$

2.2.15. Determine for which $x \in \mathbb{R}$ the limit
$$\lim_{n\to\infty} \prod_{k=0}^{n} (1 + x^{2^k})$$
exists and find its value.

2.2.16. Determine all $x \in \mathbb{R}$ such that the limit
$$\lim_{n\to\infty} \prod_{k=0}^{n} \left(1 + \frac{2}{x^{2^k} + x^{-2^k}}\right)$$
exists and find its value.

2.2.17. Establish for which $x \in \mathbb{R}$ the limit
$$\lim_{n\to\infty} \prod_{k=1}^{n} (1 + x^{3^k} + x^{2\cdot 3^k})$$
exists and find its value.

2.2. Limits. Properties of Convergent Sequences

2.2.18. Calculate
$$\lim_{n\to\infty} \frac{1\cdot 1! + 2\cdot 2! + \ldots + n\cdot n!}{(n+1)!}.$$

2.2.19. For which $x \in \mathbb{R}$ does the equality
$$\lim_{n\to\infty} \frac{n^{1999}}{n^x - (n-1)^x} = \frac{1}{2000}$$
hold?

2.2.20. Given a and b such that $a \geq b > 0$, define the sequence $\{a_n\}$ by setting
$$a_1 = a + b, \quad a_n = a_1 - \frac{ab}{a_{n-1}}, \quad n \geq 2.$$
Determine the nth term of the sequence and compute $\lim_{n\to\infty} a_n$.

2.2.21. Define the sequence $\{a_n\}$ by setting
$$a_1 = 0, \quad a_2 = 1 \quad \text{and} \quad a_{n+1} - 2a_n + a_{n-1} = 2 \quad \text{for} \quad n \geq 2.$$
Determine its nth term and calculate $\lim_{n\to\infty} a_n$.

2.2.22. For $a > 0$ and $b > 0$, consider the sequence $\{a_n\}$ defined by
$$a_1 = \frac{ab}{\sqrt{a^2 + b^2}} \quad \text{and}$$
$$a_n = \frac{aa_{n-1}}{\sqrt{a^2 + a_{n-1}^2}}, \quad n \geq 2.$$
Determine its nth term and find $\lim_{n\to\infty} a_n$.

2.2.23. Let $\{a_n\}$ be a recursive sequence defined as follows:
$$a_1 = 0, \quad a_n = \frac{a_{n-1} + 3}{4}, \quad n \geq 2.$$
Find the formula for the nth term of the sequence and find its limit.

2.2.24. Study the convergence of the sequence given by
$$a_1 = a, \quad a_n = 1 + ba_{n-1}, \quad n \geq 2.$$

2.2.25. The *Fibonacci* sequence $\{a_n\}$ is defined as follows:
$$a_1 = a_2 = 1, \quad a_{n+2} = a_n + a_{n+1}, \quad n \geq 1.$$
Show that
$$a_n = \frac{\alpha^n - \beta^n}{\alpha - \beta},$$
where α and β are roots of $x^2 = x + 1$. Compute $\lim_{n \to \infty} \sqrt[n]{a_n}$.

2.2.26. Define the sequences $\{a_n\}$ and $\{b_n\}$ by setting
$$a_1 = a, \quad b_1 = b,$$
$$a_{n+1} = \frac{a_n + b_n}{2}, \quad b_{n+1} = \frac{a_{n+1} + b_n}{2}.$$
Show that $\lim_{n \to \infty} a_n = \lim_{n \to \infty} b_n$.

2.2.27. Given $a \in \{1, 2, ..., 9\}$, compute
$$\lim_{n \to \infty} \frac{a + aa + ... + \overbrace{aa...a}^{n \text{ digits}}}{10^n}.$$

2.2.28. Calculate
$$\lim_{n \to \infty} (\sqrt[n]{n} - 1)^n.$$

2.2.29. Suppose that the sequence $\{a_n\}$ converges to zero. Find $\lim_{n \to \infty} a_n^n$.

2.2.30. Given positive $p_1, p_2, ..., p_k$ and $a_1, a_2, ..., a_k$, find
$$\lim_{n \to \infty} \frac{p_1 a_1^{n+1} + p_2 a_2^{n+1} + ... + p_k a_k^{n+1}}{p_1 a_1^n + p_2 a_2^n + ... + p_k a_k^n}.$$

2.2.31. Suppose that $\lim_{n \to \infty} \left| \frac{a_{n+1}}{a_n} \right| = q$. Show that

(a) if $q < 1$, then $\lim_{n \to \infty} a_n = 0$,

(b) if $q > 1$, then $\lim_{n \to \infty} |a_n| = \infty$.

2.2. Limits. Properties of Convergent Sequences

2.2.32. Suppose that $\lim\limits_{n\to\infty} \sqrt[n]{|a_n|} = q$. Show that

(a) if $q < 1$, then $\lim\limits_{n\to\infty} a_n = 0$,

(b) if $q > 1$, then $\lim\limits_{n\to\infty} |a_n| = \infty$.

2.2.33. Given a real number α and $x \in (0,1)$, calculate

$$\lim_{n\to\infty} n^\alpha x^n.$$

2.2.34. Calculate

$$\lim_{n\to\infty} \frac{m(m-1)\cdot\ldots\cdot(m-n+1)}{n!} x^n, \quad \text{for} \quad m \in \mathbb{N} \text{ and } |x| < 1.$$

2.2.35. Assume that $\lim\limits_{n\to\infty} a_n = 0$ and $\{b_n\}$ is a bounded sequence. Show that $\lim\limits_{n\to\infty} a_n b_n = 0$.

2.2.36. Show that if $\lim\limits_{n\to\infty} a_n = a$ and $\lim\limits_{n\to\infty} b_n = b$, then

$$\lim_{n\to\infty} \max\{a_n, b_n\} = \max\{a, b\}.$$

2.2.37. Let $a_n \geq -1$ for $n \in \mathbb{N}$ and let $\lim\limits_{n\to\infty} a_n = 0$. For $p \in \mathbb{N}$, find

$$\lim_{n\to\infty} \sqrt[p]{1+a_n}.$$

2.2.38. Assume that a positive sequence $\{a_n\}$ converges to zero. For natural $p \geq 2$, determine

$$\lim_{n\to\infty} \frac{\sqrt[p]{1+a_n} - 1}{a_n}.$$

2.2.39. For positive $a_1, a_2, ..., a_p$, find

$$\lim_{n\to\infty} \left(\sqrt[p]{(n+a_1)(n+a_2)\cdot\ldots\cdot(n+a_p)} - n \right).$$

2.2.40. Calculate
$$\lim_{n\to\infty}\left(\frac{1}{\sqrt{n^2+1}}+\frac{1}{\sqrt{n^2+2}}+\ldots+\frac{1}{\sqrt{n^2+n+1}}\right).$$

2.2.41. For positive $a_1, a_2, \ldots, a_p$, find
$$\lim_{n\to\infty}\sqrt[n]{\frac{a_1^n+a_2^n+\ldots+a_p^n}{p}}.$$

2.2.42. Compute
$$\lim_{n\to\infty}\sqrt[n]{2\sin^2\frac{n^{1999}}{n+1}+\cos^2\frac{n^{1999}}{n+1}}.$$

2.2.43. Find
$$\lim_{n\to\infty}(n+1+n\cos n)^{\frac{1}{2n+n\sin n}}.$$

2.2.44. Calculate
$$\lim_{n\to\infty}\sum_{k=1}^{n}\left(\sqrt{1+\frac{k}{n^2}}-1\right).$$

2.2.45. Determine
$$\lim_{n\to\infty}\sum_{k=1}^{n}\left(\sqrt[3]{1+\frac{k^2}{n^3}}-1\right).$$

2.2.46. For positive a_k, $k=1,2,\ldots,p$, find
$$\lim_{n\to\infty}\left(\frac{1}{p}\sum_{k=1}^{p}\sqrt[n]{a_k}\right)^p.$$

2.2.47. Given $\alpha\in(0,1)$, compute
$$\lim_{n\to\infty}\sum_{k=0}^{n-1}\left(\alpha+\frac{1}{n}\right)^k.$$

2.2. Limits. Properties of Convergent Sequences

2.2.48. Given real $x \geq 1$, show that
$$\lim_{n\to\infty} (2\sqrt[n]{x} - 1)^n = x^2.$$

2.2.49. Show that
$$\lim_{n\to\infty} \frac{(2\sqrt[n]{n} - 1)^n}{n^2} = 1.$$

2.2.50. Which of the following sequences are Cauchy sequences?

(a) $\quad a_n = \dfrac{\tan 1}{2} + \dfrac{\tan 2}{2^2} + ... + \dfrac{\tan n}{2^n},$

(b) $\quad a_n = 1 + \dfrac{1}{4} + \dfrac{2^2}{4^2} + ... + \dfrac{n^2}{4^n},$

(c) $\quad a_n = 1 + \dfrac{1}{2} + \dfrac{1}{3} + ... + \dfrac{1}{n},$

(d) $\quad a_n = \dfrac{1}{1 \cdot 2} - \dfrac{1}{2 \cdot 3} + ... + (-1)^{n-1} \dfrac{1}{n(n+1)},$

(e) $\quad a_n = \alpha_1 q^1 + \alpha_2 q^2 + ... + \alpha_n q^n,$
  for $|q| < 1$, $|\alpha_k| \leq M$, $k = 1, 2, ...,$

(f) $\quad a_n = \dfrac{1}{2^2} + \dfrac{2}{3^2} + ... + \dfrac{n}{(n+1)^2}.$

2.2.51. Suppose that a sequence $\{a_n\}$ satisfies the condition
$$|a_{n+1} - a_{n+2}| < \lambda |a_n - a_{n+1}|.$$
with a $\lambda \in (0,1)$. Prove that $\{a_n\}$ converges.

2.2.52. Given a sequence $\{a_n\}$ of positive integers, define
$$S_n = \frac{1}{a_1} + \frac{1}{a_2} + ... + \frac{1}{a_n}$$
and
$$\sigma_n = \left(1 + \frac{1}{a_1}\right)\left(1 + \frac{1}{a_2}\right) \cdot ... \cdot \left(1 + \frac{1}{a_n}\right).$$
Prove that if $\{S_n\}$ converges, then $\{\ln \sigma_n\}$ also converges.

2.2.53. Show that the sequence $\{R_n\}$ of convergents to an irrational number x (defined in Problem 1.1.20) is a Cauchy sequence.

2.2.54. For an arithmetic progression $\{a_n\}$ whose terms are different from zero, compute

$$\lim_{n\to\infty}\left(\frac{1}{a_1 a_2}+\frac{1}{a_2 a_3}+\ldots+\frac{1}{a_n a_{n+1}}\right).$$

2.2.55. For an arithmetic progression $\{a_n\}$ with positive terms, calculate

$$\lim_{n\to\infty}\frac{1}{\sqrt{n}}\left(\frac{1}{\sqrt{a_1}+\sqrt{a_2}}+\frac{1}{\sqrt{a_2}+\sqrt{a_3}}+\ldots+\frac{1}{\sqrt{a_n}+\sqrt{a_{n+1}}}\right).$$

2.2.56. Find

(a) $\lim\limits_{n\to\infty} n(\sqrt[n]{e}-1)$, \quad (b) $\lim\limits_{n\to\infty}\dfrac{e^{\frac{1}{n}}+e^{\frac{2}{n}}+\ldots+e^{\frac{n}{n}}}{n}$.

2.2.57. Let $\{a_n\}$ be a sequence defined as follows:

$$a_1=a,\quad a_2=b,\quad a_{n+1}=pa_{n-1}+(1-p)a_n,\quad n=2,3,\ldots$$

Determine for which values a,b and p the sequence converges.

2.2.58. Let $\{a_n\}$ and $\{b_n\}$ be defined by setting

$$a_1=3,\quad b_1=2,\quad a_{n+1}=a_n+2b_n\quad\text{and}\quad b_{n+1}=a_n+b_n.$$

Moreover, let

$$c_n=\frac{a_n}{b_n},\quad n\in\mathbb{N}.$$

(a) Show that $|c_{n+1}-\sqrt{2}|<\frac{1}{2}|c_n-\sqrt{2}|,\ n\in\mathbb{N}$.
(b) Calculate $\lim\limits_{n\to\infty}c_n$.

2.3. The Toeplitz transformation, the Stolz theorem and their applications

2.3.1. Prove the following *Toeplitz theorem on regular transformation of sequences into sequences*.

Let $\{c_{n,k} : 1 \leq k \leq n,\ n \geq 1\}$ be an array of real numbers such that:

(i) $c_{n,k} \xrightarrow[n \to \infty]{} 0$ for each $k \in \mathbb{N}$,

(ii) $\sum_{k=1}^{n} c_{n,k} \xrightarrow[n \to \infty]{} 1$,

(iii) there exists $C > 0$ such that for all positive integers n:
$$\sum_{k=1}^{n} |c_{n,k}| \leq C.$$

Then for any convergent sequence $\{a_n\}$ the transformed sequence $\{b_n\}$ given by $b_n = \sum_{k=1}^{n} c_{n,k} a_k$, $n \geq 1$, is also convergent and $\lim_{n \to \infty} b_n = \lim_{n \to \infty} a_n$.

2.3.2. Show that if $\lim_{n \to \infty} a_n = a$, then
$$\lim_{n \to \infty} \frac{a_1 + a_2 + \ldots + a_n}{n} = a.$$

2.3.3.

(a) Show that the assumption (iii) in the Toeplitz theorem (Problem 2.3.1) can be omitted if all the numbers $c_{n,k}$ are nonnegative.

(b) Let $\{b_n\}$ be the transformed sequence defined in the Toeplitz theorem with $c_{n,k} > 0$, $1 \leq k \leq n$, $n \geq 1$. Show that if $\lim_{n \to \infty} a_n = +\infty$, then $\lim_{n \to \infty} b_n = +\infty$.

2.3.4. Show that if $\lim_{n \to \infty} a_n = +\infty$, then
$$\lim_{n \to \infty} \frac{a_1 + a_2 + \ldots + a_n}{n} = +\infty.$$

2.3.5. Prove that if $\lim\limits_{n\to\infty} a_n = a$, then

$$\lim_{n\to\infty} \frac{na_1 + (n-1)a_2 + \ldots + 1\cdot a_n}{n^2} = \frac{a}{2}.$$

2.3.6. Show that if a positive sequence $\{a_n\}$ converges to a, then $\lim\limits_{n\to\infty} \sqrt[n]{a_1 \cdot \ldots \cdot a_n} = a$.

2.3.7. For a positive sequence $\{a_n\}$, show that if $\lim\limits_{n\to\infty} \frac{a_{n+1}}{a_n} = a$, then $\lim\limits_{n\to\infty} \sqrt[n]{a_n} = a$.

2.3.8. Let $\lim\limits_{n\to\infty} a_n = a$ and $\lim\limits_{n\to\infty} b_n = b$. Show that

$$\lim_{n\to\infty} \frac{a_1 b_n + a_2 b_{n-1} + \ldots + a_n b_1}{n} = ab.$$

2.3.9. Let $\{a_n\}$ and $\{b_n\}$ be two sequences such that

(i) $b_n > 0$, $n \in \mathbb{N}$, and $\lim\limits_{n\to\infty}(b_1 + b_2 + \ldots + b_n) = +\infty$,

(ii) $\lim\limits_{n\to\infty} \dfrac{a_n}{b_n} = g$.

Prove that
$$\lim_{n\to\infty} \frac{a_1 + a_2 + \ldots + a_n}{b_1 + b_2 + \ldots + b_n} = g.$$

2.3.10. Let $\{a_n\}$ and $\{b_n\}$ be two sequences for which

(i) $b_n > 0$, $n \in \mathbb{N}$, and $\lim\limits_{n\to\infty}(b_1 + b_2 + \ldots + b_n) = +\infty$,

(ii) $\lim\limits_{n\to\infty} a_n = a$.

Show that
$$\lim_{n\to\infty} \frac{a_1 b_1 + a_2 b_2 + \ldots + a_n b_n}{b_1 + b_2 + \ldots + b_n} = a.$$

2.3. Toeplitz Transformation and Stolz Theorem

2.3.11. Using the result in the foregoing problem, prove *the Stolz theorem*.

Let $\{x_n\}$, $\{y_n\}$ be two sequences that satisfy the conditions:

(i) $\quad\quad\quad\quad \{y_n\}$ strictly increases to $+\infty$,

(ii) $\quad\quad\quad\quad \lim\limits_{n\to\infty} \dfrac{x_n - x_{n-1}}{y_n - y_{n-1}} = g.$

Then
$$\lim_{n\to\infty} \frac{x_n}{y_n} = g.$$

2.3.12. Calculate

(a) $\quad \lim\limits_{n\to\infty} \dfrac{1}{\sqrt{n}}\left(1 + \dfrac{1}{\sqrt{2}} + \ldots + \dfrac{1}{\sqrt{n}}\right),$

(b) $\quad \lim\limits_{n\to\infty} \dfrac{n}{a^{n+1}}\left(a + \dfrac{a^2}{2} + \ldots + \dfrac{a^n}{n}\right),\ a > 1,$

(c) $\quad \lim\limits_{n\to\infty} \dfrac{1}{n^{k+1}}\left(k! + \dfrac{(k+1)!}{1!} + \ldots + \dfrac{(k+n)!}{n!}\right),\ k \in \mathbb{N},$

(d) $\quad \lim\limits_{n\to\infty} \dfrac{1}{\sqrt{n}}\left(\dfrac{1}{\sqrt{n}} + \dfrac{1}{\sqrt{n+1}} + \ldots + \dfrac{1}{\sqrt{2n}}\right),$

(e) $\quad \lim\limits_{n\to\infty} \dfrac{1^k + 2^k + \ldots + n^k}{n^{k+1}},\quad k \in \mathbb{N},$

(f) $\quad \lim\limits_{n\to\infty} \dfrac{1 + 1\cdot a + 2\cdot a^2 + \ldots + n\cdot a^n}{n\cdot a^{n+1}},\ a > 1,$

(g) $\quad \lim\limits_{n\to\infty} \left[\dfrac{1}{n^k}(1^k + 2^k + \ldots + n^k) - \dfrac{n}{k+1}\right],\ k \in \mathbb{N}.$

2.3.13. Assume that $\lim\limits_{n\to\infty} a_n = a$. Find
$$\lim_{n\to\infty} \frac{1}{\sqrt{n}}\left(a_1 + \frac{a_2}{\sqrt{2}} + \frac{a_3}{\sqrt{3}} + \ldots + \frac{a_n}{\sqrt{n}}\right).$$

2.3.14. Prove that if $\{a_n\}$ is a sequence for which
$$\lim_{n\to\infty} (a_{n+1} - a_n) = a,$$
then
$$\lim_{n\to\infty} \frac{a_n}{n} = a.$$

2.3.15. Let $\lim_{n\to\infty} a_n = a$. Determine
$$\lim_{n\to\infty} \left(\frac{a_n}{1} + \frac{a_{n-1}}{2} + \ldots + \frac{a_1}{2^{n-1}} \right).$$

2.3.16. Suppose that $\lim_{n\to\infty} a_n = a$. Find

(a) $\quad \lim_{n\to\infty} \left(\dfrac{a_n}{1\cdot 2} + \dfrac{a_{n-1}}{2\cdot 3} + \ldots + \dfrac{a_1}{n\cdot(n+1)} \right),$

(b) $\quad \lim_{n\to\infty} \left(\dfrac{a_n}{1} - \dfrac{a_{n-1}}{2^1} + \ldots + (-1)^{n-1}\dfrac{a_1}{2^{n-1}} \right).$

2.3.17. Let k be an arbitrarily fixed integer greater than 1. Calculate
$$\lim_{n\to\infty} \sqrt[n]{\binom{nk}{n}}.$$

2.3.18. For a positive arithmetic progression $\{a_n\}$, find
$$\lim_{n\to\infty} \frac{n(a_1 \cdot \ldots \cdot a_n)^{\frac{1}{n}}}{a_1 + \ldots + a_n}.$$

2.3.19. Suppose that $\{a_n\}$ is such that the sequence $\{b_n\}$ with $b_n = 2a_n + a_{n-1}$, $n \geq 2$, converges to b. Study the convergence of $\{a_n\}$.

2.3.20. Suppose that $\{a_n\}$ is a sequence such that $\lim_{n\to\infty} n^x a_n = a$ for some real x. Prove that
$$\lim_{n\to\infty} n^x (a_1 \cdot a_2 \cdot \ldots \cdot a_n)^{\frac{1}{n}} = ae^x.$$

2.3. Toeplitz Transformation and Stolz Theorem

2.3.21. Calculate

(a) $$\lim_{n\to\infty} \frac{1 + \frac{1}{2} + \ldots + \frac{1}{n-1} + \frac{1}{n}}{\ln n},$$

(b) $$\lim_{n\to\infty} \frac{1 + \frac{1}{3} + \frac{1}{5} + \ldots + \frac{1}{2n-1}}{\ln n}.$$

2.3.22. Assume that $\{a_n\}$ tends to a. Show that

$$\lim_{n\to\infty} \frac{1}{\ln n} \left(\frac{a_1}{1} + \frac{a_2}{2} + \ldots + \frac{a_n}{n} \right) = a.$$

2.3.23. Find

(a) $\lim_{n\to\infty} \left(\frac{n!}{n^n e^{-n}} \right)^{\frac{1}{n}},$

(b) $\lim_{n\to\infty} \left(\frac{(n!)^3}{n^{3n} e^{-n}} \right)^{\frac{1}{n}},$

(c) $\lim_{n\to\infty} \left(\frac{(n!)^2}{n^{2n}} \right)^{\frac{1}{n}},$

(d) $\lim_{n\to\infty} \left(\frac{n^{3n}}{(n!)^3} \right)^{\frac{1}{n}},$

(e) $\lim_{n\to\infty} \frac{\sqrt[k]{n}}{\sqrt[n]{n!}}, \; k \in \mathbb{N}.$

2.3.24. Show that if $\lim_{n\to\infty} a_n = a$, then

$$\lim_{n\to\infty} \frac{1}{\ln n} \sum_{k=1}^{n} \frac{a_k}{k} = a.$$

2.3.25. For a sequence $\{a_n\}$, consider the sequence $\{A_n\}$ of arithmetic means, i.e. $A_n = \frac{a_1 + a_2 + \ldots + a_n}{n}$. Show that if $\lim_{n\to\infty} A_n = A$, then also

$$\lim_{n\to\infty} \frac{1}{\ln n} \sum_{k=1}^{n} \frac{a_k}{k} = A.$$

2.3.26. Prove the converse to the Toeplitz theorem stated in 2.3.1:

Let $\{c_{n,k} : 1 \leq k \leq n,\ n \geq 1\}$ be an array of real numbers. If for any convergent sequence $\{a_n\}$ the transformed sequence $\{b_n\}$ given by setting
$$b_n = \sum_{k=1}^{n} c_{n,k} a_k, \quad n \geq 1$$
is convergent to the same limit, then

(i) $c_{n,k} \underset{n \to \infty}{\longrightarrow} 0$ for each $k \in \mathbb{N}$,

(ii) $\sum_{k=1}^{n} c_{n,k} \underset{n \to \infty}{\longrightarrow} 1$,

(iii) there exists $C > 0$ such that for all positive integers n
$$\sum_{k=1}^{n} |c_{n,k}| \leq C.$$

2.4. Limit Points. Limit Superior and Limit Inferior

2.4.1. Let $\{a_n\}$ be a sequence whose subsequences $\{a_{2k}\}$, $\{a_{2k+1}\}$ and $\{a_{3k}\}$ are convergent.

(a) Prove that the sequence $\{a_n\}$ is convergent.

(b) Does the convergence of any two of these subsequences imply the convergence of the sequence $\{a_n\}$?

2.4.2. Does the convergence of every subsequence of $\{a_n\}$ of the form $\{a_{s \cdot n}\}$, $s > 1$, imply the convergence of the sequence $\{a_n\}$?

2.4.3. Let $\{a_{p_n}\}, \{a_{q_n}\}, \ldots, \{a_{s_n}\}$ be subsequences of $\{a_n\}$ such that the sequences $\{p_n\}, \{q_n\}, \ldots, \{s_n\}$ are pairwise disjoint and form the sequence $\{n\}$. Show that, if $\mathbf{S}, \mathbf{S_p}, \mathbf{S_q}, \ldots, \mathbf{S_s}$ are the sets of all the limit points of the sequences $\{a_n\}, \{a_{p_n}\}, \{a_{q_n}\}, \ldots, \{a_{s_n}\}$, respectively, then
$$\mathbf{S} = \mathbf{S_p} \cup \mathbf{S_q} \cup \ldots \cup \mathbf{S_s}.$$

2.4. Limit Points. Limit Superior and Limit Inferior

Conclude that, if every subsequence $\{a_{p_n}\}, \{a_{q_n}\}, ..., \{a_{s_n}\}$ converges to a, then the sequence $\{a_n\}$ also converges to a.

2.4.4. Is the above theorem (Problem 2.4.3) true in the case of infinitely many subsequences?

2.4.5. Prove that, if every subsequence $\{a_{n_k}\}$ of a sequence $\{a_n\}$ contains a subsequence $\{a_{n_{k_l}}\}$ converging to a, then the sequence $\{a_n\}$ also converges to a.

2.4.6. Determine the set of limit points of the sequence $\{a_n\}$, where

(a) $\quad a_n = \sqrt[n]{4^{(-1)^n} + 2}$,

(b) $\quad a_n = \frac{1}{2}\left(n - 2 - 3\left[\frac{n-1}{3}\right]\right)\left(n - 3 - 3\left[\frac{n-1}{3}\right]\right)$,

(c) $\quad a_n = \frac{(1-(-1)^n)2^n + 1}{2^n + 3}$,

(d) $\quad a_n = \frac{(1+\cos n\pi)\ln 3n + \ln n}{\ln 2n}$,

(e) $\quad a_n = \left(\cos \frac{n\pi}{3}\right)^n$,

(f) $\quad a_n = \frac{2n^2}{7} - \left[\frac{2n^2}{7}\right]$.

2.4.7. Find the set of all the limit points of the sequence $\{a_n\}$ defined by

(a) $\quad a_n = n\alpha - [n\alpha], \quad \alpha \in \mathbb{Q}$,
(b) $\quad a_n = n\alpha - [n\alpha], \quad \alpha \notin \mathbb{Q}$,
(c) $\quad a_n = \sin \pi n \alpha, \quad \alpha \in \mathbb{Q}$,
(d) $\quad a_n = \sin \pi n \alpha, \quad \alpha \notin \mathbb{Q}$.

2.4.8. Let $\{a_k\}$ be a sequence arising by an arbitrary one-to-one indexing of the elements of the matrix $\{\sqrt[3]{n} - \sqrt[3]{m}\}$, $n, m \in \mathbb{N}$. Show that every real number is a limit point of this sequence.

2.4.9. Assume that $\{a_n\}$ is a bounded sequence. Prove that the set of its limit points is closed and bounded.

2.4.10. Determine $\varlimsup\limits_{n\to\infty} a_n$ and $\varliminf\limits_{n\to\infty} a_n$, where

(a) $\quad a_n = \dfrac{2n^2}{7} - \left[\dfrac{2n^2}{7}\right],$

(b) $\quad a_n = \dfrac{n-1}{n+1}\cos\dfrac{n\pi}{3},$

(c) $\quad a_n = (-1)^n n,$

(d) $\quad a_n = n^{(-1)^n n},$

(e) $\quad a_n = 1 + n\sin\dfrac{n\pi}{2},$

(f) $\quad a_n = \left(1 + \dfrac{1}{n}\right)^n (-1)^n + \sin\dfrac{n\pi}{4},$

(g) $\quad a_n = \sqrt[n]{1 + 2^{n(-1)^n}},$

(h) $\quad a_n = \left(2\cos\dfrac{2n\pi}{3}\right)^n,$

(i) $\quad a_n = \dfrac{\ln n - (1 + \cos n\pi)n}{\ln 2n}.$

2.4.11. Find the limit superior and the limit inferior of the following sequences:

(a) $\quad a_n = n\alpha - [n\alpha], \quad \alpha \in \mathbb{Q},$

(b) $\quad a_n = n\alpha - [n\alpha], \quad \alpha \notin \mathbb{Q},$

(c) $\quad a_n = \sin\pi n\alpha, \quad \alpha \in \mathbb{Q},$

(d) $\quad a_n = \sin\pi n\alpha, \quad \alpha \notin \mathbb{Q}.$

2.4.12. For an arbitrary sequence $\{a_n\}$, prove that

(a) if there exists $k \in \mathbb{N}$ such that for any n greater than k the inequality $a_n \leq A$ holds, then $\varlimsup\limits_{n\to\infty} a_n \leq A$,

2.4. Limit Points. Limit Superior and Limit Inferior

(b) if for any $k \in \mathbb{N}$ there exists n_k greater than k such that $a_{n_k} \leq A$, then $\varliminf_{n\to\infty} a_n \leq A$,

(c) if there exists $k \in \mathbb{N}$ such that for any n greater than k the inequality $a_n \geq a$ holds, then $\varlimsup_{n\to\infty} a_n \geq a$,

(d) if for any $k \in \mathbb{N}$ there exists n_k greater than k such that $a_{n_k} \geq a$, then $\varlimsup_{n\to\infty} a_n \geq a$.

2.4.13. Assume that for a sequence $\{a_n\}$ the limit inferior and the limit superior are both finite. Prove that

(a) $L = \varlimsup_{n\to\infty} a_n$ if and only if

(i) for every $\varepsilon > 0$ there exists $k \in \mathbb{N}$ such that $a_n < L + \varepsilon$ if $n > k$

and

(ii) for every $\varepsilon > 0$ and $k \in \mathbb{N}$ there is $n_k > k$ such that $L - \varepsilon < a_{n_k}$.

(b) $l = \varliminf_{n\to\infty} a_n$ if and only if

(i) for every $\varepsilon > 0$ there exists $k \in \mathbb{N}$ such that $l - \varepsilon < a_n$ if $n > k$

and

(ii) for every $\varepsilon > 0$ and $k \in \mathbb{N}$ there is $n_k > k$ such that $a_{n_k} < l + \varepsilon$.

Formulate the corresponding statements for infinite limit inferior and limit superior.

2.4.14. Assume that there is an integer n_0 such that for $n \geq n_0$ the inequality $a_n \leq b_n$ holds. Prove that

(a) $\qquad \varliminf_{n\to\infty} a_n \leq \varliminf_{n\to\infty} b_n,$

(b) $\qquad \varlimsup_{n\to\infty} a_n \leq \varlimsup_{n\to\infty} b_n.$

2.4.15. Prove that (excluding the indeterminate forms of the type $+\infty - \infty$ and $-\infty + \infty$) the following inequalities hold:

$$\varliminf_{n\to\infty} a_n + \varliminf_{n\to\infty} b_n \leq \varliminf_{n\to\infty}(a_n + b_n) \leq \varliminf_{n\to\infty} a_n + \varlimsup_{n\to\infty} b_n$$
$$\leq \varlimsup_{n\to\infty}(a_n + b_n) \leq \varlimsup_{n\to\infty} a_n + \varlimsup_{n\to\infty} b_n.$$

Give examples of sequences for which "$\leq$" in the above inequalities is replaced by "$<$".

2.4.16. Do the inequalities

$$\varliminf_{n\to\infty} a_n + \varliminf_{n\to\infty} b_n \leq \varliminf_{n\to\infty}(a_n + b_n),$$
$$\varlimsup_{n\to\infty}(a_n + b_n) \leq \varlimsup_{n\to\infty} a_n + \varlimsup_{n\to\infty} b_n$$

remain valid in the case of infinitely many sequences?

2.4.17. Let $\{a_n\}$ and $\{b_n\}$ be sequences of nonnegative numbers. Prove that (excluding the indeterminate forms of the type $0 \cdot (+\infty)$ and $(+\infty) \cdot 0$) the following inequalities hold:

$$\varliminf_{n\to\infty} a_n \cdot \varliminf_{n\to\infty} b_n \leq \varliminf_{n\to\infty}(a_n \cdot b_n) \leq \varliminf_{n\to\infty} a_n \cdot \varlimsup_{n\to\infty} b_n$$
$$\leq \varlimsup_{n\to\infty}(a_n \cdot b_n) \leq \varlimsup_{n\to\infty} a_n \cdot \varlimsup_{n\to\infty} b_n.$$

Give examples of sequences for which "$\leq$" in the above inequalities is replaced by "$<$".

2.4.18. Prove that a necessary and sufficient condition for the convergence of a sequence $\{a_n\}$ is that both the limit inferior and the limit superior are finite and

$$\varliminf_{n\to\infty} a_n = \varlimsup_{n\to\infty} a_n.$$

Prove that the analogous theorem is also true for sequences which properly diverge to $-\infty$ or $+\infty$.

2.4. Limit Points. Limit Superior and Limit Inferior

2.4.19. Show that, if $\lim\limits_{n\to\infty} a_n = a$, $a \in \mathbb{R}$, then
$$\varliminf_{n\to\infty} (a_n + b_n) = a + \varliminf_{n\to\infty} b_n,$$
$$\varlimsup_{n\to\infty} (a_n + b_n) = a + \varlimsup_{n\to\infty} b_n.$$

2.4.20. Show that, if $\lim\limits_{n\to\infty} a_n = a$, $a \in \mathbb{R}$, $a > 0$, and there exists a positive integer n_0 such that $b_n \geq 0$ for $n \geq n_0$, then
$$\varliminf_{n\to\infty} (a_n \cdot b_n) = a \cdot \varliminf_{n\to\infty} b_n,$$
$$\varlimsup_{n\to\infty} (a_n \cdot b_n) = a \cdot \varlimsup_{n\to\infty} b_n.$$

2.4.21. Prove that
$$\varliminf_{n\to\infty} (-a_n) = -\varlimsup_{n\to\infty} a_n, \qquad \varlimsup_{n\to\infty} (-a_n) = -\varliminf_{n\to\infty} a_n.$$

2.4.22. Prove that for any positive sequence $\{a_n\}$,
$$\varliminf_{n\to\infty} \frac{1}{a_n} = \frac{1}{\varlimsup\limits_{n\to\infty} a_n},$$
$$\varlimsup_{n\to\infty} \frac{1}{a_n} = \frac{1}{\varliminf\limits_{n\to\infty} a_n}.$$

(Here $\frac{1}{+\infty} = 0$, $\frac{1}{0^+} = +\infty$).

2.4.23. Prove that, if $\{a_n\}$ is a positive sequence such that
$$\varlimsup_{n\to\infty} a_n \cdot \varlimsup_{n\to\infty} \frac{1}{a_n} = 1,$$
then the sequence $\{a_n\}$ is convergent.

2.4.24. Show that, if $\{a_n\}$ is a sequence such that for any sequence $\{b_n\}$,
$$\varliminf_{n\to\infty} (a_n + b_n) = \varliminf_{n\to\infty} a_n + \varliminf_{n\to\infty} b_n$$
or
$$\varlimsup_{n\to\infty} (a_n + b_n) = \varlimsup_{n\to\infty} a_n + \varlimsup_{n\to\infty} b_n,$$
then $\{a_n\}$ is convergent.

2.4.25. Show that, if $\{a_n\}$ is a positive sequence such that for any positive sequence $\{b_n\}$,

$$\lim_{n\to\infty}(a_n\cdot b_n)=\varliminf_{n\to\infty}a_n\cdot\varliminf_{n\to\infty}b_n$$

or

$$\varlimsup_{n\to\infty}(a_n\cdot b_n)=\varlimsup_{n\to\infty}a_n\cdot\varlimsup_{n\to\infty}b_n,$$

then $\{a_n\}$ is convergent.

2.4.26. Prove that for any positive sequence $\{a_n\}$,

$$\varliminf_{n\to\infty}\frac{a_{n+1}}{a_n}\le\varliminf_{n\to\infty}\sqrt[n]{a_n}\le\varlimsup_{n\to\infty}\sqrt[n]{a_n}\le\varlimsup_{n\to\infty}\frac{a_{n+1}}{a_n}.$$

2.4.27. For a given sequence $\{a_n\}$, define $\{b_n\}$ by setting

$$b_n=\frac{1}{n}(a_1+a_2+\ldots+a_n),\quad n\in\mathbb{N}.$$

Prove that

$$\varliminf_{n\to\infty}a_n\le\varliminf_{n\to\infty}b_n\le\varlimsup_{n\to\infty}b_n\le\varlimsup_{n\to\infty}a_n.$$

2.4.28. Prove that

(a) $\quad\varlimsup_{n\to\infty}(\max\{a_n,b_n\})=\max\left\{\varlimsup_{n\to\infty}a_n,\varlimsup_{n\to\infty}b_n\right\},$

(b) $\quad\varliminf_{n\to\infty}(\min\{a_n,b_n\})=\min\left\{\varliminf_{n\to\infty}a_n,\varliminf_{n\to\infty}b_n\right\}.$

Are the equalities

(c) $\quad\varlimsup_{n\to\infty}(\min\{a_n,b_n\})=\min\left\{\varlimsup_{n\to\infty}a_n,\varlimsup_{n\to\infty}b_n\right\},$

(d) $\quad\varliminf_{n\to\infty}(\max\{a_n,b_n\})=\max\left\{\varliminf_{n\to\infty}a_n,\varliminf_{n\to\infty}b_n\right\}$

also true?

2.4.29. Prove that every sequence of real numbers contains a monotonic subsequence.

2.5. Miscellaneous Problems

2.4.30. Use the result in the foregoing exercise to deduce *the Bolzano-Weierstrass theorem:*

Every bounded sequence of real numbers contains a convergent subsequence.

2.4.31. Prove that for every positive sequence $\{a_n\}$,

$$\varlimsup_{n\to\infty} \frac{a_1 + a_2 + \ldots + a_n + a_{n+1}}{a_n} \geq 4.$$

Show that 4 is an optimal estimate.

2.5. Miscellaneous Problems

2.5.1. Show that, if $\lim\limits_{n\to\infty} a_n = +\infty$ or $\lim\limits_{n\to\infty} a_n = -\infty$, then

$$\lim_{n\to\infty} \left(1 + \frac{1}{a_n}\right)^{a_n} = e.$$

2.5.2. For $x \in \mathbb{R}$, show that

$$\lim_{n\to\infty} \left(1 + \frac{x}{n}\right)^n = e^x.$$

2.5.3. For $x > 0$, establish the inequality

$$\frac{x}{x+2} < \ln(x+1) < x.$$

Prove also (applying differentiation) that the left inequality can be strengthened to the following:

$$\frac{x}{x+1} < \frac{2x}{x+2} < \ln(x+1), \quad x > 0.$$

2.5.4. Prove that

(a) $\qquad \lim\limits_{n\to\infty} n(\sqrt[n]{a} - 1) = \ln a, \quad a > 0,$

(b) $\qquad \lim\limits_{n\to\infty} n(\sqrt[n]{n} - 1) = +\infty.$

2.5.5. Let $\{a_n\}$ be a positive sequence with terms different from 1. Show that if $\lim\limits_{n\to\infty} a_n = 1$, then
$$\lim_{n\to\infty} \frac{\ln a_n}{a_n - 1} = 1.$$

2.5.6. Let
$$a_n = 1 + \frac{1}{1!} + \frac{1}{2!} + \ldots + \frac{1}{n!}, \quad n \in \mathbb{N}.$$
Show that
$$\lim_{n\to\infty} a_n = e \quad \text{and} \quad 0 < e - a_n < \frac{1}{nn!}.$$

2.5.7. Prove that
$$\lim_{n\to\infty} \left(1 + \frac{x}{1!} + \frac{x^2}{2!} + \ldots + \frac{x^n}{n!}\right) = e^x.$$

2.5.8. Show that

(a) $\quad \lim\limits_{n\to\infty} \left(\dfrac{1}{n} + \dfrac{1}{n+1} + \ldots + \dfrac{1}{2n}\right) = \ln 2,$

(b)
$$\lim_{n\to\infty} \left(\frac{1}{\sqrt{n(n+1)}} + \frac{1}{\sqrt{(n+1)(n+2)}} + \ldots + \frac{1}{\sqrt{2n(2n+1)}}\right) = \ln 2.$$

2.5.9. Find the limit of the sequence $\{a_n\}$, where
$$a_n = \left(1 + \frac{1}{n^2}\right)\left(1 + \frac{2}{n^2}\right) \cdot \ldots \cdot \left(1 + \frac{n}{n^2}\right), \quad n \in \mathbb{N}.$$

2.5.10. Let $\{a_n\}$ be the recursive sequence defined by
$$a_1 = 1, \quad a_n = n(a_{n-1} + 1) \quad \text{for} \quad n = 2, 3, \ldots.$$
Determine
$$\lim_{n\to\infty} \prod_{k=1}^{n} \left(1 + \frac{1}{a_k}\right).$$

2.5.11. Prove that $\lim\limits_{n\to\infty} (n!e - [n!e]) = 0.$

2.5. Miscellaneous Problems

2.5.12. Given positive a and b, show that
$$\lim_{n\to\infty} \left(\frac{\sqrt[n]{a} + \sqrt[n]{b}}{2}\right)^n = \sqrt{ab}.$$

2.5.13. Let $\{a_n\}$ and $\{b_n\}$ be positive sequences such that
$$\lim_{n\to\infty} a_n^n = a, \quad \lim_{n\to\infty} b_n^n = b, \quad \text{where} \quad a, b > 0,$$
and suppose that positive numbers p and q satisfy $p + q = 1$. Prove that
$$\lim_{n\to\infty} (pa_n + qb_n)^n = a^p b^q.$$

2.5.14. Given two real numbers a and b, define the recursive sequence $\{a_n\}$ as follows:
$$a_1 = a, \quad a_2 = b, \quad a_{n+1} = \frac{n-1}{n}a_n + \frac{1}{n}a_{n-1}, \quad n \geq 2.$$
Find $\lim_{n\to\infty} a_n$.

2.5.15. Let $\{a_n\}$ be the recursive sequence defined by
$$a_1 = 1, \quad a_2 = 2, \quad a_{n+1} = n(a_n + a_{n-1}), \quad n \geq 2.$$
Find an explicit formula for the general term of the sequence.

2.5.16. Given a and b, define $\{a_n\}$ recursively by setting
$$a_1 = a, \quad a_2 = b, \quad a_{n+1} = \frac{1}{2n}a_{n-1} + \frac{2n-1}{2n}a_n, \quad n \geq 2.$$
Determine $\lim_{n\to\infty} a_n$.

2.5.17. Let
$$a_n = 3 - \sum_{k=1}^{n} \frac{1}{k(k+1)(k+1)!}, \quad n \in \mathbb{N}.$$

(a) Show that $\lim_{n\to\infty} a_n = e$.
(b) Show also that $0 < a_n - e < \frac{1}{(n+1)(n+1)!}$.

2.5.18. Calculate $\lim\limits_{n\to\infty} n\sin(2\pi n! e)$.

2.5.19. Suppose that $\{a_n\}$ is a sequence such that $a_n < n$, $n = 1, 2, \ldots$, and $\lim\limits_{n\to\infty} a_n = +\infty$. Study the convergence of the sequence
$$\left(1 - \frac{a_n}{n}\right)^n, \quad n = 1, 2, \ldots.$$

2.5.20. Suppose that a positive sequence $\{b_n\}$ diverges to $+\infty$. Study the convergence of the sequence
$$\left(1 + \frac{b_n}{n}\right)^n, \quad n = 1, 2, \ldots.$$

2.5.21. Given the recursive sequence $\{a_n\}$ defined by setting
$$0 < a_1 < 1, \quad a_{n+1} = a_n(1 - a_n), \quad n \geq 1,$$
prove that

(a) $\quad\lim\limits_{n\to\infty} na_n = 1$,

(b) $\quad\lim\limits_{n\to\infty} \dfrac{n(1 - na_n)}{\ln n} = 1$.

2.5.22. The sequence $\{a_n\}$ is defined inductively as follows:
$$0 < a_1 < \pi, \quad a_{n+1} = \sin a_n, \quad n \geq 1.$$
Prove that $\lim\limits_{n\to\infty} \sqrt{n}\, a_n = \sqrt{3}$.

2.5.23. Let
$$a_1 = 1, \quad a_{n+1} = a_n + \frac{1}{\sum\limits_{k=1}^{n} a_k}, \quad n \geq 1.$$
Prove that
$$\lim\limits_{n\to\infty} \frac{a_n}{\sqrt{2\ln n}} = 1.$$

2.5. Miscellaneous Problems

2.5.24. For $\{a_n\}$ defined inductively by
$$a_1 > 0, \quad a_{n+1} = \arctan a_n, \quad n \geq 1,$$
determine $\lim\limits_{n\to\infty} a_n$.

2.5.25. Prove that the recursive sequence defined by
$$0 < a_1 < 1, \quad a_{n+1} = \cos a_n, \quad n \geq 1,$$
converges to the unique root of the equation $x = \cos x$.

2.5.26. Define the sequence $\{a_n\}$ inductively as follows:
$$a_1 = 0, \quad a_{n+1} = 1 - \sin(a_n - 1), \quad n \geq 1.$$
Find
$$\lim_{n\to\infty} \frac{1}{n}\sum_{k=1}^{n} a_k.$$

2.5.27. Let $\{a_n\}$ be the sequence of consecutive roots of the equation $\tan x = x$, $x > 0$. Find $\lim\limits_{n\to\infty}(a_{n+1} - a_n)$.

2.5.28. For $|a| \leq \frac{\pi}{2}$ and $a_1 \in \mathbb{R}$, consider the recursive sequence defined by
$$a_{n+1} = a \sin a_n, \quad n \geq 1.$$
Study the convergence of the sequence.

2.5.29. Given $a_1 > 0$, consider the sequence $\{a_n\}$ defined by setting
$$a_{n+1} = \ln(1 + a_n), \quad n \geq 1.$$
Prove that

(a) $\quad\lim\limits_{n\to\infty} na_n = 2,$

(b) $\quad\lim\limits_{n\to\infty} \dfrac{n(na_n - 2)}{\ln n} = \dfrac{2}{3}.$

2.5.30. Define the recursive sequence $\{a_n\}$ by putting
$$a_1 = 0 \quad \text{and} \quad a_{n+1} = \left(\frac{1}{4}\right)^{a_n}, \quad n \geq 1.$$
Study the convergence of the sequence.

2.5.31. Given $a_1 > 0$, define the sequence $\{a_n\}$ as follows:
$$a_{n+1} = 2^{1-a_n}, \quad n \geq 1.$$
Study the convergence of the sequence.

2.5.32. Find the limit of the sequence defined by
$$a_1 = \sqrt{2}, \quad a_{n+1} = 2^{\frac{a_n}{2}}, \quad n \geq 1.$$

2.5.33. Prove that if $\lim_{n\to\infty}(a_n - a_{n-2}) = 0$, then
$$\lim_{n\to\infty} \frac{a_n - a_{n-1}}{n} = 0.$$

2.5.34. Show that if for a positive sequence $\{a_n\}$ the limit
$$\lim_{n\to\infty} n\left(1 - \frac{a_{n+1}}{a_n}\right)$$
exists (finite or infinite), then
$$\lim_{n\to\infty} \frac{\ln \frac{1}{a_n}}{\ln n}$$
also exists and both limits are equal.

2.5.35. Given $a_1, b_1 \in (0, 1)$, prove that the sequences $\{a_n\}$ and $\{b_n\}$ defined by
$$a_{n+1} = a_1(1 - a_n - b_n) + a_n, \quad b_{n+1} = b_1(1 - a_n - b_n) + b_n, \quad n \geq 1,$$
converge and find their limits.

2.5.36. For positive a and a_1, consider the sequence $\{a_n\}$ defined by setting
$$a_{n+1} = a_n(2 - aa_n), \quad n = 1, 2, \ldots.$$
Study the convergence of the sequence.

2.5. Miscellaneous Problems

2.5.37. Show that if a_1 and a_2 are positive and
$$a_{n+2} = \sqrt{a_n} + \sqrt{a_{n+1}}, \quad n = 1, 2, ...,$$
then the sequence $\{a_n\}$ converges. Find its limit.

2.5.38. Assume that $f : \mathbb{R}_+^k \to \mathbb{R}_+$ is a function increasing with respect to every variable and there exists $a > 0$ such that
$$f(x, x, ..., x) > x \quad \text{for} \quad 0 < x < a,$$
$$f(x, x, ..., x) < x \quad \text{for} \quad x > a.$$
Given positive $a_1, a_2, ..., a_k$, define the recursive sequence $\{a_n\}$ by
$$a_n = f(a_{n-1}, a_{n-2}, ..., a_{n-k}) \quad \text{for} \quad n > k.$$
Prove that $\lim\limits_{n\to\infty} a_n = a$.

2.5.39. Let a_1 and a_2 be given positive numbers. Study the convergence of the sequence $\{a_n\}$ defined by the recursive relation
$$a_{n+1} = a_n e^{a_n - a_{n-1}} \quad \text{for} \quad n > 1.$$

2.5.40. Given $a > 1$ and $x > 0$, define $\{a_n\}$ by setting $a_1 = a^x$, $a_{n+1} = a^{a_n}$, $n \in \mathbb{N}$. Study the convergence of the sequence.

2.5.41. Show that
$$\underbrace{\sqrt{2 + \sqrt{2 + ... + \sqrt{2}}}}_{n \text{ roots}} = 2\cos\frac{\pi}{2^{n+1}}.$$
Use this relation to find the limit of the recursive sequence given by setting
$$a_1 = \sqrt{2}, \quad a_{n+1} = \sqrt{2 + a_n}, \quad n \geq 1.$$

2.5.42. Let $\{\varepsilon_n\}$ be a sequence whose terms are equal to one of the three values -1, 0, 1. Establish the formula
$$\varepsilon_1\sqrt{2 + \varepsilon_2\sqrt{2 + ... + \varepsilon_n\sqrt{2}}} = 2\sin\left(\frac{\pi}{4}\sum_{k=1}^{n}\frac{\varepsilon_1\varepsilon_2...\varepsilon_k}{2^{k-1}}\right), \quad n \in \mathbb{N},$$

and show that the sequence
$$a_n = \varepsilon_1\sqrt{2+\varepsilon_2\sqrt{2+\ldots+\varepsilon_n\sqrt{2}}}$$
converges.

2.5.43. Calculate
$$\lim_{n\to\infty}\left(\arctan\frac{1}{2}+\arctan\frac{1}{2\cdot 2^2}+\ldots+\arctan\frac{1}{2n^2}\right).$$

2.5.44. Find $\lim_{n\to\infty}\sin(\pi\sqrt{n^2+n})$.

2.5.45. Study the convergence of the recursive sequence defined as follows:
$$a_1=\sqrt{2},\quad a_2=\sqrt{2+\sqrt{3}},\quad a_{n+2}=\sqrt{2+\sqrt{3+a_n}}\quad\text{for}\quad n\geq 1.$$

2.5.46. Show that
$$\lim_{n\to\infty}\sqrt{1+2\sqrt{1+3\sqrt{1+\ldots\sqrt{1+(n-1)\sqrt{1+n}}}}}=3.$$

2.5.47. Given $a>0$, define the recursive sequence $\{a_n\}$ by putting
$$a_1<0,\quad a_{n+1}=\frac{a}{a_n}-1\quad\text{for}\quad n\in\mathbb{N}.$$
Show that the sequence converges to the negative root of the equation $x^2+x=a$.

2.5.48. Given $a>0$, define the recursive sequence $\{a_n\}$ by setting
$$a_1>0,\quad a_{n+1}=\frac{a}{a_n+1}\quad\text{for}\quad n\in\mathbb{N}.$$
Show that the sequence converges to the positive root of the equation $x^2+x=a$.

2.5. Miscellaneous Problems

2.5.49. Let $\{a_n\}$ be the sequence defined by the recursive formula
$$a_1 = 1, \quad a_{n+1} = \frac{2 + a_n}{1 + a_n} \quad \text{for } n \in \mathbb{N}.$$
Show that the sequence is Cauchy and find its limit.

2.5.50. Show that the sequence defined by
$$a_1 > 0, \quad a_{n+1} = 2 + \frac{1}{a_n}, \quad n \in \mathbb{N},$$
is Cauchy and find its limit.

2.5.51. Given $a > 0$, define $\{a_n\}$ as follows:
$$a_1 = 0, \quad a_{n+1} = \frac{a}{2 + a_n} \quad \text{for } n \in \mathbb{N}.$$
Study the convergence of the sequence.

2.5.52. Assume that
$$a_1 \in \mathbb{R} \quad \text{and} \quad a_{n+1} = |a_n - 2^{1-n}| \quad \text{for } n \in \mathbb{N}.$$
Study the convergence of the sequence and in case of convergence find its limit.

2.5.53. Show that

(a) if $0 < a < 1$, then
$$\lim_{n \to \infty} \sum_{j=1}^{n-1} \frac{ja^j}{n-j} = 0,$$

(b) if $0 < a < 1$, then
$$\lim_{n \to \infty} na^n \sum_{j=1}^{n} \frac{1}{ja^j} = \frac{1}{1-a},$$

(c) if $b > 1$, then
$$\lim_{n \to \infty} \frac{n}{b^n} \sum_{j=1}^{n} \frac{b^{j-1}}{j} = \frac{1}{b-1}.$$

2.5.54. Calculate
$$\lim_{n\to\infty} \left(\sin\frac{\pi}{n+1} + \sin\frac{\pi}{n+2} + \ldots + \sin\frac{\pi}{2n}\right).$$

2.5.55. Find

(a) $\quad\displaystyle\lim_{n\to\infty} \prod_{k=1}^{n}\left(1+\frac{k^2}{cn^3}\right), \quad$ where $\quad c > 0$,

(b) $\quad\displaystyle\lim_{n\to\infty} \prod_{k=1}^{n}\left(1-\frac{k^2}{cn^3}\right), \quad$ where $\quad c > 1$.

2.5.56. Determine
$$\lim_{n\to\infty} \frac{\sqrt{n^{3n}}}{n!} \prod_{k=1}^{n} \sin\frac{k}{n\sqrt{n}}.$$

2.5.57. For the sequence $\{a_n\}$ defined by
$$a_n = \sum_{k=0}^{n} \binom{n}{k}^{-1}, \quad n \geq 1,$$
show that $\lim_{n\to\infty} a_n = 2$.

2.5.58. Determine for which values of α the sequence
$$a_n = \left(1-\left(\frac{1}{n}\right)^{\alpha}\right)\left(1-\left(\frac{2}{n}\right)^{\alpha}\right)\cdot\ldots\cdot\left(1-\left(\frac{n-1}{n}\right)^{\alpha}\right), \quad n \geq 2,$$
converges.

2.5.59. For $x \in \mathbb{R}$, define $\{x\} = x - [x]$. Find $\lim_{n\to\infty} \{(2+\sqrt{3})^n\}$.

2.5.60. Let $\{a_n\}$ be a positive sequence and let $S_n = a_1 + a_2 + \ldots + a_n$, $n \geq 1$. Suppose that
$$a_{n+1} \leq \frac{1}{S_{n+1}}\left((S_n - 1)a_n + a_{n-1}\right), \quad n > 1.$$
Determine $\lim_{n\to\infty} a_n$.

2.5. Miscellaneous Problems

2.5.61. Let $\{a_n\}$ be a positive sequence such that

$$\lim_{n\to\infty} \frac{a_n}{n} = 0, \quad \overline{\lim_{n\to\infty}} \frac{a_1 + a_2 + \ldots + a_n}{n} < \infty.$$

Find

$$\lim_{n\to\infty} \frac{a_1^2 + a_2^2 + \ldots + a_n^2}{n^2}.$$

2.5.62. Consider two positive sequences $\{a_n\}$ and $\{b_n\}$ such that

$$\lim_{n\to\infty} \frac{a_n}{a_1 + a_2 + \ldots + a_n} = 0 \quad \text{and} \quad \lim_{n\to\infty} \frac{b_n}{b_1 + b_2 + \ldots + b_n} = 0.$$

Define the sequence $\{c_n\}$ by setting

$$c_n = a_1 b_n + a_2 b_{n-1} + \ldots + a_n b_1, \quad n \in \mathbb{N}.$$

Show that

$$\lim_{n\to\infty} \frac{c_n}{c_1 + c_2 + \ldots + c_n} = 0.$$

2.5.63. Find

$$\lim_{n\to\infty} \left(1 + \frac{1}{n}\right)^{n^2} e^{-n}.$$

2.5.64. Suppose that a sequence $\{a_n\}$ bounded above satisfies the condition

$$a_{n+1} - a_n > -\frac{1}{n^2}, \quad n \in \mathbb{N}.$$

Establish the convergence of $\{a_n\}$.

2.5.65. Suppose that a bounded sequence $\{a_n\}$ satisfies the condition

$$a_{n+1} \sqrt[2^n]{2} \geq a_n, \quad n \in \mathbb{N}.$$

Establish the convergence of $\{a_n\}$.

2.5.66. Let l and L denote the limit inferior and the limit superior of the sequence $\{a_n\}$, respectively. Prove that if $\lim_{n\to\infty}(a_{n+1} - a_n) = 0$, then each number in the open interval (l, L) is a limit point of $\{a_n\}$.

2.5.67. Let l and L denote the limit inferior and the limit superior of the sequence $\{a_n\}$, respectively. Assume that for any n, $a_{n+1} - a_n > -\alpha_n$, where $\alpha_n > 0$ and $\lim\limits_{n\to\infty} \alpha_n = 0$. Prove that each number in the open interval (l, L) is a limit point of $\{a_n\}$.

2.5.68. Let $\{a_n\}$ be a positive and monotonically increasing sequence. Prove that the set of all limit points of the sequence

$$\frac{a_n}{n + a_n}, \quad n \in \mathbb{N},$$

is an interval (which is reduced to a singleton in case of convergence).

2.5.69. Given $a_1 \in \mathbb{R}$, consider the sequence $\{a_n\}$ defined by

$$a_{n+1} = \begin{cases} \dfrac{a_n}{2} & \text{for even } n, \\ \dfrac{1+a_n}{2} & \text{for odd } n. \end{cases}$$

Find the limit points of this sequence.

2.5.70. Is zero a limit point of the sequence $\{\sqrt{n}\sin n\}$?

2.5.71. Prove that for a positive sequence $\{a_n\}$,

$$\varlimsup_{n\to\infty} \left(\frac{a_1 + a_{n+1}}{a_n}\right)^n \geq e.$$

2.5.72. Prove the following generalization of the foregoing result: For a positive integer p and a positive sequence $\{a_n\}$,

$$\varlimsup_{n\to\infty} \left(\frac{a_1 + a_{n+p}}{a_n}\right)^n \geq e^p.$$

2.5.73. Prove that for a positive sequence $\{a_n\}$,

$$\varlimsup_{n\to\infty} n \left(\frac{1 + a_{n+1}}{a_n} - 1\right) \geq 1.$$

Show that 1 is the best possible constant.

2.5. Miscellaneous Problems

2.5.74. Let
$$a_n = \underbrace{\sqrt{1 + \sqrt{1 + \ldots + \sqrt{1}}}}_{n \text{ roots}}.$$
Find $\lim_{n \to \infty} a_n$.

2.5.75. Let $\{a_n\}$ be a sequence whose terms are greater than 1 and suppose that
$$\lim_{n \to \infty} \frac{\ln \ln a_n}{n} = \alpha.$$
Consider the sequence $\{b_n\}$ defined by
$$b_n = \sqrt{a_1 + \sqrt{a_2 + \ldots + \sqrt{a_n}}}, \quad n \in \mathbb{N}.$$
Prove that if $\alpha < \ln 2$, then $\{b_n\}$ converges; and if $\alpha > \ln 2$, then it diverges to $+\infty$.

2.5.76. Assume that the terms of the sequence $\{a_n\}$ satisfy the condition
$$0 \leq a_{n+m} \leq a_n + a_m \quad \text{for} \quad n, m \in \mathbb{N}.$$
Show that the limit $\lim_{n \to \infty} \frac{a_n}{n}$ exists.

2.5.77. Assume that the terms of the sequence $\{a_n\}$ satisfy the condition
$$0 \leq a_{n+m} \leq a_n \cdot a_m \quad \text{for} \quad n, m \in \mathbb{N}.$$
Show that the limit $\lim_{n \to \infty} \sqrt[n]{a_n}$ exists.

2.5.78. Assume that the terms of the sequence $\{a_n\}$ satisfy the conditions
$$|a_n| \leq 1,$$
$$a_n + a_m - 1 \leq a_{n+m} \leq a_n + a_m + 1$$
for $n, m \in \mathbb{N}$.
 (a) Show that the limit $\lim_{n \to \infty} \frac{a_n}{n}$ exists.
 (b) Show that if $\lim_{n \to \infty} \frac{a_n}{n} = g$, then
$$ng - 1 \leq a_n \leq ng + 1 \quad \text{for} \quad n \in \mathbb{N}.$$

2.5.79. Assume that $\{a_n\}$ is a positive and monotonically increasing sequence that satisfies the condition
$$a_{n\cdot m} \geq na_m \quad \text{for} \quad n,m \in \mathbb{N}.$$
Prove that if $\sup\left\{\frac{a_n}{n} : n \in \mathbb{N}\right\} < +\infty$, then the sequence $\left\{\frac{a_n}{n}\right\}$ converges.

2.5.80. Given two positive numbers a_1, a_2, prove that the recursive sequence $\{a_n\}$ defined by
$$a_{n+2} = \frac{2}{a_{n+1} + a_n} \quad \text{for} \quad n \in \mathbb{N}$$
converges.

2.5.81. For $b_1 \geq a_1 > 0$, consider the two sequences $\{a_n\}$ and $\{b_n\}$ defined recursively by setting
$$a_{n+1} = \frac{a_n + b_n}{2}, \quad b_{n+1} = \sqrt{a_{n+1}b_n} \quad \text{for} \quad n \in \mathbb{N}.$$
Show that both sequences converge to the same limit.

2.5.82. Let $a_{k,n}$, $b_{k,n}$, $n \in \mathbb{N}$, $k = 1, 2, ..., n$, be two triangular arrays of real numbers with $b_{k,n} \neq 0$. Suppose that $\frac{a_{k,n}}{b_{k,n}} \xrightarrow[n\to\infty]{} 1$ uniformly with respect to k, that is, for any $\varepsilon > 0$ there is a positive integer n_0 such that
$$\left|\frac{a_{k,n}}{b_{k,n}} - 1\right| < \varepsilon$$
for each $n > n_0$ and $k = 1, 2, ..., n$. Show that if $\lim\limits_{n\to\infty} \sum\limits_{k=1}^{n} b_{k,n}$ exists, then
$$\lim_{n\to\infty} \sum_{k=1}^{n} a_{k,n} = \lim_{n\to\infty} \sum_{k=1}^{n} b_{k,n}.$$

2.5.83. Given $a \neq 0$, find
$$\lim_{n\to\infty} \sum_{k=1}^{n} \sin \frac{(2k-1)a}{n^2}.$$

2.5. Miscellaneous Problems

2.5.84. For $a > 0$, determine
$$\lim_{n\to\infty} \sum_{k=1}^{n} \left(a^{\frac{k}{n^2}} - 1\right).$$

2.5.85. Find
$$\lim_{n\to\infty} \prod_{k=1}^{n} \left(1 + \frac{k}{n^2}\right).$$

2.5.86. For $p \neq 0$ and $q > 0$, determine
$$\lim_{n\to\infty} \sum_{k=1}^{n} \left(\left(1 + \frac{k^{q-1}}{n^q}\right)^{\frac{1}{p}} - 1\right).$$

2.5.87. Given positive numbers a, b and d such that $b > a$, calculate
$$\lim_{n\to\infty} \frac{a(a+d)...(a+nd)}{b(b+d)...(b+nd)}.$$

Chapter 3

Series of Real Numbers

3.1. Summation of Series

3.1.1. Find the infinite series and its sum if the sequence $\{S_n\}$ of its partial sums is given by setting

(a) $S_n = \dfrac{n+1}{n}, \ n \in \mathbb{N}$,
(b) $S_n = \dfrac{2^n - 1}{2^n}, \ n \in \mathbb{N}$,

(c) $S_n = \arctan n, \ n \in \mathbb{N}$,
(d) $S_n = \dfrac{(-1)^n}{n}, \ n \in \mathbb{N}$.

3.1.2. Find the sum of the series

(a) $\displaystyle\sum_{n=1}^{\infty} \dfrac{2n+1}{n^2(n+1)^2}$,
(b) $\displaystyle\sum_{n=1}^{\infty} \dfrac{n}{(2n-1)^2(2n+1)^2}$,

(c) $\displaystyle\sum_{n=1}^{\infty} \dfrac{n - \sqrt{n^2-1}}{\sqrt{n(n+1)}}$,
(d) $\displaystyle\sum_{n=1}^{\infty} \dfrac{1}{4n^2-1}$,

(e) $\displaystyle\sum_{n=1}^{\infty} \dfrac{1}{(\sqrt{n} + \sqrt{n+1})\sqrt{n(n+1)}}$.

3.1.3. Compute the following sums:

(a) $$\ln\frac{1}{4} + \sum_{n=1}^{\infty} \ln\frac{(n+1)(3n+1)}{n(3n+4)},$$

(b) $$\sum_{n=1}^{\infty} \ln\frac{(2n+1)n}{(n+1)(2n-1)}.$$

3.1.4. Find the sum of the series

(a) $$\sum_{n=1}^{\infty} \frac{1}{n(n+1)\ldots(n+m)}, \quad m \in \mathbb{N},$$

(b) $$\sum_{n=1}^{\infty} \frac{1}{n(n+m)}, \quad m \in \mathbb{N},$$

(c) $$\sum_{n=1}^{\infty} \frac{n^2}{(n+1)(n+2)(n+3)(n+4)}.$$

3.1.5. Compute

(a) $$\sum_{n=1}^{\infty} \sin\frac{n!\pi}{720},$$ (b) $$\sum_{n=1}^{\infty} \frac{1}{n}\left[\frac{\ln n}{n - \ln n}\right].$$

3.1.6. Calculate

$$\sum_{n=1}^{\infty} \sin\frac{1}{2^{n+1}} \cos\frac{3}{2^{n+1}}.$$

3.1.7. Find

$$\sum_{n=0}^{\infty} \frac{1}{n!(n^4 + n^2 + 1)}.$$

3.1.8. Show that

$$\sum_{n=1}^{\infty} \frac{n}{3 \cdot 5 \cdot \ldots \cdot (2n+1)} = \frac{1}{2}.$$

3.1. Summation of Series

3.1.9. Suppose that $\{a_n\}$ is a sequence satisfying
$$\lim_{n\to\infty} ((a_1+1)(a_2+1)\ldots(a_n+1)) = g, \quad 0 < g \le +\infty.$$
Prove that
$$\sum_{n=1}^{\infty} \frac{a_n}{(a_1+1)(a_2+1)\ldots(a_n+1)} = 1 - \frac{1}{g},$$
where we assume that $\frac{1}{\infty} = 0$.

3.1.10. Using the result in the foregoing problem, find the sum of the series

(a) $\displaystyle\sum_{n=1}^{\infty} \frac{n-1}{n!},$

(b) $\displaystyle\sum_{n=1}^{\infty} \frac{2n-1}{2\cdot 4\cdot 6\cdot\ldots\cdot 2n},$

(c) $\displaystyle\sum_{n=2}^{\infty} \frac{\frac{1}{n^2}}{\left(1-\frac{1}{2^2}\right)\left(1-\frac{1}{3^2}\right)\ldots\left(1-\frac{1}{n^2}\right)}.$

3.1.11. Let $\{a_n\}$ be a recursive sequence given by setting
$$a_1 > 2, \quad a_{n+1} = a_n^2 - 2 \quad \text{for} \quad n \in \mathbb{N}.$$
Show that
$$\sum_{n=1}^{\infty} \frac{1}{a_1 \cdot a_2 \cdot \ldots \cdot a_n} = \frac{a_1 - \sqrt{a_1^2 - 4}}{2}.$$

3.1.12. For $b > 2$, verify that
$$\sum_{n=1}^{\infty} \frac{n!}{b(b+1)\ldots(b+n-1)} = \frac{1}{b-2}.$$

3.1.13. For $a > 0$ and $b > a+1$, establish the equality
$$\sum_{n=1}^{\infty} \frac{a(a+1)\ldots(a+n-1)}{b(b+1)\ldots(b+n-1)} = \frac{a}{b-a-1}.$$

3.1.14. For $a > 0$ and $b > a + 2$, verify the following claim:
$$\sum_{n=1}^{\infty} n \frac{a(a+1)\ldots(a+n-1)}{b(b+1)\ldots(b+n-1)} = \frac{a(b-1)}{(b-a-1)(b-a-2)}.$$

3.1.15. Let $\sum_{n=1}^{\infty} \frac{1}{a_n}$ be a divergent series with positive terms. For $b > 0$, find the sum
$$\sum_{n=1}^{\infty} \frac{a_1 \cdot a_2 \cdot \ldots \cdot a_n}{(a_2+b)(a_3+b)\ldots(a_{n+1}+b)}.$$

3.1.16. Compute
$$\sum_{n=0}^{\infty} (-1)^n \frac{\cos^3 3^n x}{3^n}.$$

3.1.17. Given nonzero constants a, b and c, suppose that the functions f and g satisfy the condition $f(x) = af(bx) + cg(x)$.
(a) Show that, if $\lim\limits_{n \to \infty} a^n f(b^n x) = L(x)$ exists, then
$$\sum_{n=0}^{\infty} a^n g(b^n x) = \frac{f(x) - L(x)}{c}.$$
(b) Show that, if $\lim\limits_{n \to \infty} a^{-n} f(b^{-n} x) = M(x)$ exists, then
$$\sum_{n=0}^{\infty} a^{-n} g(b^{-n} x) = \frac{M(x) - af(bx)}{c}.$$

3.1.18. Applying the identity $\sin x = 3\sin\frac{x}{3} - 4\sin^3\frac{x}{3}$, show that

(a) $\quad \sum_{n=0}^{\infty} 3^n \sin^3 \frac{x}{3^{n+1}} = \frac{x - \sin x}{4},$

(b) $\quad \sum_{n=0}^{\infty} \frac{1}{3^n} \sin^3 \frac{x}{3^{-n+1}} = \frac{3}{4} \sin \frac{x}{3}.$

3.1. Summation of Series

3.1.19. Applying the identity $\cot x = 2\cot(2x) + \tan x$ for $x \neq k\frac{\pi}{2}$, $k \in \mathbb{Z}$, show that

$$\sum_{n=0}^{\infty} \frac{1}{2^n} \tan \frac{x}{2^n} = \frac{1}{x} - 2\cot(2x).$$

3.1.20. Using the identity $\arctan x = \arctan(bx) + \arctan \frac{(1-b)x}{1+bx^2}$, establish the formulas

(a) $\sum_{n=0}^{\infty} \arctan \frac{(1-b)b^n x}{1+b^{2n+1}x^2} = \arctan x$ for $0 < b < 1$,

(b) $\sum_{n=0}^{\infty} \arctan \frac{(b-1)b^n x}{1+b^{2n+1}x^2} = \operatorname{arccot} x$ for $x \neq 0$ and $b > 1$.

3.1.21. Let $\{a_n\}$ be the Fibonacci sequence defined by setting

$$a_0 = a_1 = 1, \quad a_{n+1} = a_n + a_{n-1}, \quad n \geq 1,$$

and put $S_n = \sum_{k=0}^{n} a_k^2$. Find

$$\sum_{n=0}^{\infty} \frac{(-1)^n}{S_n}.$$

3.1.22. For the Fibonacci sequence $\{a_n\}$ defined in the foregoing problem, calculate

$$\sum_{n=0}^{\infty} \frac{(-1)^n}{a_n a_{n+2}}.$$

3.1.23. For the Fibonacci sequence $\{a_n\}$ defined in 3.1.21, determine the sum of the series

$$\sum_{n=1}^{\infty} \arctan \frac{1}{a_{2n}}.$$

3.1.24. Find the sum:

(a) $\sum_{n=1}^{\infty} \arctan \dfrac{2}{n^2}$,

(b) $\sum_{n=1}^{\infty} \arctan \dfrac{1}{n^2+n+1}$,

(c) $\sum_{n=1}^{\infty} \arctan \dfrac{8n}{n^4 - 2n^2 + 5}$.

3.1.25. Let $\{a_n\}$ be a positive sequence that diverges to infinity. Show that
$$\sum_{n=1}^{\infty} \arctan \dfrac{a_{n+1} - a_n}{1 + a_n a_{n+1}} = \arctan \dfrac{1}{a_1}.$$

3.1.26. Prove that any rearrangement of the terms of an infinite positive series does not change its sum.

3.1.27. Establish the identity
$$\sum_{n=1}^{\infty} \dfrac{1}{(2n-1)^2} = \dfrac{3}{4} \sum_{n=1}^{\infty} \dfrac{1}{n^2}.$$

3.1.28. Prove that

(a) $\sum_{n=1}^{\infty} \dfrac{1}{n^2} = \dfrac{\pi^2}{6}$,

(b) $\sum_{n=1}^{\infty} \dfrac{1}{n^4} = \dfrac{\pi^4}{90}$,

(c) $\sum_{n=0}^{\infty} (-1)^n \dfrac{1}{2n+1} = \dfrac{\pi}{4}$.

3.1.29. For the sequence $\{a_n\}$ defined recursively by
$$a_1 = 2, \quad a_{n+1} = a_n^2 - a_n + 1 \quad \text{for } n \geq 1,$$
find $\sum_{n=1}^{\infty} \dfrac{1}{a_n}$.

3.1. Summation of Series

3.1.30. Let $\{a_n\}$ be defined as follows:
$$a_1 > 0, \quad a_{n+1} = \ln \frac{e^{a_n} - 1}{a_n} \quad \text{for } n \geq 1,$$
and set $b_n = a_1 \cdot a_2 \cdot \ldots \cdot a_n$. Find $\sum_{n=1}^{\infty} b_n$.

3.1.31. Let $\{a_n\}$ be defined by setting
$$a_1 = 1, \quad a_{n+1} = \frac{1}{a_1 + a_2 + \ldots + a_n} - \sqrt{2} \quad \text{for } n \geq 1.$$
Determine the sum of the series $\sum_{n=1}^{\infty} a_n$.

3.1.32. Find the sum of the following series:

(a) $\sum_{n=1}^{\infty} (-1)^{n-1} \frac{1}{n}$,

(b) $\sum_{n=1}^{\infty} (-1)^{n-1} \frac{2n+1}{n(n+1)}$,

(c) $\sum_{n=1}^{\infty} \left(\frac{1}{x+2n-1} + \frac{1}{x+2n} - \frac{1}{x+n} \right), \quad x \neq -1, -2, \ldots.$

3.1.33. Calculate
$$\sum_{n=1}^{\infty} (-1)^{n-1} \ln \left(1 + \frac{1}{n} \right).$$

3.1.34. Compute
$$\sum_{n=1}^{\infty} (-1)^{n-1} \ln \left(1 - \frac{1}{(n+1)^2} \right).$$

3.1.35. Determine the sum of the series
$$\sum_{n=1}^{\infty} \left(\frac{1}{n} - \ln \left(1 + \frac{1}{n} \right) \right).$$

3.1.36. Suppose that a function f is differentiable on $(0,\infty)$, that its derivative f' is monotonic on a subinterval $(a,+\infty)$, and $\lim_{x\to+\infty} f'(x) = 0$. Prove that the limit
$$\lim_{n\to\infty}\left(\frac{1}{2}f(1) + f(2) + f(3) + \ldots + f(n-1) + \frac{1}{2}f(n) - \int_1^n f(x)dx\right)$$
exists. Consider also the special cases of the functions $f(x) = \frac{1}{x}$ and $f(x) = \ln x$.

3.1.37. Determine the sum of the series
$$\sum_{n=1}^{\infty}(-1)^n\frac{\ln n}{n}.$$

3.1.38. Find
$$\sum_{n=1}^{\infty}\left(n\ln\frac{2n+1}{2n-1} - 1\right).$$

3.1.39. Given an integer $k \geq 2$, show that the series
$$\sum_{n=1}^{\infty}\left(\frac{1}{(n-1)k+1} + \frac{1}{(n-1)k+2} + \ldots + \frac{1}{nk-1} - \frac{x}{nk}\right)$$
converges for only one value of x. Find this value and the sum of the series.

3.1.40. For the sequence $\{a_n\}$ defined by setting
$$a_0 = 2, \quad a_{n+1} = a_n + \frac{3+(-1)^n}{2},$$
compute
$$\sum_{n=0}^{\infty}(-1)^{\left[\frac{n+1}{2}\right]}\frac{1}{a_n^2 - 1}.$$

3.1.41. Prove that the sum of the series

(a) $\sum_{n=1}^{\infty}\frac{1}{n!}$, (b) $\sum_{n=1}^{\infty}\frac{1}{(n!)^2}$

is irrational.

3.1. Summation of Series

3.1.42. Let $\{\varepsilon_n\}$ be a sequence where ε_n is either 1 or -1. Show that the sum of the series $\sum\limits_{n=1}^{\infty} \frac{\varepsilon_n}{n!}$ is an irrational number.

3.1.43. Show that for any positive integer k the sum of the series

$$\sum_{n=1}^{\infty} \frac{(-1)^n}{(n!)^k}$$

is irrational.

3.1.44. Suppose that $\{n_k\}$ is a monotonically increasing sequence of positive integers such that

$$\lim_{k\to\infty} \frac{n_k}{n_1 n_2 \cdot \ldots \cdot n_{k-1}} = +\infty.$$

Prove that $\sum\limits_{i=1}^{\infty} \frac{1}{n_i}$ is irrational.

3.1.45. Prove that, if $\{n_k\}$ is a sequence of positive integers such that

$$\varlimsup_{k\to\infty} \frac{n_k}{n_1 n_2 \cdot \ldots \cdot n_{k-1}} = +\infty \quad \text{and} \quad \lim_{k\to\infty} \frac{n_k}{n_{k-1}} > 1,$$

then $\sum\limits_{i=1}^{\infty} \frac{1}{n_i}$ is irrational.

3.1.46. Suppose that $\{n_k\}$ is a monotonically increasing sequence of positive integers for which $\lim\limits_{k\to\infty} \sqrt[2^k]{n_k} = \infty$. Prove that $\sum\limits_{k=1}^{\infty} \frac{1}{n_k}$ is irrational.

3.1.47. Let $\sum\limits_{n=1}^{\infty} \frac{p_n}{q_n}$, where $p_n, q_n \in \mathbb{N}$, be a convergent series and let

$$\frac{p_n}{q_n - 1} - \frac{p_{n+1}}{q_{n+1} - 1} \geq \frac{p_n}{q_n}.$$

Denote by **A** the set of all n for which the above inequality is sharp. Prove that $\sum\limits_{n=1}^{\infty} \frac{p_n}{q_n}$ is irrational if and only if the set **A** is infinite.

3.1.48. Prove that for any strictly increasing sequence $\{n_k\}$ of positive integers the sum of the series $\sum\limits_{k=1}^{\infty} \frac{2^{n_k}}{n_k!}$ is irrational.

3.2. Series of Nonnegative Terms

3.2.1. Determine whether the following series converge or diverge:

(a) $\sum_{n=1}^{\infty} \left(\sqrt{n^2+1} - \sqrt[3]{n^3+1} \right),$
(b) $\sum_{n=1}^{\infty} \left(\frac{n^2+1}{n^2+n+1} \right)^{n^2},$

(c) $\sum_{n=2}^{\infty} \frac{(2n-3)!!}{(2n-2)!!},$
(d) $\sum_{n=1}^{\infty} \left(\frac{n}{n+1} \right)^{n(n+1)},$

(e) $\sum_{n=1}^{\infty} \left(1 - \cos \frac{1}{n} \right),$
(f) $\sum_{n=1}^{\infty} (\sqrt[n]{n} - 1)^n,$

(g) $\sum_{n=1}^{\infty} (\sqrt[n]{a} - 1), \quad a > 1.$

3.2.2. Test the following series for convergence:

(a) $\sum_{n=1}^{\infty} \frac{1}{n} \ln \left(1 + \frac{1}{n} \right),$
(b) $\sum_{n=2}^{\infty} \frac{1}{\sqrt{n}} \ln \frac{n+1}{n-1},$

(c) $\sum_{n=1}^{\infty} \frac{1}{n^2 - \ln n},$
(d) $\sum_{n=2}^{\infty} \frac{1}{(\ln n)^{\ln n}},$

(e) $\sum_{n=2}^{\infty} \frac{1}{(\ln n)^{\ln \ln n}}.$

3.2.3. Let $\sum_{n=1}^{\infty} a_n, \sum_{n=1}^{\infty} b_n$ be series of positive terms satisfying

$$\frac{a_{n+1}}{a_n} \leq \frac{b_{n+1}}{b_n} \quad \text{for} \quad n \geq n_0.$$

Prove that if $\sum_{n=1}^{\infty} b_n$ converges, then $\sum_{n=1}^{\infty} a_n$ also converges.

3.2.4. Test these series for convergence:

(a) $\sum_{n=1}^{\infty} \frac{n^{n-2}}{e^n n!},$
(b) $\sum_{n=1}^{\infty} \frac{n^n}{e^n n!}.$

3.2. Series of Nonnegative Terms

3.2.5. Determine for which values of α the given series converges.

(a) $\displaystyle\sum_{n=1}^{\infty}(\sqrt[n]{a}-1)^{\alpha}$, $a>1$,　　(b) $\displaystyle\sum_{n=1}^{\infty}(\sqrt[n]{n}-1)^{\alpha}$,

(c) $\displaystyle\sum_{n=1}^{\infty}\left(\left(1+\frac{1}{n}\right)^{n+1}-e\right)^{\alpha}$,　　(d) $\displaystyle\sum_{n=1}^{\infty}\left(1-n\sin\frac{1}{n}\right)^{\alpha}$.

3.2.6. Prove that, if a series $\displaystyle\sum_{n=1}^{\infty}a_n$ with positive terms converges, then

$$\sum_{n=1}^{\infty}(a^{a_n}-1), \quad \text{where} \quad a>1,$$

also converges.

3.2.7. Investigate the behavior (convergence or divergence) of the following series:

(a) $\displaystyle\sum_{n=1}^{\infty}-\ln\left(\cos\frac{1}{n}\right)$,　　(b) $\displaystyle\sum_{n=1}^{\infty}e^{\frac{a\ln n+b}{c\ln n+d}}$,　$a,b,c,d\in\mathbb{R}$,

(c) $\displaystyle\sum_{n=1}^{\infty}\frac{n^{2n}}{(n+a)^{n+b}(n+b)^{n+a}}$,　$a,b>0$.

3.2.8. Suppose a series $\displaystyle\sum_{n=1}^{\infty}a_n$ of nonnegative terms converges. Prove that $\displaystyle\sum_{n=1}^{\infty}\sqrt{a_n a_{n+1}}$ also converges. Show that the converse is not true. If, however, the sequence $\{a_n\}$ is monotonically decreasing, then the converse statement does hold.

3.2.9. Assume that a positive-term series $\displaystyle\sum_{n=1}^{\infty}a_n$ diverges. Study the behavior of the following series:

(a) $\displaystyle\sum_{n=1}^{\infty}\frac{a_n}{1+a_n}$,　　(b) $\displaystyle\sum_{n=1}^{\infty}\frac{a_n}{1+na_n}$,

(c) $\displaystyle\sum_{n=1}^{\infty}\frac{a_n}{1+n^2 a_n}$,　　(d) $\displaystyle\sum_{n=1}^{\infty}\frac{a_n}{1+a_n^2}$.

3.2.10. Assume that a positive term series $\sum_{n=1}^{\infty} a_n$ diverges. Denote by $\{S_n\}$ the sequence of its partial sums. Prove that

$$\sum_{n=1}^{\infty} \frac{a_n}{S_n} \quad \text{diverges}$$

and

$$\sum_{n=1}^{\infty} \frac{a_n}{S_n^2} \quad \text{converges.}$$

3.2.11. Show that under the assumptions of the preceding proposition the series

$$\sum_{n=2}^{\infty} \frac{a_n}{S_n S_{n-1}^{\beta}}$$

converges for each $\beta > 0$.

3.2.12. Prove that under the assumptions of 3.2.10 the series

$$\sum_{n=1}^{\infty} \frac{a_n}{S_n^{\alpha}}$$

converges if $\alpha > 1$ and diverges if $\alpha \leq 1$.

3.2.13. Prove that, if a positive term series $\sum_{n=1}^{\infty} a_n$ converges and $r_n = \sum_{k=n+1}^{\infty} a_k$, $n \in \mathbb{N}$, denotes the sequence of its remainders, then

(a) $$\sum_{n=2}^{\infty} \frac{a_n}{r_{n-1}} \quad \text{diverges,}$$

(b) $$\sum_{n=2}^{\infty} \frac{a_n}{\sqrt{r_{n-1}}} \quad \text{converges.}$$

3.2.14. Prove that under the assumptions of the foregoing problem

$$\sum_{n=2}^{\infty} \frac{a_n}{r_{n-1}^{\alpha}}$$

converges if $\alpha < 1$ and diverges if $\alpha \geq 1$.

3.2. Series of Nonnegative Terms 75

3.2.15. Show that under the assumptions of Problem 3.2.13 the series $\sum_{n=1}^{\infty} a_{n+1} \ln^2 r_n$ converges.

3.2.16. Let $\sum_{n=1}^{\infty} a_n$ be a series of positive terms. Suppose that
$$\lim_{n \to \infty} n \ln \frac{a_n}{a_{n+1}} = g.$$
Prove that $\sum_{n=1}^{\infty} a_n$ converges if $g > 1$ and diverges if $g < 1$ (the cases $g = +\infty$ and $g = -\infty$ are included). Show that if $g = 1$, then the test is inconclusive.

3.2.17. Study the behavior of the following series:

(a) $\sum_{n=1}^{\infty} \frac{1}{2^{\sqrt{n}}}$,   (b) $\sum_{n=1}^{\infty} \frac{1}{2^{\ln n}}$,   (c) $\sum_{n=1}^{\infty} \frac{1}{3^{\ln n}}$,

(d) $\sum_{n=1}^{\infty} \frac{1}{a^{\ln n}}$, $a > 0$,   (e) $\sum_{n=2}^{\infty} \frac{1}{a^{\ln \ln n}}$, $a > 0$.

3.2.18. Discuss convergence of the series
$$\sum_{n=1}^{\infty} a^{1 + \frac{1}{2} + \ldots + \frac{1}{n}}, \quad a > 0.$$

3.2.19. Use the result of Problem 3.2.16 to prove the limit form of *the Test of Raabe*.

Let $a_n > 0$, $n \in \mathbb{N}$, and let
$$\lim_{n \to \infty} n \left(\frac{a_n}{a_{n+1}} - 1 \right) = r.$$
Prove that $\sum_{n=1}^{\infty} a_n$ converges if $r > 1$ and diverges if $r < 1$.

3.2.20. Let $\{a_n\}$ be defined recursively by setting
$$a_1 = a_2 = 1, \quad a_{n+1} = a_n + \frac{1}{n^2} a_{n-1} \quad \text{for} \quad n \geq 2.$$
Study convergence of the series $\sum_{n=1}^{\infty} \frac{1}{a_n}$.

3.2.21. Let a_1 and α be positive. Define the recursive sequence $\{a_n\}$ by putting
$$a_{n+1} = a_n e^{-a_n^\alpha} \quad \text{for} \quad n = 1, 2, \ldots.$$
Determine for which values of α and β the series $\sum_{n=1}^{\infty} a_n^\beta$ converges.

3.2.22. Determine for which values of a the series
$$\sum_{n=1}^{\infty} \frac{n!}{(a+1)(a+2) \cdot \ldots \cdot (a+n)}$$
converges.

3.2.23. Let a be an arbitrary positive number and let $\{b_n\}$ be a positive sequence converging to b. Study the convergence of the series
$$\sum_{n=1}^{\infty} \frac{n! a^n}{(a+b_1)(2a+b_2) \cdot \ldots \cdot (na+b_n)}.$$

3.2.24. Prove that, if a sequence $\{a_n\}$ of positive numbers satisfies
$$\frac{a_{n+1}}{a_n} = 1 - \frac{1}{n} - \frac{\gamma_n}{n \ln n},$$
where $\gamma_n \geq \Gamma > 1$, then $\sum_{n=1}^{\infty} a_n$ converges. On the other hand, if
$$\frac{a_{n+1}}{a_n} = 1 - \frac{1}{n} - \frac{\gamma_n}{n \ln n},$$
where $\gamma_n \leq \Gamma < 1$, then $\sum_{n=1}^{\infty} a_n$ diverges. (This is the so-called *Test of Bertrand*.)

3.2.25. Use the tests of Bertrand and Raabe to derive the following *Criterion of Gauss*.

If $\{a_n\}$ is a sequence of positive numbers satisfying
$$\frac{a_{n+1}}{a_n} = 1 - \frac{\alpha}{n} - \frac{\vartheta_n}{n^\lambda},$$
where $\lambda > 1$ and $\{\vartheta_n\}$ is a bounded sequence, then $\sum_{n=1}^{\infty} a_n$ converges when $\alpha > 1$ and diverges when $\alpha \leq 1$.

3.2. Series of Nonnegative Terms

3.2.26. Discuss the convergence of the series

$$\sum_{n=1}^{\infty} \frac{\alpha(\alpha+1)\cdot\ldots\cdot(\alpha+n-1)}{n!} \frac{\beta(\beta+1)\cdot\ldots\cdot(\beta+n-1)}{\gamma(\gamma+1)\cdot\ldots\cdot(\gamma+n-1)},$$

where α, β and γ are positive constants.

3.2.27. Determine for which values of p

$$\sum_{n=1}^{\infty} \left(\frac{(2n-1)!!}{(2n)!!} \right)^p$$

converges.

3.2.28. Prove the following *condensation test of Cauchy*.

Let $\{a_n\}$ be a monotonically decreasing sequence of nonnegative numbers. Prove that the series $\sum_{n=1}^{\infty} a_n$ converges if and only if the series $\sum_{n=1}^{\infty} 2^n a_{2^n}$ converges.

3.2.29. Test the following series for convergence:

(a) $\quad \sum_{n=2}^{\infty} \frac{1}{n(\ln n)^\alpha},\quad$ (b) $\quad \sum_{n=3}^{\infty} \frac{1}{n\cdot \ln n \cdot \ln\ln n}.$

3.2.30. Prove the following *Theorem of Schlömilch* (a generalization of the Cauchy theorem, see Problem 3.2.28).

If $\{g_k\}$ is a strictly increasing sequence of positive integers such that for some $c > 0$ and for all $k \in \mathbb{N}$, $g_{k+1} - g_k \leq c(g_k - g_{k-1})$ and if a positive sequence $\{a_n\}$ strictly decreases, then

$$\sum_{n=1}^{\infty} a_n < \infty \quad \text{if and only if} \quad \sum_{n=1}^{\infty} (g_{k+1}-g_k)a_{g_k} < \infty.$$

3.2.31. Let $\{a_n\}$ be a monotonically decreasing sequence of positive numbers. Prove that the series $\sum_{n=1}^{\infty} a_n$ converges if and only if the following series converges:

(a) $\sum_{n=1}^{\infty} 3^n a_{3^n}$, (b) $\sum_{n=1}^{\infty} na_{n^2}$, (c) $\sum_{n=1}^{\infty} n^2 a_{n^3}$.

(d) Apply the above tests to study convergence of the series in Problem 3.2.17.

3.2.32. Suppose that $\{a_n\}$ is a positive sequence. Prove that
$$\overline{\lim_{n\to\infty}} (a_n)^{\frac{1}{\ln n}} < \frac{1}{e}$$
implies the convergence of $\sum_{n=1}^{\infty} a_n$.

3.2.33. Suppose that $\{a_n\}$ is a positive sequence. Show that
$$\overline{\lim_{n\to\infty}} (na_n)^{\frac{1}{\ln \ln n}} < \frac{1}{e}$$
implies the convergence of $\sum_{n=1}^{\infty} a_n$.

3.2.34. Let $\{a_n\}$ be a monotonically decreasing sequence of positive numbers such that
$$\frac{2^n a_{2^n}}{a_n} \leq g < 1.$$
Show that $\sum_{n=1}^{\infty} a_n$ converges.

3.2.35. Let $\{a_n\}$ be a monotonically decreasing sequence of nonnegative numbers. Prove that if $\sum_{n=1}^{\infty} a_n$ converges, then $\lim_{n\to\infty} na_n = 0$. Show that this condition is not sufficient for the convergence of the series.

3.2.36. Give an example of a convergent series with positive terms for which the condition $\lim_{n\to\infty} na_n = 0$ does not hold.

3.2.37. Let $\sum_{n=1}^{\infty} a_n$ be a convergent positive series. Give necessary and sufficient conditions for the existence of a positive sequence $\{b_n\}$

3.2. Series of Nonnegative Terms

such that the series

$$\sum_{n=1}^{\infty} b_n \quad \text{and} \quad \sum_{n=1}^{\infty} \frac{a_n}{b_n}$$

both converge.

3.2.38. Does there exist a positive sequence $\{a_n\}$ such that the series

$$\sum_{n=1}^{\infty} a_n \quad \text{and} \quad \sum_{n=1}^{\infty} \frac{1}{n^2 a_n}$$

both converge.

3.2.39. Show that

$$\sum_{n=1}^{\infty} \frac{1}{n} \cdot \frac{1 + a_{n+1}}{a_n}$$

diverges for any positive sequence $\{a_n\}$.

3.2.40. Let $\{a_n\}$ and $\{b_n\}$ be monotonically decreasing to zero and such that the series $\sum_{n=1}^{\infty} a_n$ and $\sum_{n=1}^{\infty} b_n$ diverge. What can be said about the convergence of $\sum_{n=1}^{\infty} c_n$, where $c_n = \min\{a_n, b_n\}$?

3.2.41. Let $\{a_n\}$ be a monotonically decreasing sequence of nonnegative numbers such that $\sum_{n=1}^{\infty} \frac{a_n}{n}$ diverges. Assume that

$$b_n = \min\left\{a_n, \frac{1}{\ln(n+1)}\right\}.$$

Prove that $\sum_{n=1}^{\infty} \frac{b_n}{n}$ also diverges.

3.2.42. Let $\{a_n\}$ be a bounded, positive and monotonically increasing sequence. Prove that

$$\sum_{n=1}^{\infty} \left(1 - \frac{a_n}{a_{n+1}}\right) \quad \text{converges.}$$

3.2.43. Let $\{a_n\}$ be an increasing positive sequence diverging to infinity. Prove that
$$\sum_{n=1}^{\infty}\left(1-\frac{a_n}{a_{n+1}}\right) \quad \text{diverges.}$$

3.2.44. Let $\{a_n\}$ be a positive monotonically increasing sequence. Show that for any $\alpha > 0$
$$\sum_{n=1}^{\infty}\frac{a_{n+1}-a_n}{a_{n+1}a_n^{\alpha}} \quad \text{converges.}$$

3.2.45. Show that for any positive and divergent series $\sum_{n=1}^{\infty} a_n$ there exists a sequence $\{c_n\}$ monotonically decreasing to zero such that $\sum_{n=1}^{\infty} a_n c_n$ diverges.

3.2.46. Show that for any positive and convergent series $\sum_{n=1}^{\infty} a_n$ there exists a sequence $\{c_n\}$ monotonically increasing to infinity such that $\sum_{n=1}^{\infty} a_n c_n$ converges.

3.2.47. Let $\sum_{n=1}^{\infty} a_n$ be a convergent series with positive terms. Let $\{r_n\}$ denote the sequence of its remainders. Prove that if $\sum_{n=1}^{\infty} r_n$ converges, then $\lim_{n\to\infty} n a_n = 0$.

3.2.48. Let $\{a_n\}$ be a positive sequence diverging to infinity. What can be said about the convergence of the following series:

(a) $\sum_{n=1}^{\infty}\frac{1}{a_n^n}$, (b) $\sum_{n=1}^{\infty}\frac{1}{a_n^{\ln n}}$, (c) $\sum_{n=1}^{\infty}\frac{1}{a_n^{\ln \ln n}}$?

3.2.49. Study convergence of $\sum_{n=1}^{\infty} a_n$, where
$$a_1 = 1, \quad a_{n+1} = \cos a_n \quad \text{for} \quad n \in \mathbb{N}.$$

3.2. Series of Nonnegative Terms

3.2.50. Let p be an arbitrarily fixed nonnegative number. Study the convergence of $\sum_{n=1}^{\infty} a_n$, where

$$a_1 = 1, \quad a_{n+1} = n^{-p}\sin a_n \quad \text{for} \quad n \in \mathbb{N}.$$

3.2.51. Let $\{a_n\}$ be a sequence of consecutive positive solutions of the equation $\tan x = x$. Study the convergence of $\sum_{n=1}^{\infty} \frac{1}{a_n^2}$.

3.2.52. Let $\{a_n\}$ be a sequence of consecutive positive solutions of the equation $\tan \sqrt{x} = x$. Study the convergence of $\sum_{n=1}^{\infty} \frac{1}{a_n}$.

3.2.53. Let a_1 be an arbitrary positive number and let $a_{n+1} = \ln(1 + a_n)$ for $n \geq 1$. Study the convergence of $\sum_{n=1}^{\infty} a_n$.

3.2.54. Assume that $\{a_n\}$ is a positive monotonically decreasing sequence such that $\sum_{n=1}^{\infty} a_n$ diverges. Show that

$$\lim_{n \to \infty} \frac{a_1 + a_3 + \ldots + a_{2n-1}}{a_2 + a_4 + \ldots + a_{2n}} = 1.$$

3.2.55. Let $S_k = 1 + \frac{1}{2} + \ldots + \frac{1}{k}$ and let k_n denote the least of all positive integers k for which $S_k \geq n$. Find

$$\lim_{n \to \infty} \frac{k_{n+1}}{k_n}.$$

3.2.56. Let $\mathbf{A}$ be the set of all positive integers such that their decimal representations do not contain zero.
(a) Show that $\sum_{n \in \mathbf{A}} \frac{1}{n}$ converges.
(b) Determine all α such that $\sum_{n \in \mathbf{A}} \frac{1}{n^\alpha}$ converges.

3.2.57. Let $\sum_{n=1}^{\infty} a_n$ be a series of positive terms and let

$$\lim_{n \to \infty} \frac{\ln \frac{1}{a_n}}{\ln n} = g.$$

Prove that if $g > 1$, then the series converges, and if $g < 1$, then the series diverges (here g may be equal to $+\infty$ or $-\infty$.)

Give examples showing that in the case of $g = 1$ the criterion is indecisive.

3.2.58. Show that the Raabe test (see Problem 3.2.19) and the test given in Problem 3.2.16 are equivalent. Moreover, show that the criterion in the preceding problem is stronger than each of the above mentioned tests.

3.2.59. Study the convergence of $\sum_{n=1}^{\infty} a_n$ whose terms are given by

$$a_1 = \sqrt{2}, \qquad a_n = \sqrt{2 - \underbrace{\sqrt{2 + \sqrt{2 + ... + \sqrt{2}}}}_{(n-1)-\text{roots}}}, \qquad n \geq 2.$$

3.2.60. Let $\{a_n\}$ be a sequence monotonically decreasing to zero. Show that if the sequence with terms

$$(a_1 - a_n) + (a_2 - a_n) + ... + (a_{n-1} - a_n)$$

is bounded, then $\sum_{n=1}^{\infty} a_n$ must converge.

3.2.61. Find a series whose terms a_n satisfy the following conditions:

$$a_1 = \frac{1}{2}, \qquad a_n = a_{n+1} + a_{n+2} + ... \qquad \text{for} \quad n = 1, 2, 3, ...$$

3.2.62. Suppose that the terms of a convergent series $\sum_{n=1}^{\infty} a_n$ whose sum is S satisfy two conditions:

$$a_1 \geq a_2 \geq a_3 \geq ... \quad \text{and} \quad 0 < a_n \leq a_{n+1} + a_{n+2} + ..., \qquad n \in \mathbb{N}.$$

Show that it is possible to represent any number s in the half-closed interval $(0, S]$ by a finite sum of terms of the series $\sum_{n=1}^{\infty} a_n$ or by an infinite subseries $\sum_{k=1}^{\infty} a_{n_k}$, where $\{a_{n_k}\}$ is a subsequence of $\{a_n\}$.

3.2. Series of Nonnegative Terms

3.2.63. Assume that $\sum_{n=1}^{\infty} a_n$ is a series of positive and monotonically decreasing terms. Prove that if each number in $(0, S)$, where S denotes the sum of the series, can be represented by a finite sum of terms of $\{a_n\}$ or by an infinite subseries $\sum_{k=1}^{\infty} a_{n_k}$, where $\{a_{n_k}\}$ is a subsequence of $\{a_n\}$, then the following inequality holds:

$$a_n \leq a_{n+1} + a_{n+2} + \ldots \quad \text{for each} \quad n \in \mathbb{N}.$$

3.2.64. Let $\sum_{n=1}^{\infty} a_n$ be a divergent series of positive terms and let $\lim_{n \to \infty} \frac{a_n}{S_n} = 0$, where $S_n = a_1 + a_2 + \ldots + a_n$. Prove that

$$\lim_{n \to \infty} \frac{a_1 S_1^{-1} + a_2 S_2^{-1} + \ldots + a_n S_n^{-1}}{\ln S_n} = 1.$$

3.2.65. Using the preceding problem, show that

$$\lim_{n \to \infty} \frac{1 + \frac{1}{2} + \frac{1}{3} + \ldots + \frac{1}{n}}{\ln n} = 1.$$

3.2.66. Let $\sum_{n=1}^{\infty} a_n$ be a convergent series of positive terms. What can be said about the convergence of

$$\sum_{n=1}^{\infty} \frac{a_1 + a_2 + \ldots + a_n}{n}?$$

3.2.67. Prove that if $\{a_n\}$ is a positive sequence such that $\frac{1}{n} \sum_{k=1}^{n} a_k \geq \sum_{k=n+1}^{2n} a_k$ for $n \in \mathbb{N}$, then $\sum_{n=1}^{\infty} a_n \leq 2ea_1$.

3.2.68. Prove the following *Inequality of Carleman*:
If $\{a_n\}$ is a positive sequence, then

$$\sum_{n=1}^{\infty} \sqrt[n]{a_1 \cdot \ldots \cdot a_n} < e \sum_{n=1}^{\infty} a_n,$$

provided that $\sum_{n=1}^{\infty} a_n$ converges.

3.2.69. Show that if $\{a_n\}$ is a positive sequence, then, for every positive integer k,
$$\sum_{n=1}^{\infty} \sqrt[n]{a_1 \cdot \ldots \cdot a_n} \leq \frac{1}{k} \sum_{n=1}^{\infty} a_n \left(\frac{n+k}{n}\right)^n.$$

3.2.70. Let $\{a_n\}$ be a sequence of positive numbers. Prove that the convergence of $\sum_{n=1}^{\infty} \frac{1}{a_n}$ implies the convergence of
$$\sum_{n=1}^{\infty} \left(n^2 a_n \left(\sum_{k=1}^{n} a_k\right)^{-2}\right).$$

3.2.71. Let $\{a_n\}$ be a monotonically increasing sequence of positive numbers such that $\sum_{n=1}^{\infty} \frac{1}{a_n}$ diverges. Show that
$$\sum_{n=2}^{\infty} \frac{1}{na_n - (n-1)a_{n-1}}$$
is also divergent.

3.2.72. Let $\{p_n\}$ be a sequence of all consecutive prime numbers. Study convergence of $\sum_{n=1}^{\infty} \frac{1}{p_n}$.

3.2.73. Study convergence of
$$\sum_{n=2}^{\infty} \frac{1}{np_n - (n-1)p_{n-1}},$$
where p_n denotes the nth prime number.

3.2.74. Evaluate
$$\lim_{n \to \infty} \frac{\sum_{k=2}^{\infty} \frac{1}{k^{n+1}}}{\sum_{k=2}^{\infty} \frac{1}{k^n}}.$$

3.2. Series of Nonnegative Terms

3.2.75. Let $\{a_n\}$ be a sequence satisfying the following conditions:
$$0 \leq a_n \leq 1 \quad \text{for all} \quad n \in \mathbb{N} \quad \text{and} \quad a_1 \neq 0.$$
Let
$$S_n = a_1 + \ldots + a_n \quad \text{and} \quad T_n = S_1 + \ldots + S_n.$$
Determine for which values $\alpha > 0$ the series $\sum\limits_{n=1}^{\infty} \frac{a_n}{T_n^\alpha}$ converges.

3.2.76. Let k be an arbitrary positive integer. Assume that $\{a_n\}$ is a monotonically increasing sequence of positive numbers such that $\sum\limits_{n=1}^{\infty} \frac{1}{a_n}$ converges. Prove that the series
$$\sum_{n=1}^{\infty} \frac{\ln^k a_n}{a_n} \quad \text{and} \quad \sum_{n=1}^{\infty} \frac{\ln^k n}{a_n}$$
are either both convergent or both divergent.

3.2.77. Assume that $f : \mathbb{N} \to (0, \infty)$ is a decreasing function and $\varphi : \mathbb{N} \to \mathbb{N}$ is an increasing function such that $\varphi(n) > n$ for all $n \in \mathbb{N}$. Verify the following inequalities:

(1) $\sum\limits_{k=1}^{\varphi(n)-1} f(k) < \sum\limits_{k=1}^{\varphi(1)-1} f(k) + \sum\limits_{k=1}^{n-1} f(\varphi(k))(\varphi(k+1) - \varphi(k)),$

(2) $\sum\limits_{k=\varphi(1)-1}^{\varphi(n)} f(k) > \sum\limits_{k=2}^{n} f(\varphi(k))(\varphi(k) - \varphi(k-1)).$

3.2.78. Prove that under the assumptions of the foregoing problem, if there exists q such that for all $n \in \mathbb{N}$ the inequality
$$\frac{f(\varphi(n))(\varphi(n+1) - \varphi(n))}{f(n)} \leq q < 1$$
holds, then $\sum\limits_{n=1}^{\infty} f(n)$ converges. On the other hand, if
$$\frac{f(\varphi(n))(\varphi(n) - \varphi(n-1))}{f(n)} \geq 1, \quad n \in \mathbb{N},$$

then $\sum_{n=1}^{\infty} f(n)$ diverges.

3.2.79. Derive from the preceding problem the following test for convergence and divergence of positive series.

The series $\sum_{n=1}^{\infty} a_n$ whose terms are positive and monotonically decreasing is convergent when

$$\lim_{n \to \infty} \frac{a_{2n}}{a_n} = g < \frac{1}{2}$$

and divergent when

$$\lim_{n \to \infty} \frac{a_{2n}}{a_n} = g > \frac{1}{2}.$$

3.2.80. Derive from Problem 3.2.78 the following test for convergence and divergence of positive series (compare with Problem 3.2.34).

A positive series $\sum_{n=1}^{\infty} a_n$ whose terms are monotonically decreasing is convergent provided that

$$\overline{\lim_{n \to \infty}} \frac{2^n a_{2^n}}{a_n} = g < 1$$

and divergent provided that

$$\lim_{n \to \infty} \frac{2^n a_{2^n}}{a_n} > 2.$$

3.2.81. Using Problem 3.2.77, prove the criteria given in 3.2.31.

3.2.82. Prove the following *Test of Kummer*.

Let $\{a_n\}$ be a positive-valued sequence.

(1) If there are a sequence $\{b_n\}$ of positive numbers and a positive constant c such that

$$b_n \frac{a_n}{a_{n+1}} - b_{n+1} \geq c \quad \text{for all} \quad n \in \mathbb{N},$$

then $\sum_{n=1}^{\infty} a_n$ converges.

3.2. Series of Nonnegative Terms

(2) If there is a positive sequence $\{b_n\}$ such that $\sum_{n=1}^{\infty} \frac{1}{b_n}$ diverges and
$$b_n \frac{a_n}{a_{n+1}} - b_{n+1} \leq 0 \quad \text{for all} \quad n \in \mathbb{N},$$
then $\sum_{n=1}^{\infty} a_n$ diverges.

3.2.83. Show that the tests of d'Alembert (the ratio test), Raabe (3.2.19) and Bertrand (3.2.24) are special cases of the Kummer test (3.2.82).

3.2.84. Prove the following converse of the Kummer test.

Let $\{a_n\}$ be a positive sequence.

(1) If $\sum_{n=1}^{\infty} a_n$ converges, then there exist a positive sequence $\{b_n\}$ and a positive constant c such that
$$b_n \frac{a_n}{a_{n+1}} - b_{n+1} \geq c.$$

(2) If $\sum_{n=1}^{\infty} a_n$ diverges, then there exists a positive sequence $\{b_n\}$ such that $\sum_{n=1}^{\infty} \frac{1}{b_n}$ diverges and
$$b_n \frac{a_n}{a_{n+1}} - b_{n+1} \leq 0.$$

3.2.85. Prove the following tests for convergence and divergence of positive series.

(a) Let k be a positive integer and let $\lim_{n \to \infty} \frac{a_{n+k}}{a_n} = g$. If $g < 1$, then $\sum_{n=1}^{\infty} a_n$ converges, and if $g > 1$, then $\sum_{n=1}^{\infty} a_n$ diverges.

(b) Let k be a positive integer and let $\lim_{n \to \infty} n \left(\frac{a_n}{a_{n+k}} - 1 \right) = g$. If $g > k$, then $\sum_{n=1}^{\infty} a_n$ converges, and if $g < k$, then $\sum_{n=1}^{\infty} a_n$ diverges.

3.2.86. Let $\{a_n\}$ and $\{\varphi_n\}$ be sequences of positive numbers. Assume that $\varphi_n = O\left(\frac{1}{\ln n}\right)$. Prove that the convergence of $\sum_{n=2}^{\infty} a_n$ implies the convergence of $\sum_{n=2}^{\infty} a_n^{1-\varphi_n}$.

3.3. The Integral Test

3.3.1. Prove the following *integral test*.

Assume that f is a positive and decreasing function on the interval $[1, \infty)$. Then the series $\sum_{n=1}^{\infty} f(n)$ converges if and only if the sequence $\{I_n\}$, $I_n = \int_1^n f(x)dx$, is bounded.

3.3.2. Let f be a positive and differentiable function on $(0, \infty)$ such that f' decreases to zero. Show that the series

$$\sum_{n=1}^{\infty} f'(n) \quad \text{and} \quad \sum_{n=1}^{\infty} \frac{f'(n)}{f(n)}$$

either both converge or both diverge.

3.3.3. Let f be a positive and decreasing function on $[1, \infty)$. Set

$$S_N = \sum_{n=1}^{N} f(n) \quad \text{and} \quad I_N = \int_1^N f(x)dx.$$

Show that the sequence $\{S_N - I_N\}$ is monotonically decreasing and its limit belongs to the interval $[0, f(1)]$.

3.3.4. Show that the limits of the sequences

(a) $\quad 1 + \dfrac{1}{2} + \ldots + \dfrac{1}{n} - \ln n,$

(b) $\quad 1 + \dfrac{1}{2^\alpha} + \ldots + \dfrac{1}{n^\alpha} - \int_1^n \dfrac{1}{x^\alpha}dx, \quad 0 < \alpha < 1,$

both belong to the interval $(0, 1)$.

3.3. The Integral Test

3.3.5. Apply the integral test to study convergence of the series given in 3.2.29.

3.3.6. Let $\sum_{n=1}^{\infty} a_n$ be a positive divergent series and let $S_n = a_1 + a_2 + \ldots + a_n > 1$ for $n \geq 1$. Verify the following claims:

(a) $\displaystyle\sum_{n=1}^{\infty} \frac{a_{n+1}}{S_n \ln S_n}$ diverges,

(b) $\displaystyle\sum_{n=1}^{\infty} \frac{a_n}{S_n \ln^2 S_n}$ converges.

3.3.7. Let f be a positive and decreasing function on $[1,\infty)$. Assume that a function φ is strictly increasing, differentiable and such that $\varphi(x) > x$ for $x > 1$. Prove that, if there exists $q < 1$ such that $\frac{\varphi'(x)f(\varphi(x))}{f(x)} \leq q$ for sufficiently large x, then $\sum_{n=1}^{\infty} f(n)$ converges. Prove also that, if $\frac{\varphi'(x)f(\varphi(x))}{f(x)} \geq 1$ for sufficiently large x, then the series diverges.

3.3.8. Let f,g be positive continuously differentiable functions on $(0,\infty)$. Moreover, suppose that f is decreasing.

(a) Show that, if $\lim\limits_{x \to \infty} \left(-g(x)\frac{f'(x)}{f(x)} - g'(x)\right) > 0$, then $\sum_{n=1}^{\infty} f(n)$ converges.

(b) Show that, if the sequence with terms $\int_1^n \frac{1}{g(x)}dx$ is unbounded and for sufficiently large x, $-g(x)\frac{f'(x)}{f(x)} - g'(x) \leq 0$, then $\sum_{n=1}^{\infty} f(n)$ diverges.

3.3.9. Let f be a positive continuously differentiable function on $(0,\infty)$. Prove that

(a) if $\lim\limits_{x \to \infty} \left(-\frac{xf'(x)}{f(x)}\right) > 1$, then $\sum_{n=1}^{\infty} f(n)$ converges,

(b) if $-\frac{xf'(x)}{f(x)} \leq 1$ for sufficiently large x, then $\sum_{n=1}^{\infty} f(n)$ diverges.

3.3.10. Let f be a positive continuously differentiable function on $(0, \infty)$. Prove that

(a) if $\lim\limits_{x \to \infty} \left(-\frac{f'(x)}{f(x)} - \frac{1}{x} \right) x \ln x > 1$, then $\sum\limits_{n=1}^{\infty} f(n)$ converges,

(b) if $\left(-\frac{f'(x)}{f(x)} - \frac{1}{x} \right) x \ln x \leq 1$ for sufficiently large x, then $\sum\limits_{n=1}^{\infty} f(n)$ diverges.

3.3.11. Prove the following converse of the theorem stated in 3.3.8.

Let f be a positive decreasing and continuously differentiable function on $(0, \infty)$.

(a) If $\sum\limits_{n=1}^{\infty} f(n)$ converges, then there exists a positive continuously differentiable function g on $(0, \infty)$ such that

$$\lim_{x \to \infty} \left(-g(x) \frac{f'(x)}{f(x)} - g'(x) \right) > 0.$$

(b) If $\sum\limits_{n=1}^{\infty} f(n)$ diverges, then there exists a positive continuously differentiable function g on $(0, \infty)$ such that the sequence with terms

$$\int_1^n \frac{1}{g(x)} dx, \quad n = 1, 2, \ldots,$$

is unbounded and for sufficiently large x,

$$-g(x) \frac{f'(x)}{f(x)} - g'(x) \leq 0.$$

3.3.12. For $\gamma \geq 0$, study convergence of the series

$$\sum_{n=2}^{\infty} \frac{1}{(\ln n)^{(\ln n)^\gamma}}.$$

3.3.13. Study convergence of the series

$$\sum_{n=3}^{\infty} \frac{1}{n^{1+\frac{1}{\ln \ln n}} \ln n}.$$

3.3. The Integral Test

3.3.14. Let $\{\lambda_n\}$ be a positive monotonically increasing sequence and let f be a positive and increasing function satisfying the condition
$$\int_{\lambda_1}^\infty \frac{1}{tf(t)}dt < \infty.$$
Show that
$$\sum_{n=1}^\infty \left(1 - \frac{\lambda_n}{\lambda_{n+1}}\right)\frac{1}{f(\lambda_n)} < \infty.$$

3.3.15. Prove the following generalization of the integral test.

Let $\{\lambda_n\}$ be a sequence strictly increasing to infinity and let f be a positive continuous and decreasing function on $[\lambda_1, \infty)$.

(a) If there exists $M > 0$ such that $\lambda_{n+1} - \lambda_n \geq M$ for $n \in \mathbb{N}$ and if the improper integral $\int_{\lambda_1}^\infty f(t)dt$ converges, then the series $\sum_{n=1}^\infty f(\lambda_n)$ also converges.

(b) If there exists $M > 0$ such that $\lambda_{n+1} - \lambda_n \leq M$ for $n \in \mathbb{N}$ and if the improper integral $\int_{\lambda_1}^\infty f(t)dt$ diverges, then the series $\sum_{n=1}^\infty f(\lambda_n)$ also diverges.

3.3.16. Suppose that $f : (0, \infty) \to \mathbb{R}$ is a positive and differentiable function with positive derivative. Prove that $\sum_{n=1}^\infty \frac{1}{f(n)}$ converges if and only if $\sum_{n=1}^\infty \frac{f^{-1}(n)}{n^2}$ does.

3.3.17. Define $\ln_1 x = \ln x$ and $\ln_k x = \ln(\ln_{k-1} x)$ for $k > 1$ and sufficiently large x. For $n \in \mathbb{N}$, let $\varphi(n)$ be the unique positive integer such that $1 \leq \ln_{\varphi(n)} n < e$. Does the series
$$\sum_{n=3}^\infty \frac{1}{n(\ln_1 n)(\ln_2 n) \cdot ... \cdot (\ln_{\varphi(n)} n)}$$
converge or diverge?

3.4. Absolute Convergence. Theorem of Leibniz

3.4.1. For the indicated values of a, decide whether the series are absolutely convergent, conditionally convergent, or divergent:

(a) $\sum_{n=1}^{\infty} \left(\frac{an}{n+1}\right)^n$, $\quad a \in \mathbb{R}$,

(b) $\sum_{n=2}^{\infty} (-1)^n \frac{(\ln n)^a}{n}$, $\quad a \in \mathbb{R}$,

(c) $\sum_{n=1}^{\infty} (-1)^n \sin\frac{a}{n}$, $\quad a \in \mathbb{R}$,

(d) $\sum_{n=1}^{\infty} \frac{1}{n+1} \left(\frac{a^2 - 4a - 8}{a^2 + 6a - 16}\right)^n$, $\quad a \in \mathbb{R} \setminus \{-8, 2\}$,

(e) $\sum_{n=1}^{\infty} \frac{n^n}{a^{n^2}}$, $\quad a \neq 0$,

(f) $\sum_{n=1}^{\infty} (-1)^n \frac{(\ln n)^{\ln n}}{n^a}$, $\quad a > 0$.

3.4.2. For $a \in \mathbb{R}$, study convergence and absolute convergence of the series
$$\sum_{n=n_a}^{\infty} \frac{a^{n-1}}{na^{n-1} + \ln n},$$
where n_a is an index depending on a such that $na^{n-1} + \ln n \neq 0$ for $n \geq n_a$.

3.4.3. Suppose that a series $\sum_{n=1}^{\infty} a_n$ with nonzero terms converges. Study the convergence of the series
$$\sum_{n=1}^{\infty} \left(1 - \frac{\sin a_n}{a_n}\right).$$

3.4. Absolute Convergence. Theorem of Leibniz

3.4.4. Does the condition $\lim_{n\to\infty} \frac{a_n}{b_n} = 1$ imply that the convergence of $\sum_{n=1}^{\infty} a_n$ is equivalent to the convergence of $\sum_{n=1}^{\infty} b_n$?

3.4.5. Assume that a series $\sum_{n=1}^{\infty} a_n$ converges conditionally and set $p_n = \frac{|a_n|+a_n}{2}$, $q_n = \frac{|a_n|-a_n}{2}$. Show that both $\sum_{n=1}^{\infty} p_n$ and $\sum_{n=1}^{\infty} q_n$ diverge.

3.4.6. Assume that a series $\sum_{n=1}^{\infty} a_n$ converges conditionally. Let $\{P_n\}$ and $\{Q_n\}$ be the sequences of partial sums of $\sum_{n=1}^{\infty} p_n$ and $\sum_{n=1}^{\infty} q_n$, defined in the foregoing problem, respectively. Show that

$$\lim_{n\to\infty} \frac{P_n}{Q_n} = 1.$$

3.4.7. Study convergence and absolute convergence of the series

$$\sum_{n=1}^{\infty} \frac{(-1)^{\lfloor \frac{n}{3} \rfloor}}{n}.$$

3.4.8. For $a \in \mathbb{R}$, decide whether the series

$$\sum_{n=1}^{\infty} \frac{(-1)^{\lfloor \sqrt{n} \rfloor}}{n^a}$$

converges absolutely, converges conditionally, or diverges.

3.4.9. Decide whether the series

$$\sum_{n=1}^{\infty} \frac{(-1)^{\lfloor \ln n \rfloor}}{n}$$

is absolutely convergent, conditionally convergent, or divergent.

3.4.10. Let
$$\varepsilon_n = \begin{cases} +1 & \text{for } 2^{2k} \leq n < 2^{2k+1}, \\ -1 & \text{for } 2^{2k+1} \leq n < 2^{2k+2}, \end{cases}$$
where $k = 0, 1, 2, \ldots$. Discuss the convergence of the series

(a) $\sum_{n=1}^{\infty} \frac{\varepsilon_n}{n}$, (b) $\sum_{n=2}^{\infty} \frac{\varepsilon_n}{n \ln n}$.

3.4.11. Study the convergence of the series
$$\sum_{n=2}^{\infty} (-1)^n \frac{\sqrt{n}}{(-1)^n + \sqrt{n}} \sin \frac{1}{\sqrt{n}}.$$

3.4.12. Investigate the behavior (absolute convergence, conditional convergence) of the following series:

(a) $\sum_{n=1}^{\infty} (-1)^n (\sqrt[n]{n} - 1)^n,$

(b) $\sum_{n=1}^{\infty} (-1)^n (\sqrt[n]{a} - 1), \quad a > 1,$

(c) $\sum_{n=1}^{\infty} (-1)^n (\sqrt[n]{n} - 1),$

(d) $\sum_{n=1}^{\infty} (-1)^n \left(e - \left(1 + \frac{1}{n}\right)^n \right),$

(e) $\sum_{n=1}^{\infty} (-1)^n \left(\left(1 + \frac{1}{n}\right)^{n+1} - e \right).$

3.4.13. For $a, b > 0$, discuss convergence of the following series:

(a) $\sum_{n=1}^{\infty} (-1)^n \frac{(\ln n)^a}{n^b},$

(b) $\sum_{n=1}^{\infty} (-1)^n \frac{(\ln n)^{\ln n}}{n^b}.$

3.4. Absolute Convergence. Theorem of Leibniz

3.4.14. Let $\sum_{n=1}^{\infty}(-1)^{n-1}a_n$ be an alternating series which satisfies the conditions of the Leibniz test, that is, $0 < a_{n+1} \leq a_n$ for all n and $\lim_{n\to\infty} a_n = 0$. Denote by r_n the nth remainder of the series, $r_n = \sum_{k=n+1}^{\infty}(-1)^{k-1}a_k$. Show that r_n has the same sign as the term $(-1)^n a_{n+1}$ and $|r_n| < a_{n+1}$.

3.4.15. Suppose that a sequence $\{a_n\}$ tends to zero. Show that the series
$$\sum_{n=1}^{\infty} a_n \quad \text{and} \quad \sum_{n=1}^{\infty}(a_n + a_{n+1})$$
either both converge or both diverge.

3.4.16. For a sequence $\{a_n\}$ convergent to zero and for a, b, c such that $a + b + c \neq 0$, prove that the series
$$\sum_{n=1}^{\infty} a_n \quad \text{and} \quad \sum_{n=1}^{\infty}(aa_n + ba_{n+1} + ca_{n+2})$$
either both converge or both diverge.

3.4.17. Let $\{a_n\}$ be a sequence with $\lim_{n\to\infty} a_n = a \neq 0$ and with nonzero terms. Prove that the series
$$\sum_{n=1}^{\infty}(a_{n+1} - a_n) \quad \text{and} \quad \sum_{n=1}^{\infty}\left(\frac{1}{a_{n+1}} - \frac{1}{a_n}\right)$$
either both are absolutely convergent or both do not converge absolutely.

3.4.18. Show that, if a sequence $\{na_n\}$ and a series $\sum_{n=1}^{\infty} n(a_n - a_{n+1})$ both converge, then $\sum_{n=1}^{\infty} a_n$ also does.

3.4.19. For a sequence $\{a_n\}$ monotonically decreasing to zero, study the convergence of the series
$$\sum_{n=1}^{\infty}(-1)^{n+1}\frac{a_1 + a_2 + \ldots + a_n}{n}.$$

3.4.20. Decide for which values of a the series
$$\sum_{n=1}^{\infty}(-1)^n n! \sin a \sin \frac{a}{2} \cdot \ldots \cdot \sin \frac{a}{n}$$
converges absolutely and for which it diverges.

3.4.21. For positive a, b and c, study convergence of the series
$$\sum_{n=1}^{\infty}\left(\sqrt[n]{a} - \frac{\sqrt[n]{b} + \sqrt[n]{c}}{2}\right).$$

3.4.22. Discuss convergence of the following series:

(a) $\sum_{n=1}^{\infty}(\cos n)^n$, (b) $\sum_{n=1}^{\infty}(\sin n)^n$.

3.4.23. Let $\{a_n\}$ be a positive sequence. Prove that

(a) if $\lim\limits_{n\to\infty} n\left(\frac{a_n}{a_{n+1}} - 1\right) > 0$, then $\sum_{n=1}^{\infty}(-1)^n a_n$ converges,

(b) if $n\left(\frac{a_n}{a_{n+1}} - 1\right) \leq 0$, then $\sum_{n=1}^{\infty}(-1)^n a_n$ diverges (in particular, if $\overline{\lim}\limits_{n\to\infty} n\left(\frac{a_n}{a_{n+1}} - 1\right) < 0$, then the series diverges).

3.4.24. Assume that for a positive sequence $\{a_n\}$ there exist $\alpha \in \mathbb{R}$, $\varepsilon > 0$ and a bounded sequence $\{\beta_n\}$ such that
$$\frac{a_n}{a_{n+1}} = 1 + \frac{\alpha}{n} + \frac{\beta_n}{n^{1+\varepsilon}}.$$
Prove that the series $\sum_{n=1}^{\infty}(-1)^n a_n$ converges for $\alpha > 0$ and diverges for $\alpha \leq 0$.

3.4.25. Discuss the convergence of the series
$$\sum_{n=1}^{\infty}(-1)^n \frac{n! e^n}{n^{n+p}}, \quad p \in \mathbb{R}.$$

3.4. Absolute Convergence. Theorem of Leibniz

3.4.26. Assume that the series $\sum_{n=1}^{\infty} a_n$ converges and $\{p_n\}$ is a positive sequence which increases to $+\infty$. Show that

$$\lim_{n\to\infty} \frac{a_1 p_1 + a_2 p_2 + \ldots + a_n p_n}{p_n} = 0.$$

3.4.27. Let $\{a_n\}$ be a positive sequence decreasing to zero. Prove that, if the series $\sum_{n=1}^{\infty} a_n b_n$ converges, then

$$\lim_{n\to\infty} a_n (b_1 + b_2 + \ldots + b_n) = 0.$$

3.4.28. Let α be a given positive number. Prove that, if the series $\sum_{n=1}^{\infty} \frac{a_n}{n^\alpha}$ converges, then

$$\lim_{n\to\infty} \frac{a_1 + a_2 + \ldots + a_n}{n^\alpha} = 0.$$

3.4.29. Let $\{k_n\}$ be a strictly increasing sequence of natural numbers. Then the series $\sum_{n=1}^{\infty} a_{k_n}$ is called *a subseries* of the series $\sum_{n=1}^{\infty} a_n$. Show that, if all the subseries of a series converge, then the series is absolutely convergent.

3.4.30. Let k, l be integers such that $k \geq 1, l \geq 2$. Must the convergent series $\sum_{n=1}^{\infty} a_n$ be absolutely convergent if all its subseries of the form

$$\sum_{n=1}^{\infty} a_{k+(n-1)l}$$

are convergent?

3.4.31. Give an example of a convergent series $\sum_{n=1}^{\infty} a_n$ such that $\sum_{n=1}^{\infty} a_n^3$ diverges.

3.4.32. Does there exist a convergent series $\sum_{n=1}^{\infty} a_n$ such that all the series of the form $\sum_{n=1}^{\infty} a_n^k$, where k is an integer greater than or equal to 2, diverge?

3.4.33. Let $\{a_n\}$ be a monotonically decreasing and positive sequence such that the series $\sum_{n=1}^{\infty} a_n$ diverges. Suppose that the series $\sum_{n=1}^{\infty} \varepsilon_n a_n$, where ε_n is -1 or 1, converges. Prove that

$$\varlimsup_{n\to\infty} \frac{\varepsilon_1 + \varepsilon_2 + \ldots + \varepsilon_n}{n} \leq 0 \leq \varlimsup_{n\to\infty} \frac{\varepsilon_1 + \varepsilon_2 + \ldots + \varepsilon_n}{n}.$$

3.4.34. Assume that $\{a_n\}$ is a positive monotonically decreasing sequence and that the series $\sum_{n=1}^{\infty} \varepsilon_n a_n$, where ε_n is -1 or 1, converges. Show that

$$\lim_{n\to\infty} (\varepsilon_1 + \varepsilon_2 + \ldots + \varepsilon_n) a_n = 0.$$

(See 3.2.35.)

3.4.35. Suppose that the series $\sum_{n=1}^{\infty} b_n$ converges and $\{p_n\}$ is a monotonically increasing sequence for which $\lim_{n\to\infty} p_n = +\infty$ and $\sum_{n=1}^{\infty} \frac{1}{p_n} = +\infty$. Show that

$$\varlimsup_{n\to\infty} \frac{p_1 b_1 + p_2 b_2 + \ldots + p_n b_n}{n} \leq 0 \leq \varlimsup_{n\to\infty} \frac{p_1 b_1 + p_2 b_2 + \ldots + p_n b_n}{n}.$$

3.4.36. In the harmonic series $\sum_{n=1}^{\infty} \frac{1}{n}$ let us attach the sign "+", p times, consecutively, then the sign "−", q times, consecutively, then "+", p times, consecutively, etc. Show that the new series converges if and only if $p = q$.

3.4.37. Prove the following generalization of the Toeplitz theorem (see 2.3.1 and 2.3.36).

3.5. The Dirichlet and Abel Tests

Let $\{c_{n,k} : n, k \in \mathbb{N}\}$ be an array of real numbers. Then for any convergent sequence $\{a_n\}$ the transformed sequence $\{b_n\}$ given by

$$b_n = \sum_{k=1}^{\infty} c_{n,k} a_k, \quad n \geq 1,$$

is convergent to the same limit if and only if the following three conditions are satisfied:

(i) $\quad c_{n,k} \xrightarrow[n \to \infty]{} 0 \quad$ for each $k \in \mathbb{N}$,

(ii) $\quad \displaystyle\sum_{k=1}^{\infty} c_{n,k} = 1$,

(iii) there exists $C > 0$ such that for all positive integers n

$$\sum_{k=1}^{\infty} |c_{n,k}| \leq C.$$

3.5. The Dirichlet and Abel Tests

3.5.1. Using the Dirichlet and Abel tests, study convergence of the following series:

(a) $\quad \displaystyle\sum_{n=1}^{\infty} (-1)^n \frac{\sin^2 n}{n}$,

(b) $\quad \displaystyle\sum_{n=1}^{\infty} \frac{\sin n}{n} \left(1 + \frac{1}{2} + \ldots + \frac{1}{n}\right)$,

(c) $\quad \displaystyle\sum_{n=2}^{\infty} \frac{1}{\ln^2 n} \cos\left(\pi \frac{n^2}{n+1}\right)$,

(d) $\quad \displaystyle\sum_{n=1}^{\infty} \frac{\sin \frac{n\pi}{4}}{n^a + \sin \frac{n\pi}{4}}, \quad a > 0$.

3.5.2. Does the series $\displaystyle\sum_{n=2}^{\infty} \frac{\sin\left(n + \frac{1}{n}\right)}{\ln \ln n}$ converge?

3.5.3. For $a \in \mathbb{R}$, study convergence of the series

(a) $$\sum_{n=1}^{\infty} \frac{\sin(na)\sin(n^2 a)}{n},$$

(b) $$\sum_{n=1}^{\infty} \frac{\sin(na)\cos(n^2 a)}{n}.$$

3.5.4. Show that the series
$$\sum_{n=1}^{\infty} \frac{\cos n \sin(na)}{n}$$
converges for each $a \in \mathbb{R}$.

3.5.5. Determine whether the series $\sum_{n=1}^{\infty} \frac{\sin(na)}{n}$, $a \in \mathbb{R}$, converges absolutely.

3.5.6. Show that for $a \in \mathbb{R}$ and $n \in \mathbb{N}$,
$$\left| \sum_{k=1}^{n} \frac{\sin(ak)}{k} \right| < 2\sqrt{\pi}.$$

3.5.7. Prove that the series
$$\sum_{n=1}^{\infty} (-1)^n \frac{\arctan n}{\sqrt{n}}$$
converges.

3.5.8. For $x > 1$, study convergence of the series
$$\sum_{n=1}^{\infty} (-1)^n \frac{\sqrt[n]{\ln x}}{n}.$$

3.5. The Dirichlet and Abel Tests

3.5.9. Prove the following *lemma of Kronecker*.

Let $\sum_{n=1}^{\infty} a_n$ be a convergent series and let $\{b_n\}$ be a monotonically increasing sequence such that $\lim_{n \to \infty} b_n = +\infty$. Then

(a) $\displaystyle\sum_{k=n}^{\infty} \frac{a_k}{b_k} = o\left(\frac{1}{b_n}\right),$ (b) $\displaystyle\sum_{k=1}^{n} a_k b_k = o(b_n),$

where $o(b_n)$ means that $\lim_{n \to \infty} \frac{o(b_n)}{b_n} = 0$.

3.5.10. Assume that the series $\sum_{n=1}^{\infty} nc_n$ converges. Show that for every $n \in \mathbb{N}$ the series $\sum_{k=0}^{\infty} (k+1)c_{n+k}$ also converges. Moreover, show that if $l_n = \sum_{k=0}^{\infty} (k+1)c_{n+k}$, then $\lim_{n \to \infty} t_n = 0$.

3.5.11. Assume that the partial sums of the series $\sum_{n=1}^{\infty} a_n$ form a bounded sequence. Prove that if the series $\sum_{n=1}^{\infty} |b_n - b_{n+1}|$ converges and $\lim_{n \to \infty} b_n = 0$, then for every natural k the series $\sum_{n=1}^{\infty} a_n b_n^k$ also converges.

3.5.12. Prove that if $\sum_{n=1}^{\infty} (b_n - b_{n+1})$ converges absolutely and $\sum_{n=1}^{\infty} a_n$ converges, then the series $\sum_{n=1}^{\infty} a_n b_n$ also converges.

3.5.13. Using Abel's test, show that the convergence of $\sum_{n=1}^{\infty} a_n$ implies the convergence of the series $\sum_{n=1}^{\infty} a_n x^n$ for $|x| < 1$.

3.5.14. For a given sequence $\{a_n\}$ show that if the *Dirichlet series*

$$\sum_{n=1}^{\infty} \frac{a_n}{n^x}$$

converges at $x = x_0$, then it converges at every $x > x_0$.

3.5.15. Prove that the convergence of the Dirichlet series $\sum_{n=1}^{\infty} \frac{a_n}{n^x}$ implies the convergence of the series

$$\sum_{n=1}^{\infty} \frac{n! a_n}{x(x+1)\ldots(x+n)}, \quad x \neq 0, -1, -2, \ldots.$$

3.5.16. Prove that if the series $\sum_{n=1}^{\infty} a_n x^n$ converges for $|x| < 1$, then $\sum_{n=1}^{\infty} a_n \frac{x^n}{1-x^n}$ also converges.

3.5.17. Must the convergent series $\sum_{n=1}^{\infty} a_n$ be absolutely convergent if all its subseries of the form

$$\sum_{n=1}^{\infty} a_{kl^n}, \quad k \geq 1, l \geq 2,$$

converge?

3.6. Cauchy Product of Infinite Series

3.6.1. Prove the following *theorem of Mertens*.

If at least one of the two convergent series $\sum_{n=0}^{\infty} a_n$ and $\sum_{n=0}^{\infty} b_n$ converges absolutely, then their *Cauchy product* (that is, the series $\sum_{n=0}^{\infty} c_n$, where $c_n = a_0 b_n + a_1 b_{n-1} + \ldots + a_n b_0$) converges. Moreover, if $\sum_{n=0}^{\infty} a_n = A$ and $\sum_{n=0}^{\infty} b_n = B$, then $\sum_{n=0}^{\infty} c_n = AB$.

3.6.2. Find the sum of the series

(a) $\sum_{n=1}^{\infty} n x^{n-1}, \quad |x| < 1,$

(b) $\sum_{n=0}^{\infty} c_n, \quad$ where $\quad c_n = \sum_{k=0}^{n} x^k y^{n-k}, \quad |x| < 1, |y| < 1,$

3.6. Cauchy Product of Infinite Series

(c) $\sum_{n=1}^{\infty} c_n,$ where $c_n = \sum_{k=1}^{n} \dfrac{1}{k(k+1)(n-k+1)!}.$

3.6.3. Form the Cauchy product of the given series and calculate its sum.

(a) $\sum_{n=0}^{\infty} \dfrac{2^n}{n!}$ and $\sum_{n=0}^{\infty} \dfrac{1}{2^n n!},$

(b) $\sum_{n=1}^{\infty} (-1)^n \dfrac{1}{n}$ and $\sum_{n=1}^{\infty} \dfrac{1}{3^n},$

(c) $\sum_{n=0}^{\infty} (n+1)x^n$ and $\sum_{n=0}^{\infty} (-1)^n (n+1)x^n.$

3.6.4. Assume that the series $\sum_{n=0}^{\infty} a_n$ is convergent and set $A_n = a_0 + a_1 + ... + a_n$. Prove that for $|x| < 1$ the series $\sum_{n=0}^{\infty} A_n x^n$ converges and
$$\sum_{n=0}^{\infty} a_n x^n = (1-x) \sum_{n=0}^{\infty} A_n x^n.$$

3.6.5. Find the Cauchy product of the series $\sum_{n=0}^{\infty} (-1)^n \dfrac{x^{2n}}{(n!)^2},\ x \in \mathbb{R},$ with itself.
Hint. Use the equality $\sum_{k=0}^{n} \binom{n}{k}^2 = \binom{2n}{n}.$

3.6.6. For $a > 0$ and $|x| < 1$, verify the claim
$$\left(\dfrac{1}{a} + \dfrac{1}{2} \dfrac{x}{a+2} + \dfrac{1 \cdot 3}{2 \cdot 4} \dfrac{x^2}{a+4} + ... + \dfrac{1 \cdot 3 \cdot ... \cdot (2n-1)}{2 \cdot 4 \cdot ... \cdot (2n)} \dfrac{x^n}{a+2n} + ... \right)$$
$$\times \left(1 + \dfrac{1}{2}x + \dfrac{1 \cdot 3}{2 \cdot 4} x^2 + ... + \dfrac{1 \cdot 3 \cdot ... \cdot (2n-1)}{2 \cdot 4 \cdot ... \cdot (2n)} x^n + ... \right)$$
$$= \dfrac{1}{a} \left(1 + \dfrac{a+1}{a+2} x + \dfrac{(a+1)(a+3)}{(a+2)(a+4)} x^2 \right.$$
$$+ ... + \dfrac{(a+1) \cdot ... \cdot (a+2n-1)}{(a+2) \cdot ... \cdot (a+2n)} x^n + ... \bigg).$$

3.6.7. Prove the following *theorem of Abel*.

If the Cauchy product $\sum_{n=0}^{\infty} c_n$ of the two convergent series $\sum_{n=0}^{\infty} a_n = A$ and $\sum_{n=0}^{\infty} b_n = B$ converges to C, then $C = AB$.

3.6.8. Show that the series
$$\sum_{n=1}^{\infty} (-1)^{n-1} \frac{2}{n+1} \left(1 + \frac{1}{2} + \ldots + \frac{1}{n}\right)$$
is the Cauchy product of the series $\sum_{n=1}^{\infty} (-1)^{n-1} \frac{1}{n}$ with itself, and find its sum.

3.6.9. Study the convergence of the Cauchy product of the series $\sum_{n=1}^{\infty} (-1)^{n-1} \frac{1}{\sqrt{n}}$ with itself.

3.6.10. Prove that if at least one of two positive series is divergent, then their Cauchy product diverges.

3.6.11. Must the Cauchy product of two divergent series be divergent?

3.6.12. Prove that the Cauchy product of two convergent series $\sum_{n=0}^{\infty} a_n$ and $\sum_{n=0}^{\infty} b_n$ converges if and only if
$$\lim_{n \to \infty} \sum_{k=1}^{n} a_k (b_n + b_{n-1} + \ldots + b_{n-k+1}) = 0.$$

3.6.13. Suppose that $\{a_n\}$ and $\{b_n\}$ are positive sequences monotonically decreasing to zero. Show that the Cauchy product of the series $\sum_{n=0}^{\infty} (-1)^n a_n$ and $\sum_{n=0}^{\infty} (-1)^n b_n$ converges if and only if
$$\lim_{n \to \infty} a_n (b_0 + b_1 + \ldots + b_n) = 0 \quad \text{and} \quad \lim_{n \to \infty} b_n (a_0 + a_1 + \ldots + a_n) = 0.$$

3.7. Rearrangement of Series. Double Series

3.6.14. Show that the Cauchy product of

$$\sum_{n=1}^{\infty} \frac{(-1)^n}{n^\alpha} \quad \text{and} \quad \sum_{n=1}^{\infty} \frac{(-1)^n}{n^\beta}, \quad \alpha, \beta > 0,$$

converges if and only if $\alpha + \beta > 1$.

3.6.15. Assume that positive sequences $\{a_n\}$ and $\{b_n\}$ are monotonically decreasing to zero. Prove that the convergence of the series $\sum_{n=0}^{\infty} a_n b_n$ is a sufficient condition for convergence of the Cauchy product of the series $\sum_{n=0}^{\infty}(-1)^n a_n$ and $\sum_{n=0}^{\infty}(-1)^n b_n$, and that the convergence of $\sum_{n=0}^{\infty} (a_n b_n)^{1+\alpha}$ for every $\alpha > 0$ is a necessary condition for the convergence of this Cauchy product.

3.7. Rearrangement of Series. Double Series

3.7.1. Let $\{m_k\}$ be a strictly increasing sequence of positive integers, and put

$$b_1 = a_1 + a_2 + \ldots + a_{m_1}, \quad b_2 = a_{m_1+1} + a_{m_1+2} + \ldots + a_{m_2}, \ldots.$$

Show that if the series $\sum_{n=1}^{\infty} a_n$ converges, then $\sum_{n=1}^{\infty} b_n$ also converges and both series have the same sum.

3.7.2. Consider the series

$$1 - \frac{1}{2} - \frac{1}{4} + \frac{1}{3} - \frac{1}{6} - \frac{1}{8} + \frac{1}{5} - \ldots,$$

which is obtained by rearranging the terms of the series $\sum_{n=1}^{\infty} \frac{(-1)^{n-1}}{n}$ in such a way that each positive term is followed by two negative terms. Find the sum of this series.

3.7.3. Let us rearrange the terms of $\sum_{n=1}^{\infty} \frac{(-1)^{n-1}}{n}$ so that blocks of α positive terms alternate with blocks of β negative terms, that is,

$$1 + \frac{1}{3} + \ldots + \frac{1}{2\alpha - 1} - \frac{1}{2} - \frac{1}{4} - \ldots - \frac{1}{2\beta} + \frac{1}{2\alpha + 1} + \frac{1}{2\alpha + 3} + \ldots$$
$$+ \frac{1}{4\alpha - 1} - \frac{1}{2\beta + 2} - \frac{1}{2\beta + 4} - \ldots - \frac{1}{4\beta} + \ldots.$$

Find the sum of the rearranged series.

3.7.4. Show that

$$1 - \frac{1}{2} - \frac{1}{4} - \frac{1}{6} + \frac{1}{3} - \frac{1}{10} - \frac{1}{12} - \frac{1}{14} - \frac{1}{16} + \frac{1}{5} - \ldots = 0.$$

3.7.5. Find a rearrangement of the series $\sum_{n=1}^{\infty} \frac{(-1)^{n-1}}{n}$ which doubles its sum.

3.7.6. Rearrange the terms of $\sum_{n=1}^{\infty} \frac{(-1)^{n-1}}{n}$ to obtain a divergent series.

3.7.7. Study convergence of the series

$$1 + \frac{1}{\sqrt{3}} - \frac{1}{\sqrt{2}} + \frac{1}{\sqrt{5}} + \frac{1}{\sqrt{7}} - \frac{1}{\sqrt{4}} + \ldots$$

obtained by taking alternately two positive terms and one negative term of the series $\sum_{n=1}^{\infty} \frac{(-1)^{n-1}}{\sqrt{n}}$.

3.7.8. Prove that any rearrangement of an absolutely convergent series is convergent and has the same sum.

3.7.9. Assume that a function $f : (0, +\infty) \to (0, \infty)$, decreasing to zero as $x \to \infty$, is such that the sequence $\{nf(n)\}$ increases to $+\infty$. Let S denote the sum of the series $\sum_{n=1}^{\infty} (-1)^{n-1} f(n)$. Given l, find a rearrangement of this series convergent to $S + l$.

3.7. Rearrangement of Series. Double Series

3.7.10. Assume that a function $f : (0, +\infty) \to (0, \infty)$, decreasing to zero as $x \to \infty$, satisfies $\lim\limits_{n\to\infty} nf(n) = g$, $g \in (0, +\infty)$. Let S denote the sum of the series $\sum\limits_{n=1}^{\infty} (-1)^{n-1} f(n)$. Given l, find a rearrangement of this series convergent to $S + l$.

3.7.11. Rearrange the terms of $\sum\limits_{n=1}^{\infty} (-1)^{n-1} \frac{1}{n^p}$, $p \in (0, 1)$, to increase its sum by l.

3.7.12. For $\alpha > 0$, using the result in 3.7.10, find a rearrangement of $\sum\limits_{n=1}^{\infty} (-1)^{n-1} \frac{1}{n}$ whose sum is $\ln 2 + \frac{1}{2} \ln \alpha$.

3.7.13. Is it possible to accelerate by rearrangement the divergence of a divergent series with positive and monotonically decreasing terms?

3.7.14. Assume that the series $\sum\limits_{n=1}^{\infty} a_n$ with positive terms diverges and that $\lim\limits_{n\to\infty} a_n = 0$. Show that it is possible to slow down its divergence arbitrarily by rearrangement; that is: for any sequence $\{Q_n\}$ satisfying

$$0 < Q_1 < Q_2 < ... < Q_n < ..., \quad \lim\limits_{n\to\infty} Q_n = +\infty,$$

there is a rearrangement $\sum\limits_{k=1}^{\infty} a_{n_k}$ such that

$$a_{n_1} + a_{n_2} + ... + a_{n_m} \leq Q_m \quad \text{for} \quad m \in \mathbb{N}.$$

3.7.15. Let $\{r_n\}$ and $\{s_n\}$ be two strictly increasing sequences of positive integers without common terms. Assume also that every positive integer appears in one of the two sequences. Then the two subseries $\sum\limits_{n=1}^{\infty} a_{r_n}$ and $\sum\limits_{n=1}^{\infty} a_{s_n}$ are called *complementary subseries* of $\sum\limits_{n=1}^{\infty} a_n$. We say that the rearrangement shifts the two complementary subseries relative to each other if for all positive integers m and n such that $m < n$ the term a_{r_m} precedes a_{r_n} and a_{s_m} precedes a_{s_n}. Prove that the terms of a conditionally convergent series $\sum\limits_{n=1}^{\infty} a_n$ can be rearranged, by shifting the two complementary subseries of all its positive and all its negative terms, to give a conditionally convergent series whose sum is an arbitrarily preassigned number.

3.7.16. Let $\sum_{k=1}^{\infty} a_{n_k}$ be a rearrangement of a conditionally convergent series $\sum_{n=1}^{\infty} a_n$. Prove that if $\{n_k - k\}$ is a bounded sequence, then $\sum_{k=1}^{\infty} a_{n_k} = \sum_{n=1}^{\infty} a_n$. What happens if the sequence $\{n_k - k\}$ is unbounded?

3.7.17. Let $\sum_{k=1}^{\infty} a_{n_k}$ be a rearrangement of a conditionally convergent series $\sum_{n=1}^{\infty} a_n$. Prove that $\sum_{k=1}^{\infty} a_{n_k} = \sum_{n=1}^{\infty} a_n$ if and only if there exists a positive integer N such that each set $\{n_k : 1 \leq k \leq m\}$ is a union of at most N disjoint blocks of successive positive integers.

3.7.18. With an infinite matrix $\{a_{i,k}\}$, $i = 1, 2, ...$, $k = 1, 2, ...$, of real numbers we associate a *double series* $\sum_{i,k=1}^{\infty} a_{i,k}$. We say that the double series converges to $S \in \mathbb{R}$ if, given $\varepsilon > 0$, there exists $n_0 \in \mathbb{N}$ such that
$$|S_{m,n} - S| < \varepsilon \quad \text{for} \quad m, n > n_0,$$
where
$$S_{m,n} = \sum_{i=1}^{m} \sum_{k=1}^{n} a_{i,k}.$$
Then we write
$$S = \lim_{m,n \to \infty} S_{m,n} = \sum_{i,k=1}^{\infty} a_{i,k}.$$
We say that $\sum_{i,k=1}^{\infty} a_{i,k}$ converges absolutely if $\sum_{i,k=1}^{\infty} |a_{i,k}|$ converges. Note that the terms of an infinite matrix $(a_{i,k})_{i,k=1,2,...}$ can be ordered into a sequence $\{c_n\}$, and then the corresponding series $\sum_{n=1}^{\infty} c_n$ is called the *ordering of* $\sum_{i,k=1}^{\infty} a_{i,k}$ *into a single series*. Prove that if

3.7. Rearrangement of Series. Double Series

one of the orderings of a double series converges absolutely, then the double series converges (absolutely) to the same sum.

3.7.19. Prove that if a double series $\sum_{i,k=1}^{\infty} a_{i,k}$ converges absolutely, then any of its orderings $\sum_{n=1}^{\infty} c_n$ converges and

$$\sum_{i,k=1}^{\infty} a_{i,k} = \sum_{n=1}^{\infty} c_n.$$

3.7.20. Show that any absolutely convergent double series is convergent.

3.7.21. We say the *iterated series* $\sum_{i=1}^{\infty} \left(\sum_{k=1}^{\infty} a_{i,k} \right)$ is absolutely convergent if $\sum_{i=1}^{\infty} \left(\sum_{k=1}^{\infty} |a_{i,k}| \right)$ converges; similarly for $\sum_{k=1}^{\infty} \left(\sum_{i=1}^{\infty} a_{i,k} \right)$. Prove that an absolutely convergent iterated series is convergent.

3.7.22. Prove that if the double series $\sum_{i,k=1}^{\infty} a_{i,k}$ converges absolutely, then both the iterated series

$$\sum_{i=1}^{\infty} \left(\sum_{k=1}^{\infty} a_{i,k} \right) \quad \text{and} \quad \sum_{k=1}^{\infty} \left(\sum_{i=1}^{\infty} a_{i,k} \right)$$

converge absolutely and

$$\sum_{i,k=1}^{\infty} a_{i,k} = \sum_{i=1}^{\infty} \left(\sum_{k=1}^{\infty} a_{i,k} \right) = \sum_{k=1}^{\infty} \left(\sum_{i=1}^{\infty} a_{i,k} \right).$$

3.7.23. Prove that if one of the four series

$$\sum_{i,k=1}^{\infty} |a_{i,k}|, \quad \sum_{i=1}^{\infty} \left(\sum_{k=1}^{\infty} |a_{i,k}| \right), \quad \sum_{k=1}^{\infty} \left(\sum_{i=1}^{\infty} |a_{i,k}| \right),$$

$$\sum_{n=1}^{\infty} (|a_{n,1}| + |a_{n-1,2}| + |a_{n-2,3}| + \ldots + |a_{1,n}|)$$

converges, then all the series

$$\sum_{i,k=1}^{\infty} a_{i,k}, \quad \sum_{i=1}^{\infty}\left(\sum_{k=1}^{\infty} a_{i,k}\right), \quad \sum_{k=1}^{\infty}\left(\sum_{i=1}^{\infty} a_{i,k}\right),$$

$$\sum_{n=1}^{\infty}(a_{n,1}+a_{n-1,2}+a_{n-2,3}+\ldots+a_{1,n})$$

converge to the same sum.

3.7.24. Calculate
$$\sum_{n,k=0}^{\infty} \frac{1}{n!k!(n+k+1)}.$$

3.7.25. Find
$$\sum_{n,k=1}^{\infty} \frac{1}{nk(n+k+2)}.$$

3.7.26. Show that
$$\sum_{n,k=0}^{\infty} \frac{n!k!}{(n+k+2)!} = \frac{\pi^2}{6}.$$

3.7.27. For $0 < x < 1$, consider the infinite matrix

$$\begin{pmatrix} x & -x^2 & x^2 & -x^3 & x^3 & \ldots \\ x(1-x) & -x^2(1-x^2) & x^2(1-x^2) & -x^3(1-x^3) & x^3(1-x^3) & \ldots \\ x(1-x)^2 & -x^2(1-x^2)^2 & x^2(1-x^2)^2 & -x^3(1-x^3)^2 & x^3(1-x^3)^2 & \ldots \\ \ldots \end{pmatrix}$$

Prove that only one of the iterated series associated with this matrix converges (not absolutely).

3.7.28. Study convergence of the following double series:

(a) $\displaystyle\sum_{i,k=0}^{\infty} x^i y^k,$ where $|x|, |y| < 1,$

(b) $\displaystyle\sum_{i,k=1}^{\infty} \frac{1}{i^\alpha k^\beta},$ where $\alpha, \beta > 0,$

(c) $\displaystyle\sum_{i,k=1}^{\infty} \frac{1}{(i+k)^p},$ where $p > 0.$

3.7. Rearrangement of Series. Double Series

3.7.29. Find the sums of the following double series:

(a) $\quad \sum\limits_{i,k=2}^{\infty} \dfrac{1}{(p+i)^k}, \quad$ where $\quad p > -1,$

(b) $\quad \sum\limits_{i=2,k=1}^{\infty} \dfrac{1}{(2k)^i},$

(c) $\quad \sum\limits_{i,k=1}^{\infty} \dfrac{1}{(4i-1)^{2k}}.$

3.7.30. Given an infinite matrix $(b_{i,k})_{i,k=1,2,\ldots}$, prove that there is only one double series $\sum\limits_{i,k=1}^{\infty} a_{i,k}$ such that

$$S_{m,n} = \sum_{i=1}^{m} \sum_{k=1}^{n} a_{i,k} = b_{m,n}, \quad m,n = 1,2,\ldots.$$

3.7.31. Taking

$$b_{i,k} = (-1)^{i+k} \left(\frac{1}{2^i} + \frac{1}{2^k} \right), \quad i,k = 1,2,\ldots,$$

in the preceding problem, study convergence of the corresponding double series $\sum\limits_{i,k=1}^{\infty} a_{i,k}$.

3.7.32. Show that if $|x| < 1$, then the double series $\sum\limits_{i,k=1}^{\infty} x^{ik}$ converges absolutely. Using this fact, prove that

$$\sum_{i,k=1}^{\infty} x^{ik} = \sum_{k=1}^{\infty} \frac{x^k}{1-x^k} = \sum_{n=1}^{\infty} \theta(n) x^n = 2 \sum_{n=1}^{\infty} \frac{x^{n^2}}{1-x^n} + \sum_{n=1}^{\infty} x^{n^2},$$

where $\theta(n)$ denotes the number of all natural divisors of n.

3.7.33. Show that if $|x| < 1$, then the double series $\sum_{i,k=1}^{\infty} ix^{ik}$ converges absolutely. Moreover, prove that

$$\sum_{i,k=1}^{\infty} ix^{ik} = \sum_{k=1}^{\infty} \frac{kx^k}{1-x^k} = \sum_{n=1}^{\infty} \sigma(n)x^n,$$

where $\sigma(n)$ denotes the sum of all natural divisors of n.

3.7.34. Let $\zeta(p) = \sum_{n=1}^{\infty} \frac{1}{n^p}$, $p > 1$, be the *Riemann zeta function*. Set

$$S_p = \sum_{n=2}^{\infty} \frac{1}{n^p} = \zeta(p) - 1, \quad p > 1.$$

Show that

(a) $\sum_{p=2}^{\infty} S_p = 1,$ (b) $\sum_{p=2}^{\infty} (-1)^p S_p = \frac{1}{2}.$

3.7.35. Prove the following *theorem of Goldbach*.
If $\mathbf{A} = \{k^m : m, k = 2, 3, ...\}$, then $\sum_{n \in \mathbf{A}} \frac{1}{n-1} = 1$.

3.7.36. Let ζ denote the Riemann zeta function. Prove that for any integer $n \geq 2$,

$$\zeta(2)\zeta(2n-2) + \zeta(4)\zeta(2n-4) + ... + \zeta(2n-2)\zeta(2) = \left(n + \frac{1}{2}\right)\zeta(2n).$$

3.7.37. Using the result in the foregoing problem, find the sums of the series

$$\sum_{n=1}^{\infty} \frac{1}{n^6} \quad \text{and} \quad \sum_{n=1}^{\infty} \frac{1}{n^8}.$$

3.8. Infinite Products

3.8.1. Find the value of:

(a) $\prod_{n=2}^{\infty} \left(1 - \frac{1}{n^2}\right),$ (b) $\prod_{n=2}^{\infty} \frac{n^3-1}{n^3+1},$

3.8. Infinite Products

(c) $\displaystyle\prod_{n=1}^{\infty} \cos\frac{x}{2^n}$, $x \neq 2^m\left(\frac{\pi}{2}+k\pi\right)$, $m \in \mathbb{N}$, $k \in \mathbb{Z}$,

(d) $\displaystyle\prod_{n=1}^{\infty} \cosh\frac{x}{2^n}$, $x \in \mathbb{R}$, (e) $\displaystyle\prod_{n=0}^{\infty}\left(1+x^{2^n}\right)$, $|x|<1$,

(f) $\displaystyle\prod_{n=1}^{\infty}\left(1+\frac{1}{n(n+2)}\right)$, (g) $\displaystyle\prod_{n=1}^{\infty} a^{\frac{(-1)^n}{n}}$, $a>0$,

(h) $\displaystyle\prod_{n=1}^{\infty}\frac{e^{\frac{1}{n}}}{1+\frac{1}{n}}$, (i) $\displaystyle\prod_{n=1}^{\infty}\frac{9n^2}{9n^2-1}$.

3.8.2. Study the convergence of the following infinite products:

(a) $\displaystyle\prod_{n=2}^{\infty}\left(1+\frac{(-1)^n}{n}\right)$, (b) $\displaystyle\prod_{n=1}^{\infty}\left(1+\frac{1}{n}\right)$,

(c) $\displaystyle\prod_{n=2}^{\infty}\left(1-\frac{1}{n}\right)$.

3.8.3. Assume that $a_n \geq 0$, $n \in \mathbb{N}$. Prove that the infinite product $\prod_{n=1}^{\infty}(1+a_n)$ converges if and only if the series $\sum_{n=1}^{\infty} a_n$ converges.

3.8.4. Suppose that $a_n \geq 0$ and $a_n \neq 1$ for $n \in \mathbb{N}$. Show that the infinite product $\prod_{n=1}^{\infty}(1-a_n)$ converges if and only if the series $\sum_{n=1}^{\infty} a_n$ converges.

3.8.5. Set

$$a_{2n-1}=\frac{1}{\sqrt{n}}+\frac{1}{n}, \qquad a_{2n}=-\frac{1}{\sqrt{n}}, \quad n\in\mathbb{N}.$$

Show that the product $\prod_{n=1}^{\infty}(1+a_n)$ converges although the series $\sum_{n=1}^{\infty} a_n$ diverges.

3.8.6. Study convergence of the products:

(a) $\prod_{n=1}^{\infty} \cos \frac{1}{n}$,

(b) $\prod_{n=1}^{\infty} n \sin \frac{1}{n}$,

(c) $\prod_{n=1}^{\infty} \tan\left(\frac{\pi}{4} + \frac{1}{n}\right)$,

(d) $\prod_{n=1}^{\infty} n \ln\left(1 + \frac{1}{n}\right)$,

(e) $\prod_{n=1}^{\infty} \sqrt[n]{n}$,

(f) $\prod_{n=1}^{\infty} \sqrt[n^2]{n}$.

3.8.7. Assume that the series $\sum_{n=1}^{\infty} a_n$ converges. Prove that the infinite product $\prod_{n=1}^{\infty} (1 + a_n)$ converges if and only if the series $\sum_{n=1}^{\infty} a_n^2$ does. Prove also that if the series $\sum_{n=1}^{\infty} a_n^2$ diverges, then the infinite product $\prod_{n=1}^{\infty} (1 + a_n)$ diverges to zero.

3.8.8. Assume that the sequence $\{a_n\}$ decreases monotonically to zero. Show that the product $\prod_{n=1}^{\infty} (1 + (-1)^n a_n)$ converges if and only if the series $\sum_{n=1}^{\infty} a_n^2$ converges.

3.8.9. Prove that the product $\prod_{n=1}^{\infty} \left(1 + (-1)^{n+1} \frac{1}{\sqrt{n}}\right)$ diverges although the series $\sum_{n=1}^{\infty} (-1)^{n+1} \frac{1}{\sqrt{n}}$ converges.

3.8.10. Show that if the series

$$\sum_{n=1}^{\infty} \left(a_n - \frac{1}{2} a_n^2\right) \quad \text{and} \quad \sum_{n=1}^{\infty} |a_n|^3$$

both converge, then the product $\prod_{n=1}^{\infty} (1 + a_n)$ also converges.

3.8. Infinite Products

3.8.11. Does the convergence of the product $\prod_{n=1}^{\infty}(1+a_n)$ imply the convergence of the series $\sum_{n=1}^{\infty} a_n$ and $\sum_{n=1}^{\infty} a_n^2$?
Hint. Consider the product

$$\left(1-\frac{1}{2^\alpha}\right)\left(1+\frac{1}{2^\alpha}+\frac{1}{2^{2\alpha}}\right)\left(1-\frac{1}{3^\alpha}\right)\left(1+\frac{1}{3^\alpha}+\frac{1}{3^{2\alpha}}\right)\cdots,$$

where $\frac{1}{3} < \alpha \leq \frac{1}{2}$.

3.8.12. Prove the following generalization of the result in 3.8.10. For $k \geq 2$, if both series

$$\sum_{n=1}^{\infty}\left(a_n - \frac{1}{2}a_n^2 + \ldots + \frac{(-1)^{k-1}}{k}a_n^k\right) \quad \text{and} \quad \sum_{n=1}^{\infty}|a_n|^{k+1}$$

converge, then the product $\prod_{n=1}^{\infty}(1+a_n)$ also converges.

3.8.13. Prove that the convergence of $\prod_{n=1}^{\infty}(1+a_n)$ and of $\sum_{n=1}^{\infty} a_n^2$ implies the convergence of $\sum_{n=1}^{\infty} a_n$.

3.8.14. Show that if the products $\prod_{n=1}^{\infty}(1+a_n)$ and $\prod_{n=1}^{\infty}(1-a_n)$ converge, then both series $\sum_{n=1}^{\infty} a_n$ and $\sum_{n=1}^{\infty} a_n^2$ converge.

3.8.15. Assume that the sequence $\{a_n\}$ decreases monotonically to 1. Does the product

$$a_1 \cdot \frac{1}{a_2} \cdot a_3 \cdot \frac{1}{a_4} \cdot a_5 \cdot \ldots$$

always converge?

3.8.16. Assume that the products $\prod_{n=1}^{\infty} a_n$ and $\prod_{n=1}^{\infty} b_n$ with positive factors both converge. Study the convergence of

(a) $\prod_{n=1}^{\infty} (a_n + b_n)$, (b) $\prod_{n=1}^{\infty} a_n^2$,

(c) $\prod_{n=1}^{\infty} a_n b_n$, (d) $\prod_{n=1}^{\infty} \dfrac{a_n}{b_n}$?

3.8.17. Show that for $x_n \in (0, \frac{\pi}{2})$, $n \in \mathbb{N}$, the products

$$\prod_{n=1}^{\infty} \cos x_n \quad \text{and} \quad \prod_{n=1}^{\infty} \frac{\sin x_n}{x_n}$$

converge if and only if the series $\sum_{n=1}^{\infty} x_n^2$ converges.

3.8.18. Let $\sum_{n=1}^{\infty} a_n$ be a convergent series with positive terms and let S_n denote its nth partial sum. Show that

$$a_1 \prod_{n=2}^{\infty} \left(1 + \frac{a_n}{S_{n-1}}\right) = \sum_{n=1}^{\infty} a_n.$$

3.8.19. Show that if the infinite product $\prod_{n=1}^{\infty} (1 + a_n)$, $a_n > -1$, converges to P, then the series

$$\sum_{n=1}^{\infty} \frac{a_n}{(1 + a_1)(1 + a_2) \cdot \ldots \cdot (1 + a_n)}$$

also converges. Moreover, if S is its sum, then $S = 1 - \frac{1}{P}$.

3.8.20. Suppose that the infinite product

$$\prod_{n=1}^{\infty} (1 + a_n), \quad \text{where} \quad a_n > 0, \; n \in \mathbb{N},$$

3.8. Infinite Products

diverges. Prove that

$$\sum_{n=1}^{\infty} \frac{a_n}{(1+a_1)(1+a_2)\cdot\ldots\cdot(1+a_n)} = 1.$$

3.8.21. Show that

$$\sum_{n=1}^{\infty} \frac{x^n}{(1+x)(1+x^2)\cdot\ldots\cdot(1+x^n)} = 1 \quad \text{for} \quad x > 1.$$

3.8.22. Let $a_n \neq 0$ for $n \in \mathbb{N}$. Prove that the infinite product $\prod_{n=1}^{\infty} a_n$ converges if and only if the following *Cauchy criterion* is satisfied. For every $\varepsilon > 0$ there is an integer n_0 such that

$$|a_n a_{n+1} \cdot \ldots \cdot a_{n+k} - 1| < \varepsilon$$

for $n \geq n_0$ and $k \in \mathbb{N}$.

3.8.23. For $|x| < 1$, verify the following claim:

$$\prod_{n=1}^{\infty}(1+x^n) = \frac{1}{\prod_{n=1}^{\infty}(1-x^{2n-1})}.$$

3.8.24. The product $\prod_{n=1}^{\infty}(1+a_n)$ is said to be absolutely convergent if $\prod_{n=1}^{\infty}(1+|a_n|)$ converges. Show that the product $\prod_{n=1}^{\infty}(1+a_n)$ converges absolutely if and only if the series $\sum_{n=1}^{\infty} a_n$ converges absolutely.

3.8.25. Show that every absolutely convergent product $\prod_{n=1}^{\infty}(1+a_n)$ is convergent.

3.8.26. Prove that if the product $\prod_{n=1}^{\infty}(1+a_n)$ converges absolutely, then

$$\prod_{n=1}^{\infty}(1+a_n) = 1 + \sum_{n=1}^{\infty} a_n + \sum_{\substack{n_1,n_2=1 \\ n_1<n_2}}^{\infty} a_{n_1} a_{n_2}$$

$$+ \ldots + \sum_{\substack{n_1,n_2,\ldots,n_k=1 \\ n_1<n_2<\ldots<n_k}}^{\infty} a_{n_1} a_{n_2} \ldots a_{n_k} + \ldots .$$

3.8.27. Assume that the product $\prod_{n=1}^{\infty}(1+a_n)$ converges absolutely. Show that the product $\prod_{n=1}^{\infty}(1+a_n x)$ converges absolutely for each $x \in \mathbb{R}$ and it can be expanded in an absolutely convergent series. That is,

$$\prod_{n=1}^{\infty}(1+a_n x) = 1 + \sum_{k=1}^{\infty} A_k x^k,$$

where

$$A_k = \sum_{\substack{n_1,n_2,\ldots,n_k=1 \\ n_1<n_2<\ldots<n_k}}^{\infty} a_{n_1} a_{n_2} \ldots a_{n_k}.$$

3.8.28. Establish the equality

$$\prod_{n=1}^{\infty}(1+q^n x) = 1 + \sum_{n=1}^{\infty} \frac{q^{\frac{n(n+1)}{2}}}{(1-q)(1-q^2) \cdot \ldots \cdot (1-q^n)} x^n, \quad |q|<1.$$

3.8.29. Verify the identity

$$\prod_{n=1}^{\infty}(1+q^{2n-1} x) = 1 + \sum_{n=1}^{\infty} \frac{q^{n^2}}{(1-q^2)(1-q^4) \cdot \ldots \cdot (1-q^{2n})} x^n, \quad |q|<1.$$

3.8.30. Assume that the series $\sum_{n=1}^{\infty} a_n$ converges absolutely. Prove that if $x \neq 0$, then

$$\prod_{n=1}^{\infty}(1+a_n x)\left(1+\frac{a_n}{x}\right) = B_0 + \sum_{n=1}^{\infty} B_n \left(x^n + \frac{1}{x^n}\right),$$

3.8. Infinite Products

where $B_n = A_n + A_1 A_{n+1} + A_2 A_{n+2} + \ldots$, $n = 0, 1, 2, \ldots$, and

$$\prod_{n=1}^{\infty}(1 + a_n x) = A_0 + \sum_{k=1}^{\infty} A_k x^k \quad \text{(see 3.8.27)}.$$

3.8.31. For $|q| < 1$ and $x \neq 0$, establish the identity

$$\prod_{n=1}^{\infty}(1 - q^{2n}) \prod_{n=1}^{\infty}(1 + q^{2n-1}x)\left(1 + \frac{q^{2n-1}}{x}\right) = 1 + \sum_{n=1}^{\infty} q^{n^2}\left(x^n + \frac{1}{x^n}\right).$$

3.8.32. For $|q| < 1$, verify the following claims:

(a) $\displaystyle\prod_{n=1}^{\infty}(1 - q^{2n}) \prod_{n=1}^{\infty}(1 - q^{2n-1})^2 = 1 + 2\sum_{n=1}^{\infty}(-1)^n q^{n^2}$,

(b) $\displaystyle\prod_{n=1}^{\infty}(1 - q^{2n}) \prod_{n=1}^{\infty}(1 + q^{2n-1})^2 = 1 + 2\sum_{n=1}^{\infty} q^{n^2}$,

(c) $\displaystyle\prod_{n=1}^{\infty}(1 - q^{2n}) \prod_{n=1}^{\infty}(1 + q^{2n})^2 = 1 + \sum_{n=1}^{\infty} q^{n^2+n}$.

3.8.33. For $x > 0$, define the sequence $\{a_n\}$ by setting

$$a_1 = \frac{1}{1+x}, \quad a_n = \frac{n}{x+n}\prod_{k=1}^{n-1}\frac{x-k}{x+k}, \quad n > 1.$$

Show that the series $\displaystyle\sum_{n=1}^{\infty} a_n$ converges and find its sum.

3.8.34. Prove that if the infinite product $\displaystyle\prod_{n=1}^{\infty}(1 + ca_n)$ converges for two different values of $c \in \mathbb{R} \setminus \{0\}$, then it converges for each c.

3.8.35. Prove that if the series

$$\sum_{n=1}^{\infty} a_n \prod_{k=0}^{n}(x^2 - k^2)$$

converges at $x = x_0$, $x_0 \notin \mathbb{Z}$, then it converges for any value of x.

3.8.36. Let $\{p_n\}$ be the sequence of consecutive prime numbers greater than 1.

(a) Prove the following *Euler product formula*:
$$\prod_{n=1}^{\infty}\left(1-\frac{1}{p_n^x}\right)^{-1}=\sum_{n=1}^{\infty}\frac{1}{n^x} \quad \text{for} \quad x>1.$$

(b) Prove that the series $\sum_{n=1}^{\infty}\frac{1}{p_n}$ diverges (compare with 3.2.72).

3.8.37. Using DeMoivre's law, establish the identities

(a) $$\sin x = x\prod_{n=1}^{\infty}\left(1-\frac{x^2}{n^2\pi^2}\right),$$

(b) $$\cos x = \prod_{n=1}^{\infty}\left(1-\frac{4x^2}{(2n-1)^2\pi^2}\right).$$

3.8.38. Applying the result in the foregoing problem prove the *Wallis formula*
$$\lim_{n\to\infty}\frac{(2n)!!}{(2n-1)!!\sqrt{n}}=\sqrt{\pi}.$$

3.8.39. Study convergence of the products

(a) $$\prod_{n=1}^{\infty}\left(1+\frac{x}{n}\right)e^{-\frac{x}{n}}, \quad x>-1,$$

(b) $$\prod_{n=1}^{\infty}\frac{\left(1+\frac{1}{n}\right)^x}{1+\frac{x}{n}}, \quad x>-1.$$

3.8.40. Prove that the infinite product $\prod_{n=1}^{\infty}(1+a_n)$ converges absolutely if and only if any rearrangement of its factors does not change its value.

3.8. Infinite Products   121

3.8.41. Find the value of the product
$$\left(1+\frac{1}{2}\right)\left(1+\frac{1}{4}\right)\cdots\left(1+\frac{1}{2\alpha}\right)$$
$$\times \left(1-\frac{1}{3}\right)\cdots\left(1-\frac{1}{2\beta+1}\right)\left(1+\frac{1}{2\alpha+2}\right)\cdots$$
obtained by rearranging the factors of $\prod_{n=2}^{\infty}\left(1+\frac{(-1)^n}{n}\right)$ in such a way that blocks of α factors greater than 1 alternate with blocks of β factors smaller than 1.

3.8.42. Prove that the convergent but not absolutely convergent infinite product $\prod_{n=1}^{\infty}(1+a_n)$, $a_n > -1$, can be rearranged to give a product whose value is an arbitrarily preassigned positive number, or to give a product that diverges to zero or to infinity. (Compare with 3.7.15).

Solutions

Chapter 1

Real Numbers

1.1. Supremum and Infimum of Sets of Real Numbers. Continued Fractions

1.1.1. Set $\mathbf{A} = \{x \in \mathbb{Q} : x > 0,\ x^2 < 2\}$ and $s = \sup \mathbf{A}$. We may assume that $s > 1$. We will now show that for any positive integer n,

$$(1) \qquad \left(s - \frac{1}{n}\right)^2 \leq 2 \leq \left(s + \frac{1}{n}\right)^2.$$

Since $s - \frac{1}{n}$ is not an upper bound of $\mathbf{A}$, there exists $x^\star \in \mathbf{A}$ such that $s - \frac{1}{n} < x^\star$. Hence

$$\left(s - \frac{1}{n}\right)^2 < (x^\star)^2 < 2.$$

Assume that $\left(s + \frac{1}{n}\right)^2 < 2$. If s were rational, then $s + \frac{1}{n} \in \mathbf{A}$ and $s + \frac{1}{n} > s$, which would contradict the fact that $s = \sup \mathbf{A}$. If s were irrational, then $w = \frac{[(n+1)s]}{n+1} + \frac{1}{n+1}$ is a rational number such that $s < w < s + \frac{1}{n}$. Hence $w^2 < \left(s + \frac{1}{n}\right)^2 < 2$ and $w \in \mathbf{A}$, a contradiction. So, we have proved that $\left(s + \frac{1}{n}\right)^2 \geq 2$. By the left-hand side of (1), $s^2 - \frac{2s}{n} < s^2 - \frac{2s}{n} + \frac{1}{n^2} \leq 2$, which gives $\frac{s^2 - 2}{2s} < \frac{1}{n}$. Letting $n \to \infty$, we get $s^2 - 2 \leq 0$.

As above, the right-hand side of (1) gives $\frac{s^2 - 2}{3s} \geq -\frac{1}{n}$, which implies $s^2 - 2 \geq 0$. Therefore $s^2 = 2$.

1.1.2. Suppose **A** is bounded below and set $a = \inf \mathbf{A}$. Then

(1) $x \geq a$ for all $x \in \mathbf{A}$,
(2)   for any $\varepsilon > 0$ there is $x^\star \in \mathbf{A}$ such that $x^\star < a + \varepsilon$.

Multiplying the inequalities in (1) and (2) by -1, we get

(1′) $x \leq -a$ for all $x \in (-\mathbf{A})$,
(2′) for any $\varepsilon > 0$ there is $x^\star \in (-\mathbf{A})$ such that $x^\star > -a - \varepsilon$.

Hence $-a = \sup(-\mathbf{A})$. If **A** is not bounded below, then $-\mathbf{A}$ is not bounded above and therefore $\sup(-\mathbf{A}) = -\inf \mathbf{A} = +\infty$. The other equality can be established similarly.

1.1.3. Suppose **A** and **B** are bounded above, and put $a = \sup \mathbf{A}$ and $b = \sup \mathbf{B}$. Then a is an upper bound of **A** and b is an upper bound of **B**. Thus, $a + b$ is an upper bound of $\mathbf{A} + \mathbf{B}$. Moreover, for any $\varepsilon > 0$ there are $x^\star \in \mathbf{A}$ and $y^\star \in \mathbf{B}$ such that $x^\star > a - \frac{\varepsilon}{2}$ and $y^\star > b - \frac{\varepsilon}{2}$. Therefore, $x^\star + y^\star > a + b - \varepsilon$. Since $z^\star = x^\star + y^\star \in \mathbf{A} + \mathbf{B}$, the equality $a + b = \sup(\mathbf{A} + \mathbf{B})$ is proved. If **A** or **B** is unbounded above, then $\mathbf{A} + \mathbf{B}$ is also unbounded above, and by the definition of the supremun $\sup(\mathbf{A} + \mathbf{B}) = \sup \mathbf{A} + \sup \mathbf{B} = +\infty$.

The second equality is an immediate consequence of the first one and of the foregoing problem. Indeed, we have

$$\sup(\mathbf{A} - \mathbf{B}) = \sup(\mathbf{A} + (-\mathbf{B})) = \sup \mathbf{A} + \sup(-\mathbf{B}) = \sup \mathbf{A} - \inf \mathbf{B}.$$

Similar arguments can be applied to prove the equalities

$$\inf(\mathbf{A} + \mathbf{B}) = \inf \mathbf{A} + \inf \mathbf{B},$$
$$\inf(\mathbf{A} - \mathbf{B}) = \inf \mathbf{A} - \sup \mathbf{B}.$$

1.1.4. Suppose that both sets are bounded above, and put $a = \sup \mathbf{A}$ and $b = \sup \mathbf{B}$. Since elements of **A** and **B** are positive numbers, $xy \leq ab$ for any $x \in \mathbf{A}$ and $y \in \mathbf{B}$. We will now prove that ab is the least upper bound of $\mathbf{A} \cdot \mathbf{B}$. Let $\varepsilon > 0$ be arbitrarily fixed. There exist $x^\star \in \mathbf{A}$ and $y^\star \in \mathbf{B}$ such that $x^\star > a - \varepsilon$ and $y^\star > b - \varepsilon$. Thus $x^\star y^\star > ab - \varepsilon(a + b - \varepsilon)$. Since $\varepsilon(a + b - \varepsilon)$ can be made arbitrarily small if ε is small enough, we see that any number less than ab

1.1. Supremum and Infimum. Continued Fractions 127

cannot be an upper bound of $\mathbf{A} \cdot \mathbf{B}$. Therefore $ab = \sup(\mathbf{A} \cdot \mathbf{B})$. If $\mathbf{A}$ or $\mathbf{B}$ is not bounded above, then $\mathbf{A} \cdot \mathbf{B}$ is not either. Therefore $\sup(\mathbf{A} \cdot \mathbf{B}) = \sup \mathbf{A} \cdot \sup \mathbf{B} = +\infty$.

The task is now to prove $\sup\left(\frac{1}{\mathbf{A}}\right) = \frac{1}{\inf \mathbf{A}}$ if $a' = \inf \mathbf{A} > 0$. Then, for any $x \in \mathbf{A}$, the inequality $x \geq a'$ is equivalent to $\frac{1}{x} \leq \frac{1}{a'}$. So, $\frac{1}{a'}$ is an upper bound of $\frac{1}{\mathbf{A}}$. Moreover, for any $\varepsilon > 0$ there is $x^\star \in \mathbf{A}$ such that $x^\star < a' + \varepsilon$. Hence

$$\frac{1}{x^\star} > \frac{1}{a' + \varepsilon} = \frac{1}{a'} - \frac{\varepsilon}{a'(a' + \varepsilon)}.$$

Since $\frac{\varepsilon}{a'(a'+\varepsilon)}$ can be made arbitrarily small, $\frac{1}{a'}$ is the least upper bound of $\frac{1}{\mathbf{A}}$. We now turn to the case $a' = 0$. Then the set $\frac{1}{\mathbf{A}}$ is unbounded (indeed, for any $\varepsilon > 0$ there is $x^\star \in \frac{1}{\mathbf{A}}$ for which $x^\star > \frac{1}{\varepsilon}$). Therefore, $\sup \frac{1}{\mathbf{A}} = +\infty$.

Assume now that $\mathbf{A}$, $\mathbf{B}$ are bounded sets of real numbers (positive or nonpositive) and put $a = \sup \mathbf{A}$, $b = \sup \mathbf{B}$, $a' = \inf \mathbf{A}$ and $b' = \inf \mathbf{B}$. If a' and b' are nonnegative, the desired equality follows from the above. If $a' < 0$ and a, $b' > 0$, then $xy \leq ab$ for any $x \in \mathbf{A}$ and $y \in \mathbf{B}$. Take $\varepsilon > 0$ so small that $a - \varepsilon > 0$. Then there is a positive number $x^\star$ in $\mathbf{A}$ for which $x^\star > a - \varepsilon$. Moreover, there is $y^\star \in \mathbf{B}$ such that $y^\star > b - \varepsilon$. Hence

$$x^\star y^\star > x^\star(b - \varepsilon) > (a - \varepsilon)(b - \varepsilon) = ab - \varepsilon(a + b - \varepsilon).$$

So, in this case we have $\sup(\mathbf{A} \cdot \mathbf{B}) = ab$.

We now consider the case where a', $b' < 0$ and $a, b > 0$. Then, for any $x \in \mathbf{A}$ and $y \in \mathbf{B}$, we have

$$xy \leq \max\{ab,\, a'b'\}.$$

Assume first that $\max\{ab,\, a'b'\} = a'b'$. By the definition of the greatest lower bound, for sufficiently small $\varepsilon > 0$, there exist $x^\star \in \mathbf{A}$ and $y^\star \in \mathbf{B}$ for which $x^\star < a' + \varepsilon < 0$ and $y^\star < b' + \varepsilon < 0$. This gives

$$x^\star y^\star > x^\star(b' + \varepsilon) > (a' + \varepsilon)(b' + \varepsilon) = a'b' + \varepsilon(a' + b' + \varepsilon).$$

Notice that $a' + b' + \varepsilon$ is negative. Therefore $a'b'$ is the least upper bound of $\mathbf{A} \cdot \mathbf{B}$. In the case where $\max\{ab,\, a'b'\} = ab$ similar reasoning yields $\sup(\mathbf{A} \cdot \mathbf{B}) = ab$. All other cases can be proved analogously.

1.1.5. Suppose first that **A** and **B** are bounded above. Put $a = \sup \mathbf{A}$ and $b = \sup \mathbf{B}$. Of course we may assume that $a \leq b$. Then, for any $x \in \mathbf{A} \cup \mathbf{B}$, $x \leq b$. Moreover, for any $\varepsilon > 0$ there is $x^* \in \mathbf{B}$ such that $x^* > b - \varepsilon$. It is obvious that x^* belongs to $\mathbf{A} \cup \mathbf{B}$. Therefore, the first equality is valid. If **A** or **B** is not bounded above, then $\mathbf{A} \cup \mathbf{B}$ is not either. So, $\sup(\mathbf{A} \cup \mathbf{B}) = +\infty$, and we assume that $\max\{+\infty, c\} = \max\{+\infty, +\infty\} = +\infty$ for any real c. The proof of the second equality is similar.

1.1.6. We have
$$\mathbf{A}_1 = \left\{-3, -\frac{11}{2}, 5\right\}$$
$$\cup \left\{\frac{3}{4k}, -\frac{3}{4k+1}, -4 - \frac{3}{4k+2}, 4 + \frac{3}{4k+3} \; ; \; k \in \mathbb{N}\right\},$$
$$\mathbf{A}_2 = \left\{\frac{3k-1}{3k+1}, -\frac{3k-2}{6k}, -\frac{3k-3}{2(3k-1)} \; ; \; k \in \mathbb{N}\right\}.$$

Therefore $\inf \mathbf{A}_1 = -\frac{11}{2}$, $\sup \mathbf{A}_1 = 5$ and $\inf \mathbf{A}_2 = -\frac{1}{2}$, $\sup \mathbf{A}_2 = 1$.

1.1.7. $\sup \mathbf{A} = \frac{2}{9}$, $\inf \mathbf{A} = 0.2$, $\sup \mathbf{B} = \frac{1}{9}$, $\inf \mathbf{B} = 0$.

1.1.8. One can show by induction that for $n \geq 11$, $2^n > (n+1)^3$. Hence
$$0 < \frac{(n+1)^2}{2^n} < \frac{(n+1)^2}{(n+1)^3} = \frac{1}{n+1} \quad \text{for} \quad n \geq 11.$$
Therefore 0 is the greatest lower bound of our set.

It is also easy to show that $2^n > (n+1)^2$ for $n \geq 6$. Hence $\frac{(n+1)^2}{2^n} < 1$ for $n \geq 6$. The numbers $2, \frac{9}{4}, \frac{25}{16}, \frac{36}{32}$ (greater than 1) also belong to our set. Thus the least upper bound of the set is $\frac{9}{4}$.

1.1.9. It follows from the foregoing problem that the greatest lower bound of this set is equal to 0. By the inequality mentioned in the preceding solution, $2^{nm} > (nm+1)^2$ for $nm \geq 6$. Since $nm + 1 \geq n + m$, for $n, m \in \mathbb{N}$, we have
$$\frac{(n+m)^2}{2^{nm}} < \frac{(n+m)^2}{(nm+1)^2} \leq \frac{(n+m)^2}{(n+m)^2} = 1 \quad \text{if} \quad nm \geq 6.$$

1.1. Supremum and Infimum. Continued Fractions

For $nm < 6$ we get the following elements of our set: $1, 2, \frac{9}{4}, \frac{25}{16}, \frac{36}{32}$. Hence its least upper bound is $\frac{9}{4}$.

1.1.10.

(a) It is obvious that 2 is an upper bound for the set **A**. We will show there are no smaller upper bounds. Indeed, if $\varepsilon > 0$ is arbitrarily fixed, then for any positive integer $n^* > \left[\frac{2}{\varepsilon}\right]$, we obtain $\frac{2(n^*-1)}{n^*} > 2 - \varepsilon$. The greatest lower bound of **A** is 0, because $\frac{m}{n} > 0$ for $m, n \in \mathbb{N}$. Given $\varepsilon > 0$, there is $\widehat{n}$ such that $\frac{1}{\widehat{n}} < \varepsilon$.

(b) Clearly, $0 \leq \sqrt{n} - [\sqrt{n}] < 1$. Taking $n = k^2$, $k \in \mathbb{N}$, we see that $0 \in \mathbf{B}$. Thus $\inf \mathbf{B} = 0$. To show that $\sup \mathbf{B} = 1$, observe first that $\left[\sqrt{n^2 + 2n}\right] = n$ for each positive integer n. Suppose now $0 < \varepsilon < 1$. A simple calculation shows that the inequality

$$\sqrt{n^2 + 2n} - \left[\sqrt{n^2 + 2n}\right] = \frac{2}{\sqrt{1 + \frac{2}{n}} + 1} > 1 - \varepsilon$$

is satisfied for any integer $n > \frac{(1-\varepsilon)^2}{2\varepsilon}$.

1.1.11.

(a) $\sup\{x \in \mathbb{R} : x^2 + x + 1 > 0\} = +\infty$,

(b) $\inf\{z = x + x^{-1} : x > 0\} = 2$,

(c) $\inf\{z = 2^x + 2^{\frac{1}{x}} : x > 0\} = 4$.

The first two equalities are easily verifiable. To show the third one, observe that $\frac{a+b}{2} \geq \sqrt{ab}$ for $a, b > 0$. Therefore,

$$\frac{2^x + 2^{\frac{1}{x}}}{2} \geq \sqrt{2^{\frac{1}{x}+x}} \geq \sqrt{2^2} = 2$$

with equality if and only if $x = 1$. Thus (c) is proved.

1.1.12.

(a) Using the inequality $\frac{a+b}{2} \geq \sqrt{ab}$ for $a, b > 0$, we get

$$\frac{m}{n} + \frac{4n}{m} \geq 4$$

with equality for $m = 2n$. Therefore, $\inf \mathbf{A} = 4$. Taking $m = 1$, one can see that the set **A** is not bounded above. This means that $\sup \mathbf{A} = +\infty$.

(b) Similarly, we get
$$-\frac{1}{4} \leq \frac{mn}{4m^2+n^2} \leq \frac{1}{4},$$
with equalities for $m=-2n$ and $m=2n$, respectively. Consequently, $\inf \mathbf{B} = -\frac{1}{4}$ and $\sup \mathbf{B} = \frac{1}{4}$.

(c) We have $\inf \mathbf{C} = 0$ and $\sup \mathbf{C} = 1$. Indeed, $0 < \frac{m}{m+n} < 1$, and for any $\varepsilon > 0$ there exist positive integers n_1 and m_1 such that
$$\frac{1}{n_1+1} < \varepsilon \quad \text{and} \quad \frac{m_1}{m_1+1} > 1-\varepsilon.$$

(d) $\inf \mathbf{D} = -1$ and $\sup \mathbf{D} = 1$.

(e) One can take $m = n$ to see that the set is not bounded above. Hence $\sup \mathbf{E} = +\infty$. On the other hand, for any $m, n \in \mathbb{N}$ we have $\frac{mn}{1+m+n} \geq \frac{1}{3}$ with equality for $m = n = 1$. Therefore $\inf \mathbf{E} = \frac{1}{3}$.

1.1.13. Setting $s = a_1 + a_2 + \ldots + a_n$, we get
$$\frac{a_k}{s} \leq \frac{a_k}{a_k + a_{k+1} + a_{k+2}} \leq 1 - \frac{a_{k+1}}{s} - \frac{a_{k+2}}{s}.$$
As a result,
$$1 \leq \sum_{k=1}^{n} \frac{a_k}{a_k + a_{k+1} + a_{k+2}} \leq n-2.$$
Now, our task is to show that $\inf \sum_{k=1}^{n} \frac{a_k}{a_k+a_{k+1}+a_{k+2}} = 1$ and that $\sup \sum_{k=1}^{n} \frac{a_k}{a_k+a_{k+1}+a_{k+2}} = n-2$. To this end we take $a_k = t^k$, $t > 0$. Then
$$\sum_{k=1}^{n} \frac{a_k}{a_k + a_{k+1} + a_{k+2}} = \frac{t}{t+t^2+t^3}$$
$$+ \ldots + \frac{t^{n-2}}{t^{n-2}+t^{n-1}+t^n} + \frac{t^{n-1}}{t^{n-1}+t^n+t} + \frac{t^n}{t^n+t+t^2}$$
$$= (n-2)\frac{1}{1+t+t^2} + \frac{t^{n-2}}{t^{n-1}+t^{n-2}+1} + \frac{t^{n-1}}{t^{n-1}+t+1}.$$

1.1. Supremum and Infimum. Continued Fractions   131

Letting $t \to 0^+$, we see that $\sup \sum_{k=1}^{n} \frac{a_k}{a_k+a_{k+1}+a_{k+2}} = n-2$, and next letting $t \to +\infty$, we conclude that $\inf \sum_{k=1}^{n} \frac{a_k}{a_k+a_{k+1}+a_{k+2}} = 1$.

1.1.14. Fix $n \in \mathbb{N}$ and consider the $n+1$ real numbers
$$0,\ \alpha - [\alpha],\ 2\alpha - [2\alpha], \ldots,\ n\alpha - [n\alpha].$$
Each of them belongs to the interval $[0,1)$. Since the n intervals $[\frac{j}{n}, \frac{j+1}{n})$, $j = 0, 1, \ldots, n-1$, cover $[0, 1)$, there must be one which contains at least two of these points, say $n_1\alpha - [n_1\alpha]$ and $n_2\alpha - [n_2\alpha]$ with $0 \leq n_1 < n_2 \leq n$. So,
$$|n_2\alpha - [n_2\alpha] - n_1\alpha + [n_1\alpha]| < \frac{1}{n}.$$
Now, it is enough to take $q_n = n_2 - n_1$ and $p_n = [n_2\alpha] - [n_1\alpha]$. It follows from the above argument that $q_n \leq n$; that is, the second inequality also holds.

1.1.15. We will show that in any interval (p, q) there is at least one element of **A**. Put $0 < \varepsilon = q-p$. It follows from the preceding problem that there are p_n and q_n such that
$$\left|\alpha - \frac{p_n}{q_n}\right| < \frac{1}{q_n^2}.$$
Since α is irrational, $\lim_{n\to\infty} q_n = +\infty$. Therefore
$$|q_n\alpha - p_n| < \frac{1}{q_n} < \varepsilon$$
for almost all n. Now set $a = |q_n\alpha - p_n|$. Then at least one of the numbers ma, $m \in \mathbb{Z}$, belongs to the interval (p, q); that is, $mq_n\alpha - mp_n$ or $-mq_n\alpha + mp_n$ lies in this interval.

1.1.16. Let $t \in [-1, 1]$. Then there is an x such that $t = \cos x$. By the result in the foregoing problem, there exist integer sequences $\{m_n\}$ and $\{k_n\}$ such that $x = \lim_{n\to\infty}(k_n 2\pi + m_n)$. This and the continuity of the cosine function imply that
$$t = \cos x = \cos(\lim_{n\to\infty}(k_n 2\pi + m_n)) = \lim_{n\to\infty} \cos m_n = \lim_{n\to\infty} \cos|m_n|.$$

Hence each number of $[-1,1]$ is a limit point of the set $\{\cos n : n \in \mathbb{N}\}$. The desired result is proved.

1.1.17. It is obvious that, if there is n for which x_n is an integer, then x is rational. Assume now that $x = \frac{p}{q}$ with $p \in \mathbb{Z}$ and $q \in \mathbb{N}$. If $x - [x] \neq 0$, then $\frac{p}{q} - \left[\frac{p}{q}\right] = \frac{l}{q}$, where l is a positive integer smaller than q. Thus, the denominator of $x_1 = \frac{q}{l}$ is smaller than the denominator of x. This means that the denominators of $x_1, x_2, ...$ are successively strictly decreasing and cannot constitute an infinite sequence.

1.1.18. We will proceed by induction. It is easily verifiable that

$$R_k = \frac{p_k}{q_k} \quad \text{for } k = 0, 1, 2.$$

Assume that for an arbitrarily chosen $m \geq 2$,

$$R_m = \frac{p_m}{q_m} = \frac{p_{m-1} a_m + p_{m-2}}{q_{m-1} a_m + q_{m-2}}.$$

Note now that if we replace a_m in R_m by $a_m + \frac{1}{a_{m+1}}$, then we get the convergent R_{m+1}. Therefore

$$R_{m+1} = \frac{p_{m-1}\left(a_m + \frac{1}{a_{m+1}}\right) + p_{m-2}}{q_{m-1}\left(a_m + \frac{1}{a_{m+1}}\right) + q_{m-2}}$$

$$= \frac{(p_{m-1} a_m + p_{m-2}) a_{m+1} + p_{m-1}}{(q_{m-1} a_m + q_{m-2}) a_{m+1} + q_{m-1}} = \frac{p_{m+1}}{q_{m+1}}.$$

1.1.19. Denote

$$\Delta_k = p_{k-1} q_k - q_{k-1} p_k \quad \text{for } k = 1, 2, ..., n.$$

Then, for $k > 1$,

$$\Delta_k = p_{k-1}(q_{k-1} a_k + q_{k-2}) - q_{k-1}(p_{k-1} a_k + p_{k-2})$$
$$= -(p_{k-2} q_{k-1} - q_{k-2} p_{k-1}) = -\Delta_{k-1}.$$

Since $\Delta_1 = p_0 q_1 - q_0 p_1 = a_0 a_1 - (a_0 a_1 + 1) = -1$, we obtain $\Delta_k = (-1)^k$. This implies that p_k and q_k are relatively prime.

1.1. Supremum and Infimum. Continued Fractions

1.1.20. As in the solution of 1.1.18, we have, for $n > 1$,

$$R_n = \frac{p_n}{q_n} = \frac{p_{n-1}a_n + p_{n-2}}{q_{n-1}a_n + q_{n-2}}.$$

Analogously

$$x = \frac{p_n x_{n+1} + p_{n-1}}{q_n x_{n+1} + q_{n-1}} \quad \text{for } n = 1, 2, \ldots.$$

Hence

$$x - R_n = \frac{p_n a_{n+1} + p_{n-1}}{q_n a_{n+1} + q_{n-1}} - \frac{p_n}{q_n}$$
$$= \frac{p_{n-1}q_n - q_{n-1}p_n}{(q_n x_{n+1} + q_{n-1})q_n} = \frac{(-1)^n}{(q_n x_{n+1} + q_{n-1})q_n},$$

where the last equality follows from the result in 1.1.19. Therefore

$$x - R_n \begin{cases} > 0 & \text{for even } n, \\ < 0 & \text{for odd } n. \end{cases}$$

Thus x lies between two consecutive convergents.

1.1.21. We first prove that if α is a positive irrational, then the set $\{n - m\alpha : n, m \in \mathbb{N}\}$ is dense in $\mathbb{R}_+$. To this end take an interval (a, b), $0 < a < b$. We will show that this interval contains at least one element of our set. Put $\varepsilon = b - a > 0$. By the preceding problem there exists a convergent R_n such that

(1) $$0 < R_n - \alpha < \frac{1}{q_n^2}.$$

Indeed, take an odd n and observe that

$$(q_n x_{n+1} + q_{n-1})q_n > q_n^2.$$

Since $\lim\limits_{n \to \infty} q_n = +\infty$, for sufficiently large n we have $\frac{1}{q_n} < \varepsilon$. This and (1) imply that $0 < p_n - \alpha q_n < \frac{1}{q_n} < \varepsilon$ for sufficiently large n. Therefore there is $n_0 \in \mathbb{N}$ such that $n_0(p_n - \alpha q_n) \in (a, b)$. Now let $t \in [-1, 1]$. There is a positive x such that $t = \sin x$. It follows from the above considerations that there exist sequences of positive integers $\{m_n\}$ and $\{k_n\}$ for which $x = \lim\limits_{n\to\infty}(m_n - 2\pi k_n)$. By the continuity of the sine function,

$$t = \sin x = \sin(\lim_{n\to\infty}(m_n - 2\pi k_n)) = \lim_{n\to\infty}\sin m_n.$$

So, we have proved that any number of the interval $[-1,1]$ is a limit point of the set $\{\sin n : n \in \mathbb{N}\}$.

1.1.22. Let p_n and q_n be the integers defined in 1.1.20. Since $x_{n+1} = a_{n+1} + \frac{1}{x_{n+2}} > a_{n+1}$, we get $(q_n x_{n+1} + q_{n-1})q_n > (q_n a_{n+1} + q_{n-1})q_n = q_{n+1}q_n$. Therefore, by 1.1.20,

$$|x - R_n| < \frac{1}{q_n q_{n+1}}.$$

Since $q_{n+1} = q_n a_{n+1} + q_{n-1} > q_n a_{n+1} > q_n$, the desired inequality follows. We will show now that the sequence $\{q_n\}$ contains infinitely many odd numbers. Indeed, it follows from the result in 1.1.19 that q_n and q_{n+1} cannot be both even.

1.1.23. It is enough to apply the formula given in 1.1.19.

1.1.24. Observe first that the sequence $\{q_n\}$ is strictly increasing and $q_n \geq n$. Moreover, by Problem 1.1.20,

$$|x - R_n| = \frac{1}{(q_n x_{n+1} + q_{n-1})q_n}.$$

This and the inequality $x_{n+1} < a_{n+1} + 1$ imply that

$$|x - R_n| > \frac{1}{(q_n(a_{n+1} + 1) + q_{n-1})q_n} = \frac{1}{(q_{n+1} + q_n)q_n}.$$

Since $a_{n+2} \geq 1$, we have

$$|x - R_{n+1}| < \frac{1}{(q_{n+1}a_{n+2} + q_n)q_{n+1}} < \frac{1}{(q_{n+1} + q_n)q_n}.$$

These inequalities yield the desired result.

1.1. Supremum and Infimum. Continued Fractions

1.1.25. Let $\left|x - \frac{r}{s}\right| < |x - R_n| < |x - R_{n-1}|$. Since x lies between R_n and R_{n-1} (see Problem 1.1.20),

$$\left|\frac{r}{s} - R_{n-1}\right| < |R_{n-1} - R_n|.$$

Therefore, by the result in 1.1.23,

$$\frac{|rq_{n-1} - sp_{n-1}|}{sq_{n-1}} < \frac{1}{q_{n-1}q_n}.$$

Moreover, we have $\frac{1}{sq_{n-1}} < \frac{1}{q_{n-1}q_n}$ because $|rq_{n-1} - sp_{n-1}| \geq 1$. Hence $s > q_n$.

1.1.26. Following the algorithm given in 1.1.20, we get

$$a_0 = [\sqrt{2}] = 1, \quad x_1 = \frac{1}{\sqrt{2} - 1} = \sqrt{2} + 1.$$

Therefore, $a_1 = [x_1] = 2$. Similarly,

$$x_2 = \frac{1}{(\sqrt{2}+1) - 2} = x_1 \quad \text{and} \quad a_2 = a_1 = 2.$$

By induction,

$$\sqrt{2} = 1 + \frac{1|}{|2} + \frac{1|}{|2} + \frac{1|}{|2} + \ldots.$$

Likewise,

$$\frac{\sqrt{5}-1}{2} = \frac{1|}{|1} + \frac{1|}{|1} + \frac{1|}{|1} + \ldots.$$

1.1.27. Since $k < \sqrt{k^2 + k} < k+1$, $a_0 = [\sqrt{k^2+k}] = k$. As a result, $x_1 = \frac{\sqrt{k^2+k}+k}{k}$. Consequently, $2 < x_1 < 2 + \frac{1}{k}$ and $a_1 = 2$. Moreover,

$$x_2 = \frac{1}{\frac{1}{\sqrt{k^2+k}-k} - 2} = k + \sqrt{k^2 + k}.$$

Thus $2k < x_2 < 2k+1$ and $a_2 = 2k$. In much the same way we obtain $a_3 = 2$. Now, by induction,

$$\sqrt{k^2 + k} = k + \frac{1|}{|2} + \frac{1|}{|2k} + \frac{1|}{|2} + \frac{1|}{|2k} + \ldots.$$

1.1.28. Since $0 < x < 1$, we have $a_0 = 0$ and $x_1 = 1/x$. Therefore, $a_1 = n$ implies $[1/x] = n$. Hence, $1/x - 1 < n \leq 1/x$, which gives $1/(n+1) < x \leq 1/n$.

1.2. Some Elementary Inequalities

1.2.1. We will use induction. For $n = 1$ the inequality is obvious. Take an arbitrary positive integer n and assume that

$$(1 + a_1) \cdot (1 + a_2) \cdot \ldots \cdot (1 + a_n) \geq 1 + a_1 + a_2 + \ldots + a_n.$$

Then

$$(1 + a_1)(1 + a_2) \cdot \ldots \cdot (1 + a_n)(1 + a_{n+1})$$
$$\geq (1 + a_1 + a_2 + \ldots + a_n)(1 + a_{n+1})$$
$$= 1 + a_1 + a_2 + \ldots + a_n + a_{n+1} + a_{n+1}(a_1 + a_2 + \ldots + a_n)$$
$$\geq 1 + a_1 + a_2 + \ldots + a_n + a_{n+1}.$$

Thus the claim is established.

1.2.2. Induction will be used. For $n = 1$ our statement is clear. We suppose now that the claim holds for an arbitrarily chosen n. Without loss of generality we can assume that numbers $a_1, \ldots, a_{n+1}$ satisfying the condition $a_1 \cdot a_2 \cdot \ldots \cdot a_{n+1} = 1$ are enumerated in such a way that $a_1 \leq a_2 \leq \ldots \leq a_n \leq a_{n+1}$. Then $a_1 \leq 1$ and $a_{n+1} \geq 1$. Since $a_2 \cdot a_3 \cdot \ldots \cdot a_n \cdot (a_{n+1} \cdot a_1) = 1$, by our induction assumption, we have $a_2 + a_3 + \ldots + a_n + (a_{n+1} \cdot a_1) \geq n$. Hence

$$a_1 + a_2 + \ldots + a_n + a_{n+1} \geq n + a_{n+1} + a_1 - a_{n+1} \cdot a_1$$
$$= n + a_{n+1} \cdot (1 - a_1) + a_1 - 1 + 1$$
$$= n + 1 + (a_{n+1} - 1)(1 - a_1) \geq n + 1.$$

1.2.3. The inequalities follow from the statement proved in Problem 1.2.2. Indeed, replacing there the numbers a_j by $\frac{a_j}{\sqrt[n]{a_1 \cdot \ldots \cdot a_n}}$, we get $A_n \geq G_n$. The inequality $G_n \geq H_n$ follows from the already proved inequality $A_n \geq G_n$ provided one replaces a_j by its reciprocal $\frac{1}{a_j}$.

1.2. Some Elementary Inequalities

1.2.4. Using the arithmetic-geometric mean inequality, we have
$$\sqrt[n]{(1+nx)\cdot 1\cdot \ldots \cdot 1} \leq 1+x \qquad (n \text{ factors}).$$

1.2.5.

(a) Apply the arithmetic-harmonic mean inequality.

(b) Use the arithmetic-harmonic mean inequality.

(c) The left-hand inequality can be shown as in (a) and (b). To prove the right one, let us observe that
$$\frac{1}{3n+1}+\frac{1}{3n+2}+\ldots+\frac{1}{5n}+\frac{1}{5n+1} < \frac{1}{3n+1}+\frac{2n}{3n+2} < \frac{2}{3}.$$

(d) By the arithmetic-geometric mean inequality,
$$\frac{2}{1}+\frac{3}{2}+\frac{4}{3}+\ldots+\frac{n+1}{n} > n\sqrt[n]{n+1}.$$
Hence
$$1+1+1+\frac{1}{2}+1+\frac{1}{3}+\ldots+1+\frac{1}{n} > n\sqrt[n]{n+1}$$
and
$$1+\frac{1}{2}+\frac{1}{3}+\ldots+\frac{1}{n} > n(\sqrt[n]{n+1}-1).$$
To prove the other inequality we use the arithmetic-geometric mean inequality and get
$$\frac{1}{2}+\frac{2}{3}+\frac{3}{4}+\ldots+\frac{n}{n+1} > \frac{n}{\sqrt[n]{n+1}}.$$
This implies
$$1+\frac{1}{2}+\frac{1}{3}+\ldots+\frac{1}{n} < n\left(1-\frac{1}{\sqrt[n]{n+1}}+\frac{1}{n+1}\right).$$

1.2.6. By the inequality $G_n \leq A_n$ we get
$$x^n = \sqrt[2n+1]{1\cdot x\cdot \ldots \cdot x^{2n}} \leq \frac{1+\ldots+x^{2n}}{2n+1}.$$

1.2.7. The right-hand side of the inequality is a direct consequence of $G_n \leq A_n$. The other one can be proved by induction. This is clear for $n = 1$. Now we will prove that the inequality holds for $n + 1$ provided it does for n. To this end we show that $(a_1 a_{n+1})^{n+1} \leq (a_1 \cdot \ldots \cdot a_n \cdot a_{n+1})^2$, whenever $(a_1 a_n)^n \leq (a_1 \cdot \ldots \cdot a_n)^2$. We have

$$(a_1 a_{n+1})^{n+1} \leq a_1 \cdot a_n (a_1 \cdot \ldots \cdot a_n)^2 \cdot \left(\frac{a_{n+1}}{a_n}\right)^{n+1}.$$

Hence it is enough to show that

$$a_1 \frac{a_{n+1}^{n+1}}{a_n^n} \leq a_{n+1}^2.$$

Note that the last inequality can be rewritten as

$$a_1 \left(1 + \frac{d}{a_1 + (n-1)d}\right)^{n-1} \leq a_1 + (n-1)d,$$

where $a_n = a_1 + (n-1)d$, which is easy to prove by induction.

1.2.8. It is an immediate consequence of the foregoing result.

1.2.9. One can apply the arithmetic-harmonic mean inequality.

1.2.10.
(a) By the arithmetic-harmonic mean inequality we have

$$n \left(\sum_{k=1}^{n} \frac{1}{a_k}\right)^{-1} \leq \frac{1}{n} \sum_{k=1}^{n} a_k,$$

and consequently

$$\sum_{k=1}^{n} \frac{1}{a_k} \geq \frac{n^2}{s}.$$

Similarly, the inequality

$$n \left(\sum_{k=1}^{n} \frac{1}{s - a_k}\right)^{-1} \leq \frac{1}{n} \sum_{k=1}^{n} (s - a_k)$$

implies

$$\sum_{k=1}^{n} \frac{1}{s - a_k} \geq \frac{n^2}{s(n-1)}.$$

1.2. Some Elementary Inequalities

From this and from the equalities

$$\sum_{k=1}^{n}\frac{a_k}{s-a_k}=s\sum_{k=1}^{n}\frac{1}{s-a_k}-n \text{ and } \sum_{k=1}^{n}\frac{s-a_k}{a_k}=s\sum_{k=1}^{n}\frac{1}{a_k}-n$$

the desired result follows.

(b) See the solution of part (a).

(c) This follows by the same method as in (a).

1.2.11. Use the inequality $\frac{1+a_k}{2} \geq \sqrt{a_k}$.

1.2.12. We have

$$\sum_{k=1}^{n}a_k^2\sum_{k=1}^{n}b_k^2-\left(\sum_{k=1}^{n}a_kb_k\right)^2=\sum_{k,j=1}^{n}a_k^2b_j^2-\sum_{k,j=1}^{n}a_kb_ka_jb_j$$
$$=\frac{1}{2}\sum_{k,j=1}^{n}(a_kb_j-b_ka_j)^2\geq 0.$$

1.2.13. This inequality is equivalent to the following:

$$\sum_{k,j=1}^{n}(a_ka_j+b_kb_j)\leq\sum_{k,j=1}^{n}(a_k^2+b_k^2)^{\frac{1}{2}}(a_j^2+b_j^2)^{\frac{1}{2}},$$

which in turn is a direct consequence of the obvious inequality $a_ka_j+b_kb_j\leq (a_k^2+b_k^2)^{\frac{1}{2}}(a_j^2+b_j^2)^{\frac{1}{2}}$.

1.2.14. The claim follows from the Cauchy inequality.

1.2.15.

(a) By the Cauchy inequality,

$$\sum_{k=1}^{n}a_k\sum_{k=1}^{n}\frac{1}{a_k}\geq\left(\sum_{k=1}^{n}\sqrt{a_k\frac{1}{a_k}}\right)^2=n^2.$$

(b) By (a),

$$\sum_{k=1}^{n} a_k \sum_{k=1}^{n} \frac{1-a_k}{a_k} = \sum_{k=1}^{n} a_k \sum_{k=1}^{n} \frac{1}{a_k} - n \sum_{k=1}^{n} a_k$$

$$\geq n^2 - n \sum_{k=1}^{n} a_k = n \sum_{k=1}^{n} (1-a_k).$$

(c) By our assumption, $\log_a a_1 + \log_a a_2 + \ldots + \log_a a_n = 1$. This and the Cauchy inequality (Problem 1.2.12) give the desired result.

1.2.16. The inequality is equivalent to

$$0 \leq -4\alpha \left| \sum_{k=1}^{n} a_k b_k \right| + 4 \sum_{k=1}^{n} a_k^2 + \alpha^2 \sum_{k=1}^{n} b_k^2,$$

which holds for each real α, because

$$\Delta = 16 \left(\sum_{k=1}^{n} a_k b_k \right)^2 - 16 \sum_{k=1}^{n} a_k^2 \sum_{k=1}^{n} b_k^2 \leq 0.$$

1.2.17. Applying the Cauchy inequality, we obtain

$$\sum_{k=1}^{n} |a_k| = \sum_{k=1}^{n} 1 \cdot |a_k| \leq \sqrt{n} \left(\sum_{k=1}^{n} a_k^2 \right)^{\frac{1}{2}} \leq \sqrt{n} \sum_{k=1}^{n} |a_k|.$$

1.2.18.

(a) By the Cauchy inequality,

$$\left(\sum_{k=1}^{n} a_k b_k \right)^2 = \left(\sum_{k=1}^{n} \sqrt{k} a_k \frac{b_k}{\sqrt{k}} \right)^2 \leq \sum_{k=1}^{n} k a_k^2 \sum_{k=1}^{n} \frac{b_k^2}{k}.$$

(b) Likewise,

$$\left(\sum_{k=1}^{n} \frac{a_k}{k} \right)^2 = \left(\sum_{k=1}^{n} \frac{k^{\frac{3}{2}} a_k}{k^{\frac{5}{2}}} \right)^2 \leq \sum_{k=1}^{n} k^3 a_k^2 \sum_{k=1}^{n} \frac{1}{k^5}.$$

1.2. Some Elementary Inequalities

1.2.19. The Cauchy inequality gives

$$\left(\sum_{k=1}^{n} a_k^p\right)^2 = \left(\sum_{k=1}^{n} a_k^{\frac{p+q}{2}} a_k^{\frac{p-q}{2}}\right)^2 \leq \sum_{k=1}^{n} a_k^{p+q} \sum_{k=1}^{n} a_k^{p-q}.$$

1.2.20. By the Cauchy inequality,

$$\sum_{k=1}^{n} a_k^2 \cdot n = \sum_{k=1}^{n} a_k^2 \sum_{k=1}^{n} 1 \geq \left(\sum_{k=1}^{n} a_k\right)^2 = 1.$$

Hence $\sum_{k=1}^{n} a_k^2 \geq \frac{1}{n}$, with equality for $a_k = \frac{1}{n}$, $k = 1, 2, ..., n$. Therefore, the least value we are looking for is $\frac{1}{n}$.

1.2.21. In much the same way as in the solution of the last problem we get

$$1 = \left(\sum_{k=1}^{n} a_k\right)^2 = \left(\sum_{k=1}^{n} \sqrt{p_k} a_k \frac{1}{\sqrt{p_k}}\right)^2 \leq \sum_{k=1}^{n} p_k a_k^2 \sum_{k=1}^{n} \frac{1}{p_k}.$$

Thus,

$$\sum_{k=1}^{n} p_k a_k^2 \geq \frac{1}{\sum_{k=1}^{n} \frac{1}{p_k}}$$

with equality for $a_k = \frac{1}{p_k} \left(\sum_{k=1}^{n} \frac{1}{p_k}\right)^{-1}$. So the least value in question is $\left(\sum_{k=1}^{n} \frac{1}{p_k}\right)^{-1}$.

1.2.22. It follows from the solution of Problem 1.2.20 that

$$\left(\sum_{k=1}^{n} a_k\right)^2 \leq n \sum_{k=1}^{n} a_k^2.$$

Hence
$$\left(\sum_{k=1}^{n} a_k\right)^2 = \left((a_1 + a_2) + \sum_{k=3}^{n} a_k\right)^2 \leq (n-1)\left((a_1+a_2)^2 + \sum_{k=3}^{n} a_k^2\right)$$
$$= (n-1)\left(\sum_{k=1}^{n} a_k^2 + 2a_1 a_2\right).$$

1.2.23.

(a) By the Cauchy inequality,
$$\left(\sum_{k=1}^{n}(a_k+b_k)^2\right)^{\frac{1}{2}} = \left(\sum_{k=1}^{n}(a_k^2 + 2a_k b_k + b_k^2)\right)^{\frac{1}{2}}$$
$$\leq \left(\sum_{k=1}^{n} a_k^2 + 2\left(\sum_{k=1}^{n} a_k^2\right)^{\frac{1}{2}}\left(\sum_{k=1}^{n} b_k^2\right)^{\frac{1}{2}} + \sum_{k=1}^{n} b_k^2\right)^{\frac{1}{2}}$$
$$= \left(\sum_{k=1}^{n} a_k^2\right)^{\frac{1}{2}} + \left(\sum_{k=1}^{n} b_k^2\right)^{\frac{1}{2}}.$$

(b) By (a),
$$\left(\sum_{k=1}^{n} a_k^2\right)^{\frac{1}{2}} - \left(\sum_{k=1}^{n} b_k^2\right)^{\frac{1}{2}} \leq \left(\sum_{k=1}^{n}(a_k - b_k)^2\right)^{\frac{1}{2}}.$$
This and the inequality established in Problem 1.2.17 yield
$$\left(\sum_{k=1}^{n} a_k^2\right)^{\frac{1}{2}} - \left(\sum_{k=1}^{n} b_k^2\right)^{\frac{1}{2}} \leq \sum_{k=1}^{n} |a_k - b_k|.$$
Similarly,
$$\left(\sum_{k=1}^{n} b_k^2\right)^{\frac{1}{2}} - \left(\sum_{k=1}^{n} a_k^2\right)^{\frac{1}{2}} \leq \sum_{k=1}^{n} |a_k - b_k|$$
and the desired result is proved.

1.2. Some Elementary Inequalities   143

1.2.24. Since $\sum_{k=1}^{n} p_k a_k = 1$, we have $1 = \sum_{k=1}^{n} p_k a_k = \sum_{k=1}^{n} (p_k - \alpha) a_k + \alpha \sum_{k=1}^{n} a_k$ for any real α. Now, by the Cauchy inequality,

$$1 \leq \left(\sum_{k=1}^{n}(p_k - \alpha)^2 + \alpha^2\right)\left(\sum_{k=1}^{n} a_k^2 + \left(\sum_{k=1}^{n} a_k\right)^2\right).$$

Hence

$$\sum_{k=1}^{n} a_k^2 + \left(\sum_{k=1}^{n} a_k\right)^2 \geq \left(\sum_{k=1}^{n}(p_k - \alpha)^2 + \alpha^2\right)^{-1}.$$

Putting $\alpha = \frac{1}{n+1} \sum_{k=1}^{n} p_k$, we obtain the greatest lower bound. Therefore,

$$\sum_{k=1}^{n} a_k^2 + \left(\sum_{k=1}^{n} a_k\right)^2 \geq \frac{n+1}{(n+1)\sum_{k=1}^{n} p_k^2 - \left(\sum_{k=1}^{n} p_k\right)^2},$$

where the equality is attained for

$$a_k = \frac{(n+1)p_k - \sum_{k=1}^{n} p_k}{(n+1)\sum_{k=1}^{n} p_k^2 - \left(\sum_{k=1}^{n} p_k\right)^2}.$$

1.2.25. We will proceed by induction. For $n=1$ we get the equality $a_1 b_1 = a_1 b_1$. Moreover, if the inequality holds for n, then

$$\sum_{k=1}^{n+1} a_k \sum_{k=1}^{n+1} b_k - (n+1)\sum_{k=1}^{n+1} a_k b_k$$
$$\leq a_{n+1}\sum_{k=1}^{n} b_k + b_{n+1}\sum_{k=1}^{n} a_k - n a_{n+1} b_{n+1} - \sum_{k=1}^{n} a_k b_k$$
$$= \sum_{k=1}^{n}(b_{n+1} - b_k)(a_k - a_{n+1}) \leq 0.$$

1.2.26. We will use induction on p. For $p = 1$ the equality $a_1^p = a_1^p$ holds. Assuming the inequality to hold for p, we will prove it for $p+1$. Obviously, without loss of generality, we may suppose that the numbers a_k are enumerated in such a way that $a_1 \leq a_2 \leq \ldots \leq a_n$. Now, by our induction assumption and the result in the foregoing problem,

$$\left(\frac{1}{n}\sum_{k=1}^{n} a_k\right)^{p+1} \leq \frac{1}{n^2}\sum_{k=1}^{n} a_k^p \sum_{k=1}^{n} a_k \leq \frac{1}{n}\sum_{k=1}^{n} a_k^{p+1}.$$

1.2.27. We have

$$(1+c)a^2 + \left(1 + \frac{1}{c}\right)b^2 = a^2 + b^2 + \left(\sqrt{c}a - \frac{1}{\sqrt{c}}b\right)^2 + 2ab \geq (a+b)^2.$$

1.2.28. Clearly, $\sqrt{a^2+b^2} + \sqrt{a^2+c^2} \geq |b| + |c| \geq |b+c|$. Hence $|b^2 - c^2| \leq |b-c|\left(\sqrt{a^2+b^2} + \sqrt{a^2+c^2}\right)$, which is equivalent to the desired inequality.

1.2.29.

(a) For any real numbers a, b, c we have $a^2 + b^2 + c^2 \geq ab + bc + ca$. Thus $b^2c^2 + a^2c^2 + a^2b^2 \geq abc(a+b+c)$, which is equivalent to our claim.

(b) The desired result follows from the inequality $a^2 + b^2 + c^2 \geq ab + bc + ca$ in much the same way as in (a).

(c) This is a consequence of the arithmetic-harmonic mean inequality.

(d) We have

$$\frac{b^2 - a^2}{c+a} = \frac{b+a}{c+a}(b-a) = \frac{b+a}{c+a}\left((b+c) - (c+a)\right).$$

Putting $u = a+b$, $v = b+c$ and $z = c+a$, we obtain

1.2. Some Elementary Inequalities   145

$$\frac{b^2-a^2}{c+a}+\frac{c^2-b^2}{a+b}+\frac{a^2-c^2}{b+c} = \frac{u}{z}(v-z)+\frac{v}{u}(z-u)+\frac{z}{v}(u-v)$$

$$=\frac{u^2v^2+v^2z^2+z^2u^2-(u^2vz+v^2uz+z^2uv)}{uvz}$$

$$=\frac{u^2(v^2+z^2)+v^2(u^2+z^2)+z^2(u^2+v^2)-2(u^2vz+v^2uz+z^2uv)}{2uvz}$$

≥ 0.

(e) For $a=b$ the inequality is clear. Assume now that $0<b<a$. Then

$$\frac{a-b}{2\sqrt{a}} = \frac{(\sqrt{a}-\sqrt{b})(\sqrt{a}+\sqrt{b})}{2\sqrt{a}} < \sqrt{a}-\sqrt{b} < \frac{a-b}{2\sqrt{b}}$$

and thus

$$\frac{(a-b)^2}{4a} < (\sqrt{a}-\sqrt{b})^2 = a+b-2\sqrt{ab} < \frac{(a-b)^2}{4b}.$$

1.2.30. Let $m=\frac{a_i}{b_i}$. Then

$$m(b_1+\ldots+b_n) = \frac{a_i}{b_i}(b_1+b_2+\ldots+b_n) = \frac{a_i}{b_i}b_1+\frac{a_i}{b_i}b_2+\ldots+\frac{a_i}{b_i}b_n$$

$$\leq \frac{a_1}{b_1}b_1+\frac{a_2}{b_2}b_2+\ldots+\frac{a_n}{b_n}b_n = a_1+\ldots+a_n \leq M(b_1+\ldots+b_n).$$

1.2.31. The inequalities follow from the result in the foregoing problem and from the monotonicity of the tangent function on $(0,\pi/2)$.

1.2.32. Apply the inequality given in 1.2.30 with $a_i=\ln c_i$ and $b_i=k_i$, $i=1,2,\ldots,n$.

1.2.33. Note that

$$\frac{a_1}{b_1} \leq M, \quad \frac{a_2^2}{Mb_2^2} \leq M, \quad \ldots, \quad \frac{a_n^n}{M^{n-1}b_n^n} \leq M$$

and use the inequality proved in 1.2.30.

1.2.34. By the arithmetic-harmonic mean inequality (see, e.g. 1.2.3),

$$\frac{n}{\frac{1}{x-a_1}+\frac{1}{x-a_2}+\ldots+\frac{1}{x-a_n}} \leq \frac{(x-a_1)+(x-a_2)+\ldots+(x-a_n)}{n}$$

$$= \frac{nx-(a_1+a_2+\ldots+a_n)}{n}.$$

The desired result follows easily.

1.2.35. Observe that

$$1+c_1+c_2+\ldots+c_n = (1+1)^n = 2^n,$$

and apply the Cauchy inequality (Problem 1.2.12) with $a_k = 1$ and $b_k = \sqrt{c_k}$, $k = 1, 2, \ldots, n$.

1.2.36. Since

$$\prod_{k=0}^{n}\binom{n}{k} = \prod_{k=1}^{n-1}\binom{n}{k} \quad \text{and} \quad 2^n - 2 = \sum_{k=1}^{n-1}\binom{n}{k},$$

the claim follows at once from the arithmetic-geometric mean inequality (Problem 1.2.3).

1.2.37. By the arithmetic-geometric mean inequality (see 1.2.3),

$$A_k^{p-1}A_{k-1} \leq \frac{(p-1)A_k^p + A_{k-1}^p}{p}, \quad k = 1, 2, \ldots, n,$$

where $A_0 = 0$. It follows that

$$A_k^p - \frac{p}{p-1}A_k^{p-1}a_k = A_k^p - \frac{p}{p-1}A_k^{p-1}(kA_k - (k-1)A_{k-1})$$

$$= A_k^p\left(1 - \frac{kp}{p-1}\right) + A_k^{p-1}A_{k-1}\frac{(k-1)p}{p-1} \leq A_k^p\left(1 - \frac{kp}{p-1}\right)$$

$$+ \frac{k-1}{p-1}\left((p-1)A_k^p + A_{k-1}^p\right) = \frac{1}{p-1}\left((k-1)A_{k-1}^p - kA_k^p\right).$$

Now, adding these inequalities we get our claim.

1.2. Some Elementary Inequalities

1.2.38. Assume that $a_i = \max\{a_1, a_2, ..., a_n\}$. Then

$$\sum_{k=1}^{n-1} a_k a_{k+1} = \sum_{k=1}^{i-1} a_k a_{k+1} + \sum_{k=i}^{n-1} a_k a_{k+1} \leq a_i \sum_{k=1}^{i-1} a_k + a_i \sum_{k=i}^{n-1} a_{k+1}$$

$$= a_i(a - a_i) = \frac{a^2}{4} - \left(\frac{a}{2} - a_i\right)^2 \leq \frac{a^2}{4}.$$

1.2.39. One can apply the result in 1.2.2.

1.2.40. The left inequality follows from 1.2.1.
(a) Observe that

$$1 + a_k = \frac{1 - a_k^2}{1 - a_k} < \frac{1}{1 - a_k}.$$

Hence

$$\prod_{k=1}^{n}(1 + a_k) < \left(\prod_{k=1}^{n}(1 - a_k)\right)^{-1}.$$

Since $a_1 + a_2 + ... + a_n < 1$, applying once again the result in 1.2.1, we get

$$\prod_{k=1}^{n}(1 + a_k) < \left(1 - \sum_{k=1}^{n} a_k\right)^{-1}.$$

(b) Use the same reasoning as in (a).

1.2.41. Apply the inequality given in 1.2.15 (b), replacing a_k by $1 - a_k$.

1.2.42. Since $0 < a_k \leq 1$ for $k = 1, 2, ..., n$, the inequality

(1) $$\sum_{k=1}^{n} a_k \geq \prod_{k=1}^{n} a_k \cdot \sum_{k=1}^{n} \frac{1}{a_k}$$

holds for $n \geq 2$. Now, applying the inequality from 1.2.15 (b) with a_k replaced by $\frac{a_k}{1+a_k}$, $k = 1, 2, ..., n$, we get

$$\sum_{k=1}^{n} \frac{1}{a_k}\left(n - \sum_{k=1}^{n} \frac{1}{1 + a_k}\right) \geq n \sum_{k=1}^{n} \frac{1}{1 + a_k}.$$

Multiplying both sides of this inequality by $\prod_{k=1}^{n} a_k$ and using (1), we get our result.

1.2.43.

(a) By the arithmetic-geometric mean inequality (Problem 1.2.3),

$$\frac{\prod_{k=1}^{n}(1+a_k)}{(n+1)^n} = \frac{2a_1+a_2+\ldots+a_n}{n+1} \cdot \frac{a_1+2a_2+a_3+\ldots+a_n}{n+1}$$
$$\cdot \ldots \cdot \frac{a_1+a_2+\ldots+2a_n}{n+1} \geq \prod_{k=1}^{n} a_k.$$

(b) The proof of this part runs as in (a).

1.2.44. Observe first that if $\sum_{k=1}^{n} \frac{1}{1+a_k} = n-1$, then $\sum_{k=1}^{n} \frac{a_k}{1+a_k} = 1$. To get our result it is enough to apply the inequality given in 1.2.43 (b) with a_k replaced by $\frac{a_k}{1+a_k}$.

1.2.45. [M. S. Klamkin, Amer. Math. Monthly 82(1975), 741-742] We may assume that $a_1, a_2, \ldots, a_n$ are enumerated in such a way that $a_1 = \min\{a_1, a_2, \ldots, a_n\}$ and $a_2 = \max\{a_1, a_2, \ldots, a_n\}$, and let $A_n = 1/n$ be the arithmetic mean of $a_1, \ldots, a_n$. Define a new sequence $\{a'_k\}$ by setting $a'_1 = A_n$, $a'_2 = a_1 + a_2 - A_n$, $a'_i = a_i$ for $3 \leq i \leq n$. We will show that

(1) $$\prod_{k=1}^{n} \frac{1+a_k}{1-a_k} \geq \prod_{k=1}^{n} \frac{1+a'_k}{1-a'_k}.$$

It follows from the definition of the sequence $\{a'_k\}$ that (1) is equivalent to

$$\frac{(1+a_1)(1+a_2)}{(1-a_1)(1-a_2)} \geq \frac{(1+A_n)(1+a_1+a_2-A_n)}{(1-A_n)(1-a_1-a_2+A_n)},$$

which in turn is equivalent to

$$(A_n - a_1)(A_n - a_2) \leq 0.$$

The last inequality is an immediate consequence of our assumptions. Now, we repeat the above procedure for the sequence $\{a'_k\}$ to get the

1.2. Some Elementary Inequalities

sequence $\{a_k''\}$. At least two terms of the sequence $\{a_k''\}$ are equal to an A_n. Moreover, the sequence satisfies an inequality of type (1). If we repeat this procedure at most $n-1$ times, we get the constant sequence with each term equal to A_n. In view of (1),

$$\prod_{k=1}^n \frac{1+a_k}{1-a_k} \geq \prod_{k=1}^n \frac{1+A_n}{1-A_n} = \left(\frac{n+1}{n-1}\right)^n.$$

1.2.46. Let $a_{k_1} = \max\{a_1, a_2, ..., a_n\}$. There is a fraction on the left side of the inequality whose numerator is equal to a_{k_1}. The denominator of this fraction has two terms. Let us denote the greater one by a_{k_2}. Now, take the fraction whose numerator is a_{k_2} and denote by a_{k_3} the greater of the two terms of its denominator, etc. Note that

(1) $$\frac{a_{k_i}}{a_{k_i+1} + a_{k_i+2}} \geq \frac{a_{k_i}}{2a_{k_{i+1}}}, \quad i = 1, 2, ...$$

It follows from the above construction that there exists an l such that $a_{k_{l+1}} = a_{k_1}$. Next, observe that the numbers a_{k_i} and $a_{k_{i+1}}$ appear in our inequality as numerators of either two neighbor fractions or two fractions which are separated by only one term (we assume here that the last and the first are neighbor quotients). Moreover, $a_{k_{i+1}}$ appears as a numerator of a fraction that is to the right of the fraction with the numerator a_{k_i}. To pass from the fraction with the numerator a_{k_1} to the fraction with the numerator $a_{k_{l+1}}$, l steps are needed, where $l \geq \frac{n}{2}$. Hence, by (1) and the arithmetic-geometric mean inequality,

$$\frac{a_{k_1}}{2a_{k_2}} + \frac{a_{k_2}}{2a_{k_3}} + ... + \frac{a_{k_l}}{2a_{k_1}} \geq l\sqrt[l]{\frac{1}{2^l}} \geq \frac{n}{4}.$$

1.2.47. We have

$$\sum_{k=1}^n \frac{\sqrt{|a_k-t|}}{2^k} \geq \left(\frac{1}{2^2} + \frac{1}{2^3} + ... + \frac{1}{2^n}\right)\sqrt{|a_1-t|} + \frac{\sqrt{|a_2-t|}}{2^2}$$

$$+ ... + \frac{\sqrt{|a_n-t|}}{2^n} \geq \frac{1}{2^2}(\sqrt{|a_1-t|} + \sqrt{|a_2-t|}) + \frac{1}{2^3}(\sqrt{|a_1-t|}$$

$$+ \sqrt{|a_3-t|}) + ... + \frac{1}{2^n}(\sqrt{|a_1-t|} + \sqrt{|a_n-t|}),$$

which implies the desired inequality.

1.2.48. By the arithmetic-geometric mean inequality,
$$\sqrt[n]{\frac{a_1}{a_1+b_1}\cdot\frac{a_2}{a_2+b_2}\cdot\ldots\cdot\frac{a_n}{a_n+b_n}}+\sqrt[n]{\frac{b_1}{a_1+b_1}\cdot\frac{b_2}{a_2+b_2}\cdot\ldots\cdot\frac{b_n}{a_n+b_n}}$$
$$\leq \frac{1}{n}\left(\frac{a_1}{a_1+b_1}+\ldots+\frac{a_n}{a_n+b_n}+\frac{b_1}{a_1+b_1}+\ldots+\frac{b_n}{a_n+b_n}\right)=1.$$

1.2.49. [V. Ptak, Amer. Math. Monthly 102(1995), 820-821] First, observe that if we replace each a_k by ca_k with $c>0$, neither the left side nor the right side of the inequality is changed. Therefore, we can assume that $G=1$. Then $a_n=\frac{1}{a_1}$. Observe now that if $a_1\leq x\leq \frac{1}{a_1}$, then $x+\frac{1}{x}\leq a_1+\frac{1}{a_1}$. Hence
$$\sum_{n=1}^{n}p_k a_k + \sum_{k=1}^{n}p_k\frac{1}{a_k}\leq a_1+\frac{1}{a_1}=2A.$$
Now, to obtain our claim we may apply the arithmetic-geometric mean inequality.

1.2.50. Let us arrange all the positive divisors of n into pairs (k,l) in such a way that $kl=n$. By the arithmetic-geometric mean inequality, $\frac{k+l}{2}\geq\sqrt{kl}$. Adding these inequalities, we get
$$\frac{\sigma(n)}{2}\geq\frac{\tau(n)}{2}\sqrt{n}.$$

Chapter 2

Sequences of Real Numbers

2.1. Monotonic Sequences

2.1.1.

(a) Let $\{a_n\}$ be an increasing sequence bounded from above. Then $\sup\{a_n : n \in \mathbb{N}\} = A < \infty$. Thus for any $n \in \mathbb{N}$, $a_n \leq A$. Since for any $\varepsilon > 0$ the number $A - \varepsilon$ is not an upper bound of the set $\{a_n : n \in \mathbb{N}\}$, there is a_{n_0} such that $a_{n_0} > A - \varepsilon$. By the monotonicity of the sequence, $A \geq a_n > A - \varepsilon$ for any $n > n_0$. Hence $\lim\limits_{n \to \infty} a_n = A$.

Assume now that $\{a_n\}$ is not bounded above. Then for any M there is a_{n_0} such that $a_{n_0} > M$. Again, by the monotonicity of the sequence, $a_n > M$ for $n > n_0$, and therefore $\lim\limits_{n \to \infty} a_n = +\infty$.

(b) See the solution of (a).

2.1.2. We have
$$\frac{s_n}{s_{n-1}} \leq \frac{s_{n+1}}{s_n} \quad \text{for} \quad n \geq 2.$$
Indeed, by 1.2.19,

(1) $$s_n^2 \leq s_{n+1} s_{n-1}.$$

We will show that $\{x_n\}$ is an increasing sequence. The inequality $x_1 \leq x_2$ follows from $\left(\sum_{k=1}^{p} a_k\right)^2 \leq p \sum_{k=1}^{p} a_k^2$ (see the solution of 1.2.20). Assume now that $x_{n-1} \leq x_n$. Then

(2) $$s_{n-1} \leq s_n^{\frac{n-1}{n}}.$$

Hence, by (1) and (2),

$$x_{n+1} = \sqrt[n+1]{s_{n+1}} \geq \sqrt[n+1]{\frac{s_n^2}{s_{n-1}}} \geq \sqrt[n+1]{\frac{s_n^2}{s_n^{\frac{n-1}{n}}}} = x_n.$$

2.1.3. We have $a_{n+1} = \frac{n+1}{2n} a_n < a_n$, $n > 1$. Therefore $\{a_n\}$ is a strictly decreasing sequence. Since it is bounded below (e.g. by 0), $\lim_{n\to\infty} a_n = g$ exists. The number g satisfies the equation $g = \frac{1}{2}g$. Consequently, $g = 0$.

2.1.4. Let $b_n = a_n - \frac{1}{2^{n-1}}$. We have $b_{n+1} - b_n = a_{n+1} - a_n + \frac{1}{2^n} \geq 0$. Hence the sequence $\{b_n\}$ converges, and so does $\{a_n\}$.

2.1.5.

(a) We will show that the sequence $\{a_n\}$ is monotonically decreasing and bounded below. Indeed,

$$a_{n+1} - a_n = \frac{-1}{\sqrt{n+1}(\sqrt{n+1}+\sqrt{n})^2} < 0.$$

Moreover, by the inequality given in the hint (one can prove it by induction), $a_n > 2(\sqrt{n+1} - \sqrt{n} - 1) > -2$.

(b) The proof follows by the same method as in (a).

2.1.6. We first show by induction that $\frac{3}{2} \leq a_n \leq 2$ for $n \in \mathbb{N}$ and that the sequence $\{a_n\}$ is strictly increasing. These two facts imply the convergence of $\{a_n\}$. Let $g = \lim_{n\to\infty} a_n$. Because $a_n = \sqrt{3a_{n-1} - 2}$ we get $g = \sqrt{3g - 2}$, and consequently, $g = 2$.

2.1.7. One can establish by induction that $a_n > 2c$. Of course, $a_1 < a_2$. Moreover, if $a_n > a_{n-1}$, then

$$a_{n+1} = (a_n - c)^2 > (a_{n-1} - c)^2 = a_n.$$

2.1. Monotonic Sequences

The last inequality follows from the monotonicity of the function $f(x) = x^2$ on $\mathbb{R}_+$.

2.1.8. By the arithmetic-geometric mean inequality and by our assumptions we get
$$\frac{a_n + (1 - a_{n+1})}{2} \geq \sqrt{a_n(1 - a_{n+1})} > \frac{1}{2}.$$
Hence $a_n - a_{n+1} > 0$. Therefore the sequence $\{a_n\}$ converges to a g. Since $a_n(1 - a_{n+1}) > \frac{1}{4}$, we get $g(1 - g) \geq \frac{1}{4}$. The last inequality is equivalent to $(2g - 1)^2 \leq 0$, which gives $g = \frac{1}{2}$.

2.1.9. Obviously, $0 \leq a_n < 3$ for $n \geq 1$. Moreover, $a_{n+1}^2 - a_n^2 = -a_n^2 + a_n + 6 > 0$ for $0 \leq a_n < 3$. Thus the sequence is monotonically increasing and bounded above, so it converges. By the definition of the sequence, $\lim\limits_{n \to \infty} a_n = 3$.

2.1.10. We see at once that $0 \leq a_n < 1$ for $n \geq 1$. To prove the monotonicity of the sequence we will need the following form of the principle of induction:

$W(n)$ *is true for all natural numbers* n, *if the following two conditions hold:*

(i) $W(1)$ *is true.*

(ii) *The truth of* $W(k)$ *for* $1 \leq k \leq n$ *implies* $W(n + 1)$ *is also true.*

Assume now that $a_{n-1} \geq a_{n-2}$ and $a_n \geq a_{n-1}$. Then
$$a_{n+1} - a_n = \frac{1}{3}(a_n - a_{n-1} + a_{n-1}^3 - a_{n-2}^3) \geq 0.$$
Therefore the sequence is convergent. Let g denote its limit. Then $g = \frac{1}{3}(1 + g + g^3)$. Consequently,
$$g = 1 \quad \text{or} \quad g = \frac{-1 + \sqrt{5}}{2} \quad \text{or} \quad g = \frac{-1 - \sqrt{5}}{2}.$$
Observe that all the terms of the sequence are nonnegative and less than $\frac{-1+\sqrt{5}}{2}$. Thus $\lim\limits_{n \to \infty} a_n = \frac{-1+\sqrt{5}}{2}$.

2.1.11. We have $a_{n+1} = a_n \frac{n+1}{2n+3} < a_n$, $n \geq 1$. Therefore (see the solution of 2.1.3) we get $g = 0$.

2.1.12. Since $a_{n+1} = a_n \frac{2n+2}{2n+3} < a_n$, $n \geq 1$, the sequence is monotonically decreasing. It is bounded below by zero, so it converges.

2.1.13.

(a) Clearly, $\{a_n\}$ is monotonically increasing. We will show that it is also bounded above. Indeed,

$$a_n = 1 + \frac{1}{2^2} + \frac{1}{3^2} + \ldots + \frac{1}{n^2} < 1 + \frac{1}{1 \cdot 2} + \frac{1}{2 \cdot 3} + \ldots + \frac{1}{(n-1)n}$$
$$= 1 + \left(1 - \frac{1}{2}\right) + \left(\frac{1}{2} - \frac{1}{3}\right) + \ldots + \left(\frac{1}{n-1} - \frac{1}{n}\right) = 2 - \frac{1}{n} < 2.$$

(b) Obviously, $\{a_n\}$ is monotonically increasing. Moreover,

$$a_n = 1 + \frac{1}{2^2} + \frac{1}{3^3} + \ldots + \frac{1}{n^n} < 1 + \frac{1}{2^2} + \frac{1}{3^2} + \ldots + \frac{1}{n^2}.$$

Hence it follows from (a) that the sequence is bounded above.

2.1.14. For $n \geq 1$, we have

$$a_{n+1} - a_n = -\frac{1}{\sqrt{n(n+1)}} + \frac{1}{\sqrt{2n(2n+1)}} + \frac{1}{\sqrt{(2n+1)(2n+2)}} < 0.$$

Hence the sequence is convergent, as it is monotonically decreasing and bounded below.

2.1.15. From the arithmetic-geometric mean inequality we get

$$a_{n+1} \geq \sqrt[p]{a_n^{p-1} \frac{a}{a_n^{p-1}}} = \sqrt[p]{a}, \quad n \geq 1.$$

Therefore

$$a_{n+1} - a_n = -\frac{a_n}{p} + \frac{a}{p a_n^{p-1}} = -\frac{a_n^p - a}{p a_n^{p-1}} \leq 0, \quad n \geq 2,$$

which shows that the sequence converges and $\lim_{n \to \infty} a_n = \sqrt[p]{a}$.

2.1. Monotonic Sequences

2.1.16. Clearly, $0 < a_n < 2$ for $n \geq 1$. Moreover,
$$a_{n+1}^2 - a_n^2 = \sqrt{a_n} - \sqrt{a_{n-1}} > 0 \quad \text{provided} \quad a_n > a_{n-1}.$$
Hence the sequence converges to a g which satisfies the equation $g = \sqrt{2 + \sqrt{g}}$.

Remark. Using Cardan's formula for real roots of cubic polynomials one can show that
$$g = \frac{1}{3}\left(\sqrt[3]{\frac{1}{2}(79 + 3\sqrt{249})} + \sqrt[3]{\frac{1}{2}(79 - 3\sqrt{249})} - 1\right).$$

2.1.17. Note that $a_{n+1} = 2\left(2 - \frac{5}{a_n+3}\right)$, $n \geq 1$. Now one can establish by induction that $0 < a_n < 2$, $n \geq 1$. Moreover,
$$a_{n+1} - a_n = -\frac{(a_n+1)(a_n-2)}{a_n+3} \geq 0.$$
Hence the sequence converges and $\lim\limits_{n \to \infty} a_n = 2$.

2.1.18. One can show by induction that the sequence $\{a_n\}$ is strictly increasing. If it were bounded above then there would exist a number g such that $g = \lim\limits_{n \to \infty} a_n$. We would also have $g^2 - 2g + c = 0$. This equation has a real solution provided $c \leq 1$. So, assume that $0 < c \leq 1$. Then the sequence $\{a_n\}$ is bounded above by $1 - \sqrt{1-c}$, and $\lim\limits_{n \to \infty} a_n = 1 - \sqrt{1-c}$.

For $c > 1$, the sequence is strictly increasing and it does not converge. So, it diverges to $+\infty$.

2.1.19. Since
$$a_{n+1} = a_n\left(1 - 2\frac{a_n^2 - a}{3a_n^2 + a}\right), \quad n \geq 1,$$
we get
$$\text{if } a_n > \sqrt{a}, \quad \text{then} \quad a_{n+1} < a_n,$$
$$\text{if } a_n < \sqrt{a}, \quad \text{then} \quad a_{n+1} > a_n,$$
$$\text{if } a_n = \sqrt{a}, \quad \text{then} \quad a_{n+1} = \sqrt{a}.$$
Observe now that
$$a_n\frac{a_n^2 + 3a}{3a_n^2 + a} > \sqrt{a} \quad \text{if and only if} \quad (a_n - \sqrt{a})^3 > 0,$$

which in turn is equivalent to $a_n > \sqrt{a}$. Finally,

if $0 < a_1 < \sqrt{a}$, then $\{a_n\}$ is increasing and bounded above by $\sqrt{a}$;
if $a_1 > \sqrt{a}$, then $\{a_n\}$ is decreasing and bounded below by $\sqrt{a}$;
if $a_1 = \sqrt{a}$, then $\{a_n\}$ is a constant sequence.

In each of the above cases the sequence converges to $\sqrt{a}$.

2.1.20. One can show by induction that
$$a_n = \frac{(3^{n-1} - 1) - (3^{n-1} - 3)a_1}{(3^n - 1) - (3^n - 3)a_1} \quad \text{for} \quad n = 1, 2, 3, \ldots.$$

Therefore the sequence is not defined for $a_1 = \frac{3^{n+1}-1}{3^{n+1}-3}$ with $n \in \mathbb{N}$. Moreover, if $a_1 = 1$, then $a_n = 1$ for $n = 1, 2, 3, \ldots$. For other values of a_1, the sequence converges to $1/3$.

2.1.21. We have $a_{n+1} = (a_n - a)^2 + a_n \geq a_n$ for $n \geq 1$. Hence the sequence is monotonically increasing. Moreover, if it converges, then $\lim\limits_{n\to\infty} a_n = a$. Therefore if $a_1 > a$, then the given sequence diverges. In the case where $a-1 \leq a_1 \leq a$, we have also $a-1 \leq a_n \leq a$ for $n > 1$. Thus for such a_1 the sequence converges. Finally, if $a_1 < a-1$, then $a_2 > a$, and consequently, the sequence diverges.

2.1.22. It is obvious that the sequence may converge either to a or to b. We will consider the following cases.
1^0 $c > b$.
Then $a_2 = \frac{c^2+ab}{a+b} > c = a_1$ and by induction $a_{n+1} > a_n$. Hence $\lim\limits_{n\to\infty} a_n = +\infty$.
2^0 $c = b$.
Obviously, $a_n = b$ for $n = 1, 2, 3, \ldots$.
3^0 $a < c < b$.
One can establish inductively that the sequence $\{a_n\}$ is monotonically decreasing and bounded below by a. Hence $\lim\limits_{n\to\infty} a_n = a$.
4^0 $c = a$.
Clearly, $a_n = a$ for $n = 1, 2, 3, \ldots$.

2.1. Monotonic Sequences

5^0 $0 < c < a$.

Induction is used once again to show that $\{a_n\}$ is monotonically increasing and bounded above by a. It then follows that $\lim\limits_{n \to \infty} a_n = a$.

2.1.23. Note that $a_{n+1} = 6\left(1 - \frac{6}{a_n+7}\right)$ for $n \in \mathbb{N}$. Hence by induction

$$\text{if } a_1 < 2, \quad \text{then} \quad a_n < 2, \; n \in \mathbb{N};$$
$$\text{if } a_1 > 2, \quad \text{then} \quad a_n > 2, \; n \in \mathbb{N}.$$

Moreover,
$$a_{n+1} - a_n = -\frac{(a_n + 3)(a_n - 2)}{a_n + 7}.$$

Therefore

1^0 if $0 < a_1 < 2$, then the sequence is increasing and bounded above by 2, and $\lim\limits_{n \to \infty} a_n = 2$,

2^0 if $a_1 > 2$, then the sequence is decreasing and bounded below by 2, and $\lim\limits_{n \to \infty} a_n = 2$,

3^0 if $a_1 = 2$, then $a_n = 2$ for $n \in \mathbb{N}$.

2.1.24. Since $0 = a_1 \leq a_2$ and $a_{n+1}^2 - a_n^2 = a_n - a_{n-1}$, we see by induction that $a_{n+1} \geq a_n$ for $n \in \mathbb{N}$. The sequence is bounded above, e.g. by $\sqrt{1+4c}$. One can easily establish that $\lim\limits_{n \to \infty} a_n = \frac{1+\sqrt{1+4c}}{2}$.

2.1.25. Since $a_2 = \sqrt{2\sqrt{2}} > \sqrt{2} = a_1$ and $a_{n+1}^2 - a_n^2 = 2(a_n - a_{n-1})$, one can show inductively that $a_{n+1} \geq a_n$ for all positive integers. The sequence is bounded above by 2, and $\lim\limits_{n \to \infty} a_n = 2$.

2.1.26. For $k = 1$, we get $a_n = 5^n$, $n \in \mathbb{N}$, and therefore $\{a_n\}$ diverges to $+\infty$.

For $k > 1$,
$$a_2 = \sqrt[k]{5\sqrt[k]{5}} > \sqrt[k]{5} = a_1 \quad \text{and} \quad a_{n+1}^k - a_n^k = 5(a_n - a_{n-1}).$$

It then follows by induction that $\{a_n\}$ is strictly increasing. Moreover, $a_n < \sqrt[k-1]{5}$ for $n \in \mathbb{N}$. One can easily verify that $\lim\limits_{n \to \infty} a_n = \sqrt[k-1]{5}$.

2.1.27. We see (by induction) that $1 \leq a_n \leq 2$, $n \in \mathbb{N}$. The monotonicity of the sequence follows from the equality $a_{n+1}^2 - a_n^2 = 3(a_n - a_{n-1})$. Hence for $1 < a_1 < 2$ the sequence is monotonically increasing and its limit is 2. On the other hand, if $a_1 = 1$ or $a_1 = 2$, the sequence is constant.

2.1.28.

(a) We have $a_1 < a_2$ and $a_{n+1}^2 - a_n^2 = a_n - a_{n-1}$. It follows by induction that the sequence is monotonically increasing and bounded above by c. Obviously, $\lim\limits_{n \to \infty} a_n = c$.

(b) Since $b_2 = \sqrt{c\sqrt{c}} > \sqrt{c} = b_1$ and $b_{n+1}^2 - b_n^2 = c(b_n - b_{n-1})$, using induction we conclude that the sequence is monotonically increasing and bounded above by c, which is its limit.

2.1.29. One can establish by induction that $0 < a_n < b$, and next prove that the sequence is strictly increasing. Its limit is equal to b.

2.1.30. The sequence is strictly increasing and bounded above, e.g. by 3. Its limit is $\frac{3+\sqrt{15}}{3}$.

2.1.31. We have $a_1 < a_2 < a_3$. Moreover, we see that for any $n \in \mathbb{N}$,

$$\text{if} \quad a_n < a_{n+1} < a_{n+2}, \quad \text{then} \quad a_{n+2} < a_{n+3}.$$

It then follows from the principle of induction stated in the solution of Problem 2.1.10 that the sequence $\{a_n\}$ is strictly increasing. It is also bounded above by 4, and $\lim\limits_{n \to \infty} a_n = 4$.

2.1.32. As in the solution of the foregoing problem, one can show that the sequence $\{a_n\}$ is monotonically decreasing, bounded below by 4, and $\lim\limits_{n \to \infty} a_n = 4$.

2.1.33. By the arithmetic-geometric mean inequality, $a_n \geq b_n$. Thus

$$a_{n+1} = \frac{a_n + b_n}{2} \leq a_n, \quad n \in \mathbb{N}.$$

This means that the sequence $\{a_n\}$ is decreasing. On the other hand, the sequence $\{b_n\}$ is increasing because

$$b_{n+1} = \sqrt{b_n a_n} \geq \sqrt{b_n^2} = b_n, \quad n \in \mathbb{N}.$$

2.1. Monotonic Sequences

Moreover, $b_1 < a_n$, $b_n < a_1$. Therefore both sequences converge. Let $\alpha = \lim\limits_{n\to\infty} a_n$ and $\beta = \lim\limits_{n\to\infty} b_n$. Passage to the limit as $n \to \infty$ in $a_{n+1} = \frac{a_n+b_n}{2}$ gives $\alpha = \frac{\alpha+\beta}{2}$ or, in other words, $\alpha = \beta$.

2.1.34. Since $2(a_n^2 + b_n^2) \geq (a_n + b_n)^2$, we get $a_n \geq b_n$, $n \in \mathbb{N}$. Therefore

$$a_{n+1} = \frac{a_n^2 + b_n^2}{a_n + b_n} \leq \frac{a_n^2 + a_n b_n}{a_n + b_n} = a_n, \quad n \in \mathbb{N},$$

which means that the sequence $\{a_n\}$ is decreasing.

In much the same way we show that $\{b_n\}$ increases. Moreover, we see that $b_1 < a_n$, $b_n < a_1$, and consequently, both sequences converge.

Let $\alpha = \lim\limits_{n\to\infty} a_n$, $\beta = \lim\limits_{n\to\infty} b_n$. Letting $n \to \infty$ in $b_{n+1} = \frac{a_n+b_n}{2}$, we obtain $\beta = \frac{\alpha+\beta}{2}$, or $\alpha = \beta$.

2.1.35. By the arithmetic-harmonic mean inequality, $a_n \geq b_n$. Hence

$$a_{n+1} = \frac{a_n + b_n}{2} \leq a_n, \quad n \in \mathbb{N},$$

which means that $\{a_n\}$ decreases. On the other hand, $\{b_n\}$ increases because

$$b_{n+1} = \frac{2a_n b_n}{a_n + b_n} \geq b_n, \quad n \in \mathbb{N}.$$

Moreover, $b_1 < a_n$, $b_n < a_1$, and therefore the sequences converge.

Let $\alpha = \lim\limits_{n\to\infty} a_n$, $\beta = \lim\limits_{n\to\infty} b_n$. Passing to the limit in the equation $a_{n+1} = \frac{a_n+b_n}{2}$ yields $\alpha = \frac{\alpha+\beta}{2}$. Thus $\alpha = \beta$.

Note also that $a_{n+1}b_{n+1} = a_n b_n$, which means that all the terms of $\{a_n b_n\}$ are equal to $a_1 b_1$. It follows that $\alpha = \beta = \sqrt{a_1 b_1}$.

2.1.36. We have $a_{n+1} = \frac{n+2}{2(n+1)}(a_n + 1)$, $n \in \mathbb{N}$. Consequently,

$$a_{n+1} - a_n = \frac{-na_n + (n+2)}{2(n+1)}.$$

Now applying the inequality $na_n > n + 2$ for $n \geq 4$ (which can be established by induction), we see that the sequence is monotonically decreasing and hence is convergent. Put $\alpha = \lim\limits_{n\to\infty} a_n$. Passage to the limit in the equation $a_{n+1} = \frac{n+2}{2(n+1)}(a_n + 1)$ gives $\alpha = 1$.

2.1.37. It follows from the inequality $a_{n+2} \leq \frac{1}{3}a_{n+1} + \frac{2}{3}a_n$ that $a_{n+2} + \frac{2}{3}a_{n+1} \leq a_{n+1} + \frac{2}{3}a_n$. Hence the sequence $b_n = a_{n+1} + \frac{2}{3}a_n$ is decreasing, bounded, and therefore convergent. Let b be its limit. We will show that $\{a_n\}$ converges to $a = \frac{3}{5}b$. Let $\varepsilon > 0$ be arbitrarily fixed. Then there exists $n_0 \in \mathbb{N}$ such that $\frac{\varepsilon}{6} > |b_n - b|$ for $n \geq n_0$. Consequently,

$$\frac{\varepsilon}{6} > \left|a_{n+1} + \frac{2}{3}a_n - \frac{5}{3}a\right| \geq |a_{n+1} - a| - \frac{2}{3}|a_n - a| \quad \text{for} \quad n \geq n_0.$$

Thus $|a_{n+1} - a| < \frac{2}{3}|a_n - a| + \frac{\varepsilon}{6}$. We can see by induction that

$$|a_{n_0+k} - a| \leq \left(\frac{2}{3}\right)^k |a_{n_0} - a| + \left(\left(\frac{2}{3}\right)^{k-1} + \ldots + \frac{2}{3} + 1\right)\frac{\varepsilon}{6}$$

$$\leq \left(\frac{2}{3}\right)^k |a_{n_0} - a| + \frac{1 - \left(\frac{2}{3}\right)^k}{1 - \frac{2}{3}}\frac{\varepsilon}{6} < \left(\frac{2}{3}\right)^k |a_{n_0} - a| + \frac{\varepsilon}{2}.$$

Since $\left(\frac{2}{3}\right)^k |a_{n_0} - a| < \frac{\varepsilon}{2}$ for sufficiently large k, $|a_n - a| < \varepsilon$ for n large enough.

2.1.38.
(a) $b_n = \left(1 + \frac{1}{n}\right)^{n+1} = \left(1 + \frac{1}{n}\right)a_n > a_n$.
(b) By the geometric-arithmetic mean inequality $G_{n+1} < A_{n+1}$ (see Problem 1.2.3) with $a_1 = 1, a_2 = a_3 = \ldots = a_{n+1} = 1 + \frac{1}{n}$,

$$\sqrt[n+1]{\left(1 + \frac{1}{n}\right)^n} < 1 + \frac{1}{n+1}.$$

Hence

$$\left(1 + \frac{1}{n}\right)^n < \left(1 + \frac{1}{n+1}\right)^{n+1}, \quad n \in \mathbb{N}.$$

(c) By the harmonic-geometric mean inequality $H_{n+1} < G_{n+1}$, $n > 1$ (see Problem 1.2.3) with $a_1 = 1, a_2 = a_3 = \ldots = a_{n+1} = 1 + \frac{1}{n-1}$,

$$1 + \frac{1}{n} < \sqrt[n+1]{\left(\frac{n}{n-1}\right)^n},$$

2.1. Monotonic Sequences

which in turn gives $b_n < b_{n-1}$, $n > 1$.
To show that both sequences $\{a_n\}$ and $\{b_n\}$ converge it is enough to observe that $a_1 \leq a_n < b_n \leq b_1$, $n \in \mathbb{N}$. Moreover,
$$\lim_{n\to\infty} b_n = \lim_{n\to\infty} \left(1 + \frac{1}{n}\right) a_n = \lim_{n\to\infty} a_n.$$

2.1.39.

(a) By the geometric-arithmetic mean inequality $G_{n+1} < A_{n+1}$ (see Problem 1.2.3) with $a_1 = 1$, $a_2 = a_3 = \ldots = a_{n+1} = 1 + \frac{x}{n}$, $n \in \mathbb{N}$, we see that the sequence is strictly increasing.
If $0 < x \leq 1$, then by the preceding problem,
$$\left(1 + \frac{x}{n}\right)^n \leq \left(1 + \frac{1}{n}\right)^n < e.$$

If $x > 1$, then there exists a positive integer n_0 such that $x \leq n_0$. Consequently, the monotonicity of the sequence $\left\{\left(1 + \frac{n_0}{n}\right)^n\right\}$ and the result stated in the foregoing problem imply
$$\left(1 + \frac{x}{n}\right)^n \leq \left(1 + \frac{n_0}{n}\right)^n < \left(1 + \frac{n_0}{n_0 n}\right)^{n_0 n} < e^{n_0}.$$

(b) It is enough to apply the same reasoning as in (a) and observe that for $x \leq 0$, the sequence is bounded above, e.g. by 1.

2.1.40. Applying the geometric-harmonic mean inequality $G_{n+l+1} > H_{n+l+1}$ (see Problem 1.2.3) with $a_1 = 1$, $a_2 = a_3 = \ldots = a_{n+l+1} = 1 + \frac{x}{n}$, we get
$$\sqrt[n+l+1]{\left(1 + \frac{x}{n}\right)^{n+l}} > 1 + \frac{x(n+l)}{n^2 + nl + x + n} > 1 + \frac{x(n+l)}{(n+1)(n+l)}.$$
This shows that $b_n > b_{n+1}$, $n \in \mathbb{N}$.

2.1.41. By the inequality given in the hint,
$$a_{n+1} - a_n = \frac{1}{n} - \log \frac{n+1}{n} > 0,$$
$$b_{n+1} - b_n = \frac{1}{n+1} - \log \frac{n+1}{n} < 0.$$
Clearly, $a_1 \leq a_n < b_n \leq b_1$, $n \in \mathbb{N}$, and consequently, both sequences converge (to the same limit).

2.1.42. Monotonicity and boundedness of the sequence $\{a_n\}$ are easily verified. It follows from the equality $a_{n+1}^2 = a_n$ that its limit is 1. We now show the monotonicity of $\{c_n\}$. Assume first that $x > 1$. Then

$$c_n = 2^n(a_n - 1) = 2^n(a_{n+1}^2 - 1) = 2^n(a_{n+1} - 1)(a_{n+1} + 1)$$
$$= 2^{n+1}(a_{n+1} - 1)\frac{a_{n+1} + 1}{2} > c_{n+1}.$$

This means that for $x > 1$, the sequence $\{c_n\}$ is strictly decreasing. The same reasoning applies to the case $0 < x < 1$. For $x = 1$, the sequence is constant. The monotonicity of $\{d_n\}$ can be proved analogously.

For $x > 1$, the sequence $\{c_n\}$ converges (because it is monotonically decreasing and bounded below by 0). On the other hand, for $0 < x < 1$, the sequence $\{d_n\}$ is monotonically increasing and bounded above by 0. Now, it follows from the equality

$$d_n = \frac{c_n}{a_n}$$

that both sequences tend to same limit for all positive x different from 1. If $x = 1$, then $c_n = d_n = 0$.

2.2. Limits. Properties of Convergent Sequences

2.2.1.
(a) 1.
(b) 1.
(c) -1.
(d) We have

$$0 < (\sqrt{2} - \sqrt[3]{2})(\sqrt{2} - \sqrt[5]{2}) \cdot \ldots \cdot (\sqrt{2} - \sqrt[2n+1]{2}) < (\sqrt{2} - 1)^n.$$

Thus the limit of the sequence is equal to 0.

(e) We will first show that the sequence $a_n = \frac{n^2}{2^n}$ converges to zero. We have $a_{n+1} = a_n \frac{1}{2} \frac{(n+1)^2}{n^2} < a_n$ for $n \geq 3$. Therefore the

2.2. Limits. Properties of Convergent Sequences

sequence is monotonically decreasing. Clearly, it is bounded below by zero. Hence it is a convergent sequence and its limit g satisfies the equation $g = \frac{1}{2}g$. Thus $g = 0$. We will now find the limit of our sequence. To this end, set $k_n = [\sqrt{n}]$. Then $k_n \leq \sqrt{n} < k_n + 1$, which gives

$$0 < \frac{n}{2^{\sqrt{n}}} < 2\frac{(k_n+1)^2}{2^{k_n+1}}.$$

Therefore the limit of the given sequence is equal to zero.

(f) Let $a_n = \frac{n!}{2^{n^2}}$. Then $a_{n+1} = a_n \frac{1}{2} \frac{(n+1)}{2^{2n}} < a_n$, $n \in \mathbb{N}$, which implies (see the solution of Problem 2.1.3) that $g = 0$.

(g) Set

$$a_n = \frac{1}{\sqrt{n}}\left(\frac{1}{\sqrt{1}+\sqrt{3}} + \frac{1}{\sqrt{3}+\sqrt{5}} + ... + \frac{1}{\sqrt{2n-1}+\sqrt{2n+1}}\right).$$

Then $a_n = \frac{\sqrt{2n+1}-1}{2\sqrt{n}}$, which is a consequence of the equality $\frac{1}{\sqrt{2k-1}+\sqrt{2k+1}} = \frac{\sqrt{2k-1}-\sqrt{2k+1}}{-2}$. So, $\lim\limits_{n\to\infty} a_n = \frac{\sqrt{2}}{2}$.

(h) It follows from the inequalities

$$(1+2+...+n)\frac{1}{n^2+n} \leq \frac{1}{n^2+1} + \frac{2}{n^2+2} + ... + \frac{n}{n^2+n}$$

$$\leq (1+2+...+n)\frac{1}{n^2+1}$$

and from the squeeze law that the limit is $\frac{1}{2}$.

(i) As in (h) we show that this limit is also equal to $\frac{1}{2}$.

2.2.2. Set $a_n = \frac{n^s}{(1+p)^n}$. Then $\frac{a_{n+1}}{a_n} = \left(\frac{n+1}{n}\right)^s \frac{1}{p+1}$. Moreover, we have $\lim\limits_{n\to\infty} \left(\frac{n+1}{n}\right)^s \frac{1}{p+1} = \frac{1}{p+1}$. Consequently, the sequence $\{a_n\}$ is monotonically decreasing beginning with some value n_0 of the index n. It is also bounded below, e.g. by zero. Its limit g satisfies the equality $g = \frac{1}{p+1}g$. Therefore $g = 0$.

2.2.3. We have

$$0 < (n+1)^\alpha - n^\alpha = n^\alpha \left(\left(1+\frac{1}{n}\right)^\alpha - 1\right)$$

$$< n^\alpha \left(\left(1+\frac{1}{n}\right) - 1\right) = \frac{1}{n^{1-\alpha}}.$$

Thus the limit of the sequence is equal to zero.

2.2.4. Let $\alpha = \frac{p}{q}$, with $p \in \mathbb{Z}$ and $q \in \mathbb{N}$. For $n > q$ the number $n!\alpha\pi$ is a multiple of π, which means that the terms of the sequence, beginning with some value n_0 of the index n, are all equal to zero.

2.2.5. If the limit existed then we would get
$$0 = \lim_{n\to\infty} (\sin(n+2) - \sin n) = 2\sin 1 \lim_{n\to\infty} \cos(n+1),$$
and consequently, $\lim_{n\to\infty} \cos n = 0$. Similarly,
$$0 = \lim_{n\to\infty} (\cos(n+2) - \cos n) = -2\sin 1 \lim_{n\to\infty} \sin(n+1),$$
which is impossible because $\sin^2 n + \cos^2 n = 1$, $n \in \mathbb{N}$. Therefore the limit $\lim_{n\to\infty} \sin n$ does not exist.

2.2.6. See the solution of the foregoing problem.

2.2.7. We have
$$\lim_{n\to\infty} \frac{1}{n}\left(\left(a+\frac{1}{n}\right)^2 + \left(a+\frac{2}{n}\right)^2 + \ldots + \left(a+\frac{n-1}{n}\right)^2\right)$$
$$= \lim_{n\to\infty} \left(\frac{n-1}{n}a^2 + \frac{n(n-1)}{n^2}a + \frac{1 + 2^2 + \ldots + (n-1)^2}{n^3}\right)$$
$$= a^2 + a + \frac{1}{3}.$$

The last equality follows from the fact that $1^2 + 2^2 + \ldots + n^2 = \frac{n(n+1)(2n+1)}{6}$.

2.2.8. We have
$$a_n + a_n^2 + \ldots + a_n^k - k = (a_n - 1) + (a_n^2 - 1) + \ldots + (a_n^k - 1).$$

Moreover,
$$\lim_{n\to\infty} \frac{a_n^l - 1}{a_n - 1} = l \quad \text{for } l = 1, 2, \ldots, k.$$
Therefore the limit is equal to $1 + 2 + \ldots + k = \frac{k(k+1)}{2}$.

2.2. Limits. Properties of Convergent Sequences

2.2.9. Using the equality
$$\frac{1}{k(k+1)(k+2)} = \frac{1}{2} \cdot \frac{1}{k} - \frac{1}{k+1} + \frac{1}{2} \cdot \frac{1}{k+2}, \quad k \in \mathbb{N},$$
one can show that the limit is equal to $\frac{1}{4}$.

2.2.10. Since
$$\frac{k^3 - 1}{k^3 + 1} = \frac{(k-1)((k+1)^2 - (k+1) + 1)}{(k+1)(k^2 - k + 1)},$$
we get
$$\prod_{k=2}^{n} \frac{k^3 - 1}{k^3 + 1} = \frac{2}{3} \cdot \frac{n^2 + n + 1}{n^2 + n} \xrightarrow[n \to \infty]{} \frac{2}{3}.$$

2.2.11. $\frac{1}{6}$.

2.2.12. Since $1 - \frac{2}{k(k+1)} = \frac{(k+2)(k-1)}{k(k+1)}$, we obtain
$$\left(1 - \frac{2}{2 \cdot 3}\right)\left(1 - \frac{2}{3 \cdot 4}\right) \cdots \left(1 - \frac{2}{(n+1) \cdot (n+2)}\right) = \frac{1}{3} \cdot \frac{n+3}{n+1} \xrightarrow[n \to \infty]{} \frac{1}{3}.$$

2.2.13. We have
$$k^3 + 6k^2 + 11k + 5 = (k+1)(k+2)(k+3) - 1.$$
Hence
$$\lim_{n \to \infty} \sum_{k=1}^{n} \frac{k^3 + 6k^2 + 11k + 5}{(k+3)!} = \lim_{n \to \infty} \sum_{k=1}^{n} \left(\frac{1}{k!} - \frac{1}{(k+3)!}\right) = \frac{5}{3}.$$

2.2.14. Observe that
$$\frac{x^{2^{k-1}}}{1 - x^{2^k}} = \frac{1}{1 - x^{2^{k-1}}} - \frac{1}{1 - x^{2^k}} \quad \text{for} \quad k = 1, 2, \ldots, n.$$
Therefore
$$\lim_{n \to \infty} \sum_{k=1}^{n} \frac{x^{2^{k-1}}}{1 - x^{2^k}} = \begin{cases} \dfrac{x}{1-x} & \text{for} \quad |x| < 1, \\ \dfrac{1}{1-x} & \text{for} \quad |x| > 1. \end{cases}$$

2.2.15. For $x \neq 1$,
$$\frac{(1-x)(1+x)(1+x^2) \cdot \ldots \cdot (1+x^{2^n})}{1-x} = \frac{1-x^{2^{n+1}}}{1-x},$$
and consequently,
$$a_n = \prod_{k=0}^{n}(1+x^{2^k}) = \begin{cases} \dfrac{1-x^{2^{n+1}}}{1-x} & \text{for} \quad n = 0, 1, \ldots, \; x \neq 1, \\ 2^{n+1} & \text{for} \quad n = 0, 1, \ldots, \; x = 1. \end{cases}$$

Finally,
$$\lim_{n \to \infty} a_n = \begin{cases} -\infty & \text{for} \quad x < -1, \\ 0 & \text{for} \quad x = -1, \\ \dfrac{1}{1-x} & \text{for} \quad |x| < 1, \\ +\infty & \text{for} \quad x \geq 1. \end{cases}$$

2.2.16. For $x \neq 1$,
$$a_n = \prod_{k=0}^{n}\left(1 + \frac{2}{x^{2^k} + x^{-2^k}}\right) = \prod_{k=0}^{n} \frac{(x^{2^k}+1)^2}{x^{2^{k+1}}+1}$$
$$= \frac{(x+1)(x-1)(x+1)(x^2+1) \cdot \ldots \cdot (x^{2^n}+1)}{(x-1)(x^{2^{n+1}}+1)}$$
$$= \frac{x+1}{x-1} \cdot \frac{x^{2^{n+1}}-1}{x^{2^{n+1}}+1}.$$

Hence
$$\lim_{n \to \infty} a_n = \begin{cases} -\dfrac{x+1}{x-1} & \text{for} \quad |x| < 1, \\ \dfrac{x+1}{x-1} & \text{for} \quad |x| > 1, \\ 0 & \text{for} \quad x = -1, \\ +\infty & \text{for} \quad x = 1. \end{cases}$$

2.2.17. Let x be different from 1. Then
$$1 + x^{3^k} + x^{2 \cdot 3^k} = \frac{(1+x^{3^k}+x^{2 \cdot 3^k})(x^{3^k}-1)}{x^{3^k}-1} = \frac{x^{3^{k+1}}-1}{x^{3^k}-1}.$$
Thus
$$\prod_{k=1}^{n}(1+x^{3^k}+x^{2 \cdot 3^k}) = \frac{x^{3^{n+1}}-1}{x^3-1}.$$

2.2. Limits. Properties of Convergent Sequences

Let g denote the limit of the sequence. Then

$$g = \begin{cases} \dfrac{1}{1-x^3} & \text{for } |x| < 1, \\ +\infty & \text{for } |x| > 1, \\ 1 & \text{for } x = -1, \\ +\infty & \text{for } x = 1. \end{cases}$$

2.2.18. Clearly, $k \cdot k! = (k+1)! - k!$, $k \in \mathbb{N}$. Hence

$$\lim_{n\to\infty} \frac{1\cdot 1! + 2\cdot 2! + \ldots + n\cdot n!}{(n+1)!} = \lim_{n\to\infty} \frac{(n+1)! - 1}{(n+1)!} = 1.$$

2.2.19. Note first that the problem is meaningful for $x \neq 0$. By 2.2.3, the denominator $n^x - (n-1)^x$ tends to zero if $0 < x < 1$. Moreover, if $x < 0$, then the denominator also tends to zero. For $x = 1$ it equals 1. Therefore the sequence diverges to infinity ($+\infty$ or $-\infty$) for $x \leq 1$, $x \neq 0$. Now let $x > 1$ and set $k = [x]$. Then $k \geq 1$ and

$$1 - \left(1 - \frac{1}{n}\right)^k \leq 1 - \left(1 - \frac{1}{n}\right)^x < 1 - \left(1 - \frac{1}{n}\right)^{k+1}.$$

It follows from these inequalities that there exist α and β such that

$$\alpha < n\left(1 - \left(1 - \frac{1}{n}\right)^x\right) < \beta,$$

which in turn gives

$$\alpha n^{x-1} < n^x\left(1 - \left(1 - \frac{1}{n}\right)^x\right) < \beta n^{x-1}.$$

Hence if $x - 1 < 1999$, then the sequence diverges to $+\infty$. If $x - 1 > 1999$, the sequence converges to zero. Now let $x = 2000$. By the binomial formula,

$$\lim_{n\to\infty} \frac{n^{1999}}{n^{2000} - (n-1)^{2000}} = \frac{1}{2000}.$$

2.2.20. We have
$$a_n = \begin{cases} \dfrac{a^{n+1}-b^{n+1}}{a^n-b^n} & \text{if } a > b, \\ \dfrac{n+1}{n}a & \text{if } a = b. \end{cases}$$
Hence $\lim\limits_{n\to\infty} a_n = a$.

2.2.21. It can be shown by induction that $a_n = (n-1)^2$. Consequently, $\lim\limits_{n\to\infty} a_n = +\infty$.

2.2.22. We show by induction that $a_n = \dfrac{ab}{\sqrt{a^2+nb^2}}$. Thus $\lim\limits_{n\to\infty} a_n = 0$.

2.2.23. One can show that $a_n = 1 - \left(\dfrac{1}{4}\right)^{n-1}$. Therefore $\lim\limits_{n\to\infty} a_n = 1$.

2.2.24. It is easy to verify that $a_{n+1} = 1+b+...+b^{n-1}+b^n a$. Hence
$$a_{n+1} = \begin{cases} \dfrac{1}{1-b} + \left(a - \dfrac{1}{1-b}\right)b^n & \text{for } b \neq 1, \\ n+a & \text{for } b = 1. \end{cases}$$
Thus if $b = 1$, $a \in \mathbb{R}$, the sequence diverges to $+\infty$. If $b \neq 1$ and $a = \dfrac{1}{1-b}$, the sequence converges to $\dfrac{1}{1-b}$. In the case $a \neq \dfrac{1}{1-b}$ and $|b| < 1$, it also converges to $\dfrac{1}{1-b}$. In the remaining cases the sequence is divergent. Namely, if $b \leq -1$ and $a \neq \dfrac{1}{1-b}$, the sequence has neither finite nor infinite limit. If $b > 1$ and $a > \dfrac{1}{1-b}$, the sequence diverges properly to $+\infty$. Finally, if $b > 1$ and $a < \dfrac{1}{1-b}$, the sequence diverges properly to $-\infty$.

2.2.25. The formula for the nth term of the Fibonacci sequence can be proved by induction. We may assume that $\alpha > \beta$. Then $\alpha = \dfrac{1+\sqrt{5}}{2}$ and $\beta = \dfrac{1-\sqrt{5}}{2}$. Moreover,
$$\alpha \sqrt[n]{1 - \left|\dfrac{\beta}{\alpha}\right|^n} \leq \sqrt[n]{\alpha^n - \beta^n} \leq \alpha \sqrt[n]{1 + \left|\dfrac{\beta}{\alpha}\right|^n}.$$
Since $\lim\limits_{n\to\infty} \left|\dfrac{\beta}{\alpha}\right|^n = 0$, we get $\lim\limits_{n\to\infty} \sqrt[n]{a_n} = \alpha$.

2.2. Limits. Properties of Convergent Sequences

2.2.26. Note first that $b_{n+1} = \frac{a_n + 3b_n}{4}$. From this $a_{n+1} - b_{n+1} = \frac{1}{4}(a_n - b_n)$, which means that the sequence $\{a_n - b_n\}$ is a geometric progression with the ratio $\frac{1}{4}$. Hence this sequence converges to zero. Now it is enough to show that the sequence $\{a_n\}$ converges. Assume first that $a \leq b$. Then $\{a_n\}$ monotonically increases and $a_n \leq b_n \leq b$. Therefore it converges. It follows from the above that $\{b_n\}$ also converges and $\lim\limits_{n\to\infty} a_n = \lim\limits_{n\to\infty} b_n$. The same reasoning applies to the case $a > b$.

2.2.27. We have

$$a + aa + \ldots + \overbrace{aa\ldots a}^{n \text{ digits}} = a(1 + 11 + \ldots + \overbrace{11\ldots 1}^{n \text{ digits}})$$
$$= a(10^{n-1} + 2 \cdot 10^{n-2} + \ldots + n \cdot 10^0)$$
$$= a((1 + 10 + \ldots + 10^{n-1}) + (1 + 10 + \ldots + 10^{n-2})$$
$$+ \ldots + (1 + 10) + 1)$$
$$= a\left(\frac{10^n - 1}{9} + \frac{10^{n-1} - 1}{9} + \ldots + \frac{10^2 - 1}{9} + \frac{10 - 1}{9}\right)$$
$$= a\frac{10(10^n - 1) - 9n}{81}.$$

Therefore the limit is $\frac{10a}{81}$.

2.2.28. Note that the sequence with terms $\sqrt[n]{n}$, $n \geq 3$, is monotonically decreasing and its limit is 1. Now it is easy to check that

$$(\sqrt[n]{n} - 1)^n < \left(\frac{1}{2}\right)^n \quad \text{for} \quad n \in \mathbb{N}.$$

Thus $\lim\limits_{n\to\infty} (\sqrt[n]{n} - 1)^n = 0$.

2.2.29. Since $\lim\limits_{n\to\infty} a_n = 0$, beginning with some value n_0 of the index n, $|a_n|^n < \left(\frac{1}{2}\right)^n$. Consequently, $\lim\limits_{n\to\infty} a_n^n = 0$.

2.2.30. Let $\max\{a_1, a_2, \ldots, a_k\} = a_l$. Dividing the denominator and numerator by a_l^n we show that

$$\lim_{n\to\infty} \frac{p_1 a_1^{n+1} + p_2 a_2^{n+1} + \ldots + p_k a_k^{n+1}}{p_1 a_1^n + p_2 a_2^n + \ldots + p_k a_k^n} = a_l.$$

2.2.31.

(a) Let $\varepsilon > 0$ be so small that $q + \varepsilon < 1$. Then there exists $n_0 \in \mathbb{N}$ such that
$$\left|\frac{a_{n+1}}{a_n}\right| < q + \varepsilon \quad \text{for } n \geq n_0.$$
Hence
$$|a_n| < (q + \epsilon)^{n-n_0}|a_{n_0}|, \quad n \geq n_0.$$
This implies $\lim_{n \to \infty} |a_n| = 0$, that is, $\lim_{n \to \infty} a_n = 0$.

(b) Let $\varepsilon > 0$ be so small that $q - \varepsilon > 1$. Then, beginning with some value n_1 of the index n, $|a_n| > (q - \varepsilon)^{n-n_1}|a_{n_1}|$. Since $\lim_{n \to \infty}(q-\varepsilon)^{n-n_1} = +\infty$, we get $\lim_{n \to \infty}|a_n| = +\infty$.

2.2.32.

(a) Take $\varepsilon > 0$ small enough to get $q + \varepsilon < 1$. Then there exists $n_0 \in \mathbb{N}$ such that $|a_n| < (q+\varepsilon)^n$, $n \geq n_0$. Therefore $\lim_{n \to \infty} a_n = 0$.

(b) We have $|a_n| > (q - \varepsilon)^n$ for $n > n_1$. If $\varepsilon > 0$ is small enough, then $q - \varepsilon > 1$ and therefore $\lim_{n \to \infty}(q-\varepsilon)^n = +\infty$. So, $\lim_{n \to \infty}|a_n| = +\infty$.

2.2.33. Setting $a_n = n^\alpha x^n$, we get
$$\lim_{n \to \infty} \frac{a_{n+1}}{a_n} = \lim_{n \to \infty} \left(\frac{n+1}{n}\right)^\alpha x = x, \quad \text{where} \quad 0 < x < 1.$$
Hence, by Problem 2.2.31, the sequence tends to zero.

2.2.34. Let a_n denote the nth term of the sequence. Then
$$\left|\frac{a_{n+1}}{a_n}\right| = \left|\frac{m-n}{n+1}x\right| \xrightarrow[n \to \infty]{} |x|.$$
By Problem 2.2.31 the sequence converges to zero.

2.2.35. Assume that $|b_n| < M$ for $n \in \mathbb{N}$. Since $\lim_{n \to \infty} a_n = 0$, for any $\varepsilon > 0$ there exists $n_0 \in \mathbb{N}$ such that $|a_n| < \frac{\varepsilon}{M}$ for $n > n_0$. Hence
$$|a_n b_n| < \varepsilon \quad \text{for } n > n_0.$$
This means that $\lim_{n \to \infty} a_n b_n = 0$.

2.2. Limits. Properties of Convergent Sequences 171

2.2.36. Without loss of generality we can assume that $a \leq b$. Suppose first that $a < b$. Let $\varepsilon > 0$ be so small that $a + \varepsilon < b - \varepsilon$. By the definition of the limit of a sequence, $a_n < a + \varepsilon < b - \varepsilon < b_n$ for n sufficiently large. Hence $\max\{a_n, b_n\} = b_n$, and consequently,

$$\lim_{n\to\infty} \max\{a_n, b_n\} = \lim_{n\to\infty} b_n = b = \max\{a, b\}.$$

If $a = b$, then for any $\varepsilon > 0$ there exists n_0 such that for $n > n_0$, the inequalities $|a_n - a| < \varepsilon$ and $|b_n - a| < \varepsilon$ hold. This means that

$$|\max\{a_n, b_n\} - a| < \varepsilon.$$

In this way we have proved that

$$\lim_{n\to\infty} \max\{a_n, b_n\} = \max\{a, b\}.$$

2.2.37. Since $\lim_{n\to\infty} a_n = 0$, for any $\varepsilon \in (0, 1)$ we have

$$\sqrt[p]{1-\varepsilon} < \sqrt[p]{1+a_n} < \sqrt[p]{1+\varepsilon} \quad \text{for } n \text{ sufficiently large.}$$

This implies that $\lim_{n\to\infty} \sqrt[p]{1+a_n} = 1$.

2.2.38. Put $x_n = \sqrt[p]{1+a_n}$. It follows from the foregoing problem that $\lim_{n\to\infty} x_n = 1$. Consequently,

$$\lim_{n\to\infty} \frac{\sqrt[p]{1+a_n} - 1}{a_n} = \lim_{n\to\infty} \frac{x_n - 1}{x_n^p - 1}$$
$$= \lim_{n\to\infty} \frac{x_n - 1}{(x_n - 1)(x_n^{p-1} + \ldots + 1)} = \frac{1}{p}.$$

2.2.39. By Problem 1.2.1,

(1)
$$n\left(\sqrt[p]{1 + \frac{a_1 + a_2 + \ldots + a_p}{n}} - 1\right)$$
$$\leq n\left(\sqrt[p]{\left(1+\frac{a_1}{n}\right)\left(1+\frac{a_2}{n}\right) \cdot \ldots \cdot \left(1+\frac{a_p}{n}\right)} - 1\right)$$
$$= \sqrt[p]{(n+a_1)(n+a_2) \cdot \ldots \cdot (n+a_p)} - n.$$

Moreover, by 1.2.4 we get

$$n\left(\sqrt[p]{\left(1+\frac{a_1}{n}\right)\left(1+\frac{a_2}{n}\right)\cdot\ldots\cdot\left(1+\frac{a_p}{n}\right)}-1\right)$$

(2) $$=n\left(\sqrt[p]{1+\frac{a_1+\ldots+a_p}{n}+\frac{\sum\limits_{i<j}a_ia_j}{n^2}+\ldots+\frac{a_1\cdot\ldots\cdot a_p}{n^p}}-1\right)$$

$$\leq \frac{a_1+\ldots+a_p}{p}+\frac{\sum\limits_{i<j}a_ia_j}{np}+\ldots+\frac{a_1\cdot\ldots\cdot a_p}{pn^{p-1}}.$$

Combining (1) and (2) with the result in the foregoing problem, we show that the limit is $\frac{a_1+a_2+\ldots+a_p}{p}$.

2.2.40. Note that

$$\frac{n+1}{\sqrt{n^2+n+1}} \leq \frac{1}{\sqrt{n^2+1}}+\frac{1}{\sqrt{n^2+2}}+\ldots+\frac{1}{\sqrt{n^2+n+1}} \leq \frac{n+1}{\sqrt{n^2+1}}.$$

This and the squeeze law for sequences imply that the limit is 1.

2.2.41. Let a denote the largest of the numbers $a_1, a_2, \ldots, a_p$. Then

$$\frac{a}{\sqrt[n]{p}} \leq \sqrt[n]{\frac{a_1^n+a_2^n+\ldots+a_p^n}{p}} \leq a.$$

By the squeeze law for sequences,

$$\lim_{n\to\infty}\sqrt[n]{\frac{a_1^n+a_2^n+\ldots+a_p^n}{p}}=a=\max\{a_1,a_2,\ldots,a_p\}.$$

2.2.42. Since

$$1\leq \sqrt[n]{2\sin^2\frac{n^{1999}}{n+1}+\cos^2\frac{n^{1999}}{n+1}}\leq \sqrt[n]{2},$$

it follows that

$$\lim_{n\to\infty}\sqrt[n]{2\sin^2\frac{n^{1999}}{n+1}+\cos^2\frac{n^{1999}}{n+1}}=1.$$

2.2. Limits. Properties of Convergent Sequences

2.2.43. We will apply the squeeze law for sequences. We have
$$1 < (1+n(1+\cos n))^{\frac{1}{2n+n\sin n}} < (1+2n)^{\frac{1}{2n+n\sin n}}.$$
We will now show that
$$(*) \qquad \lim_{n\to\infty} (1+2n)^{\frac{1}{2n+n\sin n}} = 1.$$
Indeed,
$$1 < (1+2n)^{\frac{1}{2n+n\sin n}} < (1+2n)^{\frac{1}{n}}.$$
Hence $(*)$ follows from the squeeze law. Thus the limit we are looking for is 1.

2.2.44. By the harmonic-geometric-arithmetic mean inequality (see 1.2.3), for $x > -1$ we have
$$1 + \frac{x}{2+x} = \frac{2}{\frac{1}{1+x}+1} \le \sqrt{(1+x)1} = \sqrt{1+x} \le \frac{1+x+1}{2} = 1 + \frac{x}{2}.$$
Now, putting $x = \frac{k}{n^2}$, $k = 1, 2, ..., n$, and adding the obtained inequalities, we get
$$(*) \qquad \sum_{k=1}^{n} \frac{\frac{k}{n^2}}{2+\frac{k}{n^2}} \le \sum_{k=1}^{n} \left(\sqrt{1+\frac{k}{n^2}}-1\right) \le \sum_{k=1}^{n} \frac{k}{2n^2}.$$
Moreover,
$$\sum_{k=1}^{n} \frac{k}{2n^2} = \frac{n(n+1)}{4n^2} \xrightarrow[n\to\infty]{} \frac{1}{4}$$
and
$$\sum_{k=1}^{n} \frac{\frac{k}{n^2}}{2+\frac{k}{n^2}} = \sum_{k=1}^{n} \frac{k}{2n^2+k} \ge \frac{1}{2n^2+n} \sum_{k=1}^{n} k = \frac{n(n+1)}{2(2n^2+n)} \xrightarrow[n\to\infty]{} \frac{1}{4}.$$
Therefore, by $(*)$ and the squeeze law,
$$\lim_{n\to\infty} \sum_{k=1}^{n} \left(\sqrt{1+\frac{k}{n^2}}-1\right) = \frac{1}{4}.$$

2.2.45. One can apply reasoning analogous to that used in the solution of the preceding problem. Let $x > -1$. By the harmonic-geometric-arithmetic mean inequality,

$$1 + \frac{x}{3+2x} = \frac{3}{\frac{1}{1+x}+1+1} \leq \sqrt[3]{(1+x)1 \cdot 1} \leq \frac{1+x+1+1}{3} = 1 + \frac{x}{3}.$$

Substituting $x = \frac{k^2}{n^3}$, we get

$$(*) \qquad \sum_{k=1}^{n} \frac{\frac{k^2}{n^3}}{3+2\frac{k^2}{n^3}} \leq \sum_{k=1}^{n} \left(\sqrt[3]{1+\frac{k^2}{n^3}} - 1 \right) \leq \sum_{k=1}^{n} \frac{k^2}{3n^3}.$$

Moreover,

$$\sum_{k=1}^{n} \frac{k^2}{3n^3} = \frac{n(n+1)(2n+1)}{18n^3} \xrightarrow[n\to\infty]{} \frac{1}{9}$$

and

$$\sum_{k=1}^{n} \frac{\frac{k^2}{n^3}}{3+2\frac{k^2}{n^3}} = \sum_{k=1}^{n} \frac{k^2}{3n^3+2k^2} \geq \frac{1}{3n^3+2n^2} \sum_{k=1}^{n} k^2$$
$$= \frac{n(n+1)(2n+1)}{6(3n^3+2n^2)} \xrightarrow[n\to\infty]{} \frac{1}{9}.$$

By the above, together with $(*)$ and the squeeze law,

$$\lim_{n\to\infty} \sum_{k=1}^{n} \left(\sqrt[3]{1+\frac{k^2}{n^3}} - 1 \right) = \frac{1}{9}.$$

2.2.46. Clearly, $\lim_{n\to\infty} \sqrt[n]{a_k} = 1$ for $k = 1, 2, ..., p$. So we find that

$$\lim_{n\to\infty} \left(\frac{1}{p} \sum_{k=1}^{p} \sqrt[n]{a_k} \right)^p = 1.$$

2.2.47. For sufficiently large n_0 and for $n > n_0$, we have $0 < \alpha + \frac{1}{n} < \alpha + \frac{1}{n_0} < 1$. Thus

$$\lim_{n\to\infty} \sum_{k=0}^{n-1} \left(\alpha + \frac{1}{n} \right)^k = \lim_{n\to\infty} \frac{1-\left(\alpha+\frac{1}{n}\right)^n}{1-\left(\alpha+\frac{1}{n}\right)} = \frac{1}{1-\alpha}.$$

2.2. Limits. Properties of Convergent Sequences

2.2.48. The equality is obvious for $x = 1$. Assume now that $x > 1$. To calculate the limit, we will apply the squeeze law for sequences. We have
$$0 < (\sqrt[n]{x} - 1)^2 = \sqrt[n]{x^2} - 2\sqrt[n]{x} + 1.$$
Hence
$$(*) \qquad (2\sqrt[n]{x} - 1)^n < (\sqrt[n]{x^2})^n = x^2.$$
Moreover,
$$(2\sqrt[n]{x} - 1)^n = x^2 \left(\frac{2}{\sqrt[n]{x}} - \frac{1}{\sqrt[n]{x^2}} \right)^n = x^2 \left(1 + \left(\frac{2}{\sqrt[n]{x}} - \frac{1}{\sqrt[n]{x^2}} - 1 \right) \right)^n.$$
Now, by the Bernoulli inequality we get
$$(**) \qquad \begin{aligned} (2\sqrt[n]{x} - 1)^n &\geq x^2 \left(1 + n \left(\frac{2}{\sqrt[n]{x}} - \frac{1}{\sqrt[n]{x^2}} - 1 \right) \right) \\ &= x^2 \left(1 - n \frac{(\sqrt[n]{x} - 1)^2}{\sqrt[n]{x^2}} \right). \end{aligned}$$
Also, by the Bernoulli inequality,
$$x = (\sqrt[n]{x} - 1 + 1)^n \geq 1 + n(\sqrt[n]{x} - 1) > n(\sqrt[n]{x} - 1).$$
Consequently,
$$(\sqrt[n]{x} - 1)^2 < \frac{x^2}{n^2}.$$
Therefore, by (**),
$$(***) \qquad (2\sqrt[n]{x} - 1)^n > x^2 \left(1 - \frac{x^2}{n\sqrt[n]{x^2}} \right).$$
Combining (*) and (***) with the squeeze law, we see that
$$\lim_{n \to \infty} (2\sqrt[n]{x} - 1)^n = x^2.$$

2.2.49. As in the solution of the foregoing problem, we may establish the inequalities
$$1 \geq \frac{(2\sqrt[n]{n} - 1)^n}{n^2} \geq 1 - n\frac{(\sqrt[n]{n} - 1)^2}{\sqrt[n]{n^2}}.$$

Now, it is enough to show that
$$\lim_{n\to\infty} n\frac{(\sqrt[n]{n}-1)^2}{\sqrt[n]{n^2}} = 0.$$
To this end, note that for $n \geq 3$,
$$n = (\sqrt[n]{n}-1+1)^n > \frac{n(n-1)(n-2)}{3!}(\sqrt[n]{n}-1)^3.$$
Hence
$$0 \leq n(\sqrt[n]{n}-1)^2 \leq n\left(\frac{3!}{(n-1)(n-2)}\right)^{\frac{2}{3}}.$$
So, $\lim_{n\to\infty} n(\sqrt[n]{n}-1)^2 = 0.$

2.2.50.

(a) We have
$$|a_{n+k} - a_n| = \frac{\arctan(n+1)}{2^{n+1}} + \ldots + \frac{\arctan(n+k)}{2^{n+k}}$$
$$< \frac{\pi}{2}\left(\frac{1}{2^{n+1}} + \ldots + \frac{1}{2^{n+k}}\right) < \frac{\pi}{2^{n+1}}.$$

For arbitrarily fixed $\varepsilon > 0$, let $n_0 = [\log_2 \frac{\pi}{\varepsilon} - 1]$. Then for any $k \in \mathbb{N}$ and $n > n_0$ we get $|a_{n+k} - a_n| < \varepsilon$. Therefore $\{a_n\}$ is a Cauchy sequence.

(b) One can show by induction that $4^n > n^4$ for all $n \geq 5$. Hence
$$|a_{n+k} - a_n| < \frac{1}{(n+1)^2} + \frac{1}{(n+2)^2} + \ldots + \frac{1}{(n+k)^2}.$$
Consequently,
$$|a_{n+k} - a_n|$$
$$< \frac{1}{n(n+1)} + \frac{1}{(n+1)(n+2)} + \ldots + \frac{1}{(n+k-1)(n+k)}$$
$$= \frac{1}{n} - \frac{1}{n+1} + \frac{1}{n+1} - \frac{1}{n+2} + \ldots + \frac{1}{n+k-1} - \frac{1}{n+k}$$
$$= \frac{1}{n} - \frac{1}{n+k} < \frac{1}{n} < \varepsilon$$
for any $k \in \mathbb{N}$ and $n > [\frac{1}{\varepsilon}]$.

2.2. Limits. Properties of Convergent Sequences

(c) Note that

$$|a_{2n} - a_n| = \frac{1}{2n} + \frac{1}{2n-1} + \ldots + \frac{1}{n+1} \geq n\frac{1}{2n} = \frac{1}{2}.$$

This means that $\{a_n\}$ is not a Cauchy sequence.

(d) We have

$$|a_{n+k} - a_n|$$
$$= \left| \frac{(-1)^{n+k-1}}{(n+k)(n+k+1)} + \frac{(-1)^{n+k-2}}{(n+k-1)(n+k)} + \ldots + \frac{(-1)^n}{(n+1)(n+2)} \right|$$
$$\leq \frac{1}{(n+k)(n+k+1)} + \frac{1}{(n+k-1)(n+k)} + \ldots + \frac{1}{(n+1)(n+2)}$$
$$= \frac{1}{n+k} - \frac{1}{n+k+1} + \frac{1}{n+k-1} - \frac{1}{n+k} + \ldots + \frac{1}{n+1} - \frac{1}{n+2}$$
$$= \frac{1}{n+1} - \frac{1}{n+k+1} < \frac{1}{n+1} < \varepsilon$$

for any $k \in \mathbb{N}$ and $n > [\frac{1}{\varepsilon} - 1]$.

(e) We have

$$|a_{n+k} - a_n| \leq M(|q|^{n+k} + |q|^{n+k-1} + \ldots + |q|^{n+1})$$
$$= M\left(\frac{|q|^{n+1}(1-|q|^k)}{1-|q|}\right) \leq \frac{M}{1-|q|}|q|^{n+1} < \varepsilon$$

for any $k \in \mathbb{N}$ and $n > n_0 = \left[\frac{\ln \frac{(1-|q|)\varepsilon}{M}}{\ln |q|} - 1\right]$.

(f) We have

$$a_{2n} - a_n = \frac{2n}{(2n+1)^2} + \frac{2n-1}{(2n)^2} + \ldots + \frac{n+1}{(n+2)^2}$$
$$\geq n\frac{2n}{(2n+1)^2} \geq \frac{2n^2}{(3n)^2} = \frac{2}{9}.$$

Therefore $\{a_n\}$ is not a Cauchy sequence.

2.2.51. By the given condition,

$$|a_{n+k} - a_n| = |a_{n+k} - a_{n+k-1} + a_{n+k-1} - a_{n+k-2} + \ldots + a_{n+1} - a_n|$$
$$< \lambda(|a_{n+k-1} - a_{n+k-2}| + |a_{n+k-2} - a_{n+k-3}| + \ldots + |a_n - a_{n-1}|)$$
$$< (\lambda^k + \lambda^{k-1} + \ldots + \lambda^2 + \lambda)|a_n - a_{n-1}|$$
$$\leq (\lambda^k + \lambda^{k-1} + \ldots + \lambda^2 + \lambda)\lambda^{n-2}|a_2 - a_1|$$
$$= \frac{\lambda^{n-1}(1-\lambda^k)}{1-\lambda}|a_2 - a_1| < \frac{\lambda^{n-1}}{1-\lambda}|a_2 - a_1|.$$

Hence for arbitrarily fixed $\varepsilon > 0$, for $n > \left[1 + \frac{\ln \frac{\varepsilon(1-\lambda)}{|a_2-a_1|}}{\ln \lambda}\right]$, and for any $k \in \mathbb{N}$, we have $|a_{n+k} - a_n| < \varepsilon$.

2.2.52. Since $\{S_n\}$ is convergent, it is a Cauchy sequence. We will show that $\{\ln \sigma_n\}$ is also a Cauchy sequence. By the inequality in 2.1.41,

$$\ln \sigma_{n+k} - \ln \sigma_n = \ln\left(1 + \frac{1}{a_{n+k}}\right) + \ldots + \ln\left(1 + \frac{1}{a_{n+1}}\right)$$
$$< \frac{1}{a_{n+k}} + \ldots + \frac{1}{a_{n+1}} < \varepsilon$$

for $k \in \mathbb{N}$ and for sufficiently large n.

2.2.53. By the result in 1.1.23,

$$R_{n+k} - R_n$$
$$= (R_{n+k} - R_{n+k-1}) + (R_{n+k-1} - R_{n+k-2}) + \ldots + (R_{n+1} - R_n)$$
$$= (-1)^n \left(\frac{(-1)^{k-1}}{q_{n+k-1}q_{n+k}} + \frac{(-1)^{k-2}}{q_{n+k-2}q_{n+k-1}} + \ldots - \frac{1}{q_{n+1}q_{n+2}} + \frac{1}{q_n q_{n+1}}\right).$$

Hence, by the monotonicity of the sequence $\{q_n\}$ and the fact that $q_n \geq n$ (see the solution of 1.1.24),

$$|R_{n+k} - R_n| \leq \frac{1}{q_{n+1}q_n} \leq \frac{1}{n^2}.$$

2.2. Limits. Properties of Convergent Sequences

2.2.54. Let d denote the common difference of the given progression. Assume first that $d \neq 0$. Then
$$\frac{1}{a_k a_{k+1}} = \left(\frac{1}{a_k} - \frac{1}{a_{k+1}}\right) \frac{1}{d}.$$
Hence
$$\lim_{n\to\infty} \left(\frac{1}{a_1 a_2} + \frac{1}{a_2 a_3} + \ldots + \frac{1}{a_n a_{n+1}}\right) = \frac{1}{a_1 d}.$$
For $d = 0$, the arithmetic progression is a constant sequence and therefore
$$\lim_{n\to\infty} \left(\frac{1}{a_1 a_2} + \frac{1}{a_2 a_3} + \ldots + \frac{1}{a_n a_{n+1}}\right) = +\infty.$$

2.2.55. Let d denote the common difference of the given progression. Assume first that $d \neq 0$. Since
$$\frac{1}{\sqrt{a_k} + \sqrt{a_{k+1}}} = \frac{\sqrt{a_{k+1}} - \sqrt{a_k}}{d},$$
we have
$$\lim_{n\to\infty} \frac{1}{\sqrt{n}} \left(\frac{1}{\sqrt{a_1}+\sqrt{a_2}} + \frac{1}{\sqrt{a_2}+\sqrt{a_3}} + \ldots + \frac{1}{\sqrt{a_n}+\sqrt{a_{n+1}}}\right) = \frac{1}{\sqrt{d}}.$$
For $d = 0$, the arithmetic progression is a constant sequence and therefore the limit is equal to $+\infty$.

2.2.56.
(a) By Problem 2.1.38,
$$\left(1 + \frac{1}{n}\right)^n < e < \left(1 + \frac{1}{n}\right)^{n+1}.$$
Thus

(*) $\quad 1 < n(\sqrt[n]{e} - 1) < n\left(\left(1 + \frac{1}{n}\right)^{1+\frac{1}{n}} - 1\right).$

Now using the Bernoulli inequality (see 1.2.4) one can show that
$$\left(1 + \frac{1}{n}\right)^{\frac{1}{n}} \leq 1 + \frac{1}{n^2}.$$

Hence
$$n\left(\left(1+\frac{1}{n}\right)^{1+\frac{1}{n}}-1\right) \leq 1+\frac{1}{n}+\frac{1}{n^2}.$$
Therefore, by (*) and by the squeeze law,
$$\lim_{n\to\infty} n(\sqrt[n]{e}-1) = 1.$$

(b) For an arbitrarily fixed n,
$$\frac{e^{\frac{1}{n}}+e^{\frac{2}{n}}+\ldots+e^{\frac{n}{n}}}{n} = \frac{(e-1)e^{\frac{1}{n}}}{n(e^{\frac{1}{n}}-1)}.$$

Hence in view of (a) we get
$$\lim_{n\to\infty}\frac{e^{\frac{1}{n}}+e^{\frac{2}{n}}+\ldots+e^{\frac{n}{n}}}{n} = e-1.$$

2.2.57. We have $a_{n+1}-a_n = -p(a_n-a_{n-1})$. Therefore
$$a_n = a+(b-a)+(a_3-a_2)+\ldots+(a_n-a_{n-1})$$
$$= a+(b-a)(1-p+p^2+\ldots+(-1)^n p^{n-2}).$$

If $b=a$, then $\{a_n\}$ is a constant sequence convergent to a. If $a \neq b$, then the sequence is convergent provided $|p|<1$, and its limit is $a+\frac{b-a}{1+p}$.

2.2.58. Observe that
$$c_{n+1} = \frac{a_n+2b_n}{a_n+b_n} = \frac{c_n+2}{c_n+1}.$$

Hence
$$|c_{n+1}-\sqrt{2}| = \frac{\sqrt{2}-1}{c_n+1}|c_n-\sqrt{2}| < (\sqrt{2}-1)|c_n-\sqrt{2}| < \frac{1}{2}|c_n-\sqrt{2}|.$$

Consequently, by induction,
$$|c_{n+1}-\sqrt{2}| < \frac{1}{2^n}|c_1-\sqrt{2}|,$$

which means that the limit of $\{c_n\}$ is $\sqrt{2}$.

2.3. The Toeplitz Transformation, the Stolz Theorem and their Applications

2.3.1. If all the terms of the sequence $\{a_n\}$ are equal to a, then by (ii), $\lim_{n\to\infty} b_n = a \lim_{n\to\infty} \sum_{k=1}^{n} c_{n,k} = a$. Thus it is enough to consider the case where the sequence converges to zero. Then, for any $m > 1$ and $n \geq m$,

$$(*) \qquad |b_n - 0| = \left| \sum_{k=1}^{n} c_{n,k} a_k \right| \leq \sum_{k=1}^{m-1} |c_{n,k}| \cdot |a_k| + \sum_{k=m}^{n} |c_{n,k}| \cdot |a_k|.$$

The convergence to zero of $\{a_n\}$ implies that for a given $\varepsilon > 0$ there exists n_1 such that

$$|a_n| < \frac{\varepsilon}{2C} \quad \text{for} \quad n \geq n_1.$$

Of course, the sequence $\{a_n\}$ is bounded, say by $D > 0$. It follows from (i) that there exists n_2 such that for $n \geq n_2$,

$$\sum_{k=1}^{n_1-1} |c_{n,k}| < \frac{\varepsilon}{2D}.$$

Next putting $m = n_1$ in $(*)$, we get

$$|b_n| \leq D \sum_{k=1}^{n_1-1} |c_{n,k}| + \frac{\varepsilon}{2C} \sum_{k=n_1}^{n} |c_{n,k}| < \frac{\varepsilon}{2} + \frac{\varepsilon}{2} = \varepsilon$$

for $n \geq \max\{n_1, n_2\}$. Hence $\lim_{n\to\infty} b_n = 0$.

2.3.2. Apply the Toeplitz theorem with $c_{n,k} = \frac{1}{n}$, $k = 1, 2, ..., n$.

2.3.3.

(a) If $c_{n,k}$ are nonnegative, then (iii) follows from (ii).

(b) By (ii) in Problem 2.3.1, $\sum_{k=1}^{n} c_{n,k} > \frac{1}{2}$ for sufficiently large n, say $n > n_0$. It follows from the divergence of $\{a_n\}$ to $+\infty$ that, given $M > 0$, there exists n_1 such that $a_n \geq 2M$ if $n > n_1$.

Without loss of generality we can assume that all the terms a_n are positive. Set $n_2 = \max\{n_0, n_1\}$. Then

$$\sum_{k=1}^{n} c_{n,k} a_k = \sum_{k=1}^{n_2} c_{n,k} a_k + \sum_{k=n_2+1}^{n} c_{n,k} a_k \geq \sum_{k=1}^{n_2} c_{n,k} a_k + M \geq M,$$

and so $\{b_n\}$ diverges to $+\infty$.

2.3.4. This is a special case of 2.3.3. Take $c_{n,k} = \frac{1}{n}$ for $k = 1, 2, ..., n$.

2.3.5. Apply the Toeplitz theorem (2.3.1) with $c_{n,k} = \frac{2(n-k+1)}{n^2}$.

2.3.6. Use the harmonic-geometric-arithmetic mean inequality (see 1.2.3), the squeeze principle for sequences and the result in 2.3.2.

2.3.7. Apply the foregoing problem to the sequence $\{\frac{a_{n+1}}{a_n}\}$.

2.3.8. If $b \neq 0$, we take $c_{n,k} = \frac{b_{n-k+1}}{nb}$ and see that condition (i) in 2.3.1 is satisfied. In view of 2.3.2 condition (ii) is also satisfied. In this case the desired result follows from the Toeplitz theorem. For $b = 0$, setting $c_{n,k} = \frac{1+b_{n-k+1}}{n}$ yields

$$\lim_{n \to \infty} \frac{a_1(1+b_n) + a_2(1+b_{n-1}) + ... + a_n(1+b_1)}{n} = a.$$

Thus, by 2.3.2,

$$\lim_{n \to \infty} \frac{a_1 b_n + a_2 b_{n-1} + ... + a_n b_1}{n} = 0.$$

2.3.9. We apply the Toeplitz theorem to the sequence $\{\frac{a_n}{b_n}\}$ with $c_{n,k} = \frac{b_k}{b_1 + .. + b_n}$.

2.3.10. One can apply the Toeplitz theorem with $c_{n,k} = \frac{b_k}{b_1 + .. + b_n}$.

2.3.11. For $n > 1$, we put

$$a_n = \frac{x_n - x_{n-1}}{y_n - y_{n-1}}, \quad b_n = y_n - y_{n-1}$$

and apply the result in the foregoing problem.

2.3. Toeplitz Transformation and Stolz Theorem 183

2.3.12.

(a) In 2.3.10 we put $x_n = 1 + \frac{1}{\sqrt{2}} + \ldots + \frac{1}{\sqrt{n}}$, $y_n = \sqrt{n}$ and show that the limit is 2.

(b) Set
$$x_n = a + \frac{a^2}{2} + \ldots + \frac{a^n}{n}, \quad y_n = \frac{a^{n+1}}{n}.$$
Beginning with some value of the index n, the sequence $\{y_n\}$ is strictly increasing. By 2.2.31 (b) we see that $\lim\limits_{n\to\infty} y_n = +\infty$. Therefore
$$\lim_{n\to\infty} \frac{n}{a^{n+1}} \left(a + \frac{a^2}{2} + \ldots + \frac{a^n}{n} \right) = \frac{1}{a-1}.$$

(c) We can apply the Stolz theorem (see 2.3.11) to the sequences
$$x_n = k! + \frac{(k+1)!}{1!} + \ldots + \frac{(k+n)!}{n!}, \quad y_n = n^{k+1}.$$
We have
$$\lim_{n\to\infty} \frac{x_n - x_{n-1}}{y_n - y_{n-1}} = \lim_{n\to\infty} \frac{(n+1)\cdot(n+2)\cdot\ldots\cdot(n+k)}{n^{k+1} - (n-1)^{k+1}}$$
$$= \lim_{n\to\infty} \frac{\left(1+\frac{1}{n}\right)\cdot\ldots\cdot\left(1+\frac{k}{n}\right)}{n\left(1 - \left(1-\frac{1}{n}\right)^{k+1}\right)} = \lim_{n\to\infty} \frac{\left(1+\frac{1}{n}\right)\cdot\ldots\cdot\left(1+\frac{k}{n}\right)}{1 + \left(1-\frac{1}{n}\right) + \ldots + \left(1-\frac{1}{n}\right)^k}$$
$$= \frac{1}{k+1}.$$

(d) Set $x_n = \frac{1}{\sqrt{n}} + \ldots + \frac{1}{\sqrt{2n}}$, $y_n = \sqrt{n}$. Then
$$\lim_{n\to\infty} \frac{x_n - x_{n-1}}{y_n - y_{n-1}} = \lim_{n\to\infty} \frac{\frac{1}{\sqrt{2n}} + \frac{1}{\sqrt{2n-1}} - \frac{1}{\sqrt{n-1}}}{\sqrt{n} - \sqrt{n-1}}$$
$$= \lim_{n\to\infty} (\sqrt{n} + \sqrt{n-1})\left(\frac{1}{\sqrt{2n}} + \frac{1}{\sqrt{2n-1}} - \frac{1}{\sqrt{n-1}} \right)$$
$$= \lim_{n\to\infty} \left(\frac{1}{\sqrt{2}} + \sqrt{\frac{n}{2n-1}} - \sqrt{\frac{n}{n-1}} + \sqrt{\frac{n-1}{2n}} + \sqrt{\frac{n-1}{2n-1}} - 1 \right)$$
$$= 2(\sqrt{2} - 1).$$
Hence by the Stolz theorem the limit is $2(\sqrt{2} - 1)$.

(e) Taking $x_n = 1^k + 2^k + \ldots + n^k$ and $y_n = n^{k+1}$, we see that

$$\frac{x_n - x_{n-1}}{y_n - y_{n-1}} = \frac{n^k}{n^{k+1} - (n-1)^{k+1}} \xrightarrow[n \to \infty]{} \frac{1}{k+1}.$$

Now it is enough to apply the Stolz theorem.

(f) By the Stolz theorem,

$$\lim_{n \to \infty} \frac{1 + 1 \cdot a + 2 \cdot a^2 + \ldots + n \cdot a^n}{n \cdot a^{n+1}} = \frac{1}{a-1}.$$

(g) One can also apply the Stolz theorem with

$$x_n = (k+1)(1^k + 2^k + \ldots + n^k) - n^{k+1} \quad \text{and} \quad y_n = (k+1)n^k.$$

Then

$$\frac{x_n - x_{n-1}}{y_n - y_{n-1}} = \frac{(k+1)n^k - n^{k+1} + (n-1)^{k+1}}{(k+1)[n^k - (n-1)^k]} \xrightarrow[n \to \infty]{} \frac{1}{2}.$$

2.3.13. Applying the Stolz theorem to

$$x_n = a_1 + \frac{a_2}{\sqrt{2}} + \ldots + \frac{a_n}{\sqrt{n}} \quad \text{and} \quad y_n = \sqrt{n},$$

we see that

$$\lim_{n \to \infty} \frac{1}{\sqrt{n}} \left(a_1 + \frac{a_2}{\sqrt{2}} + \frac{a_3}{\sqrt{3}} + \ldots + \frac{a_n}{\sqrt{n}} \right) = 2a.$$

2.3.14. In the Stolz theorem we set $x_n = a_{n+1}$ and $y_n = n$.

2.3.15. Applying the Toeplitz transformation to $\{a_n\}$ with $c_{n,k} = \frac{1}{2^{n-k+1}}$, we see that

$$\lim_{n \to \infty} \left(\frac{a_n}{1} + \frac{a_{n-1}}{2} + \ldots + \frac{a_1}{2^{n-1}} \right) = 2a.$$

2.3. Toeplitz Transformation and Stolz Theorem

2.3.16.

(a) Using the Toeplitz transformation to $\{a_n\}$ with
$$c_{n,k} = \frac{1}{(n+1-k)(n+2-k)},$$
we can show that
$$\lim_{n\to\infty}\left(\frac{a_n}{1\cdot 2} + \frac{a_{n-1}}{2\cdot 3} + \ldots + \frac{a_1}{n(n+1)}\right) = a.$$

(b) As in the proof of (a), we can apply the Toeplitz theorem to $\{a_n\}$ with $c_{n,k} = \frac{3}{2}\frac{(-1)^{n-k}}{2^{n-k}}$ and show that the limit is $\frac{2}{3}a$.

2.3.17. Set $a_n = \binom{nk}{n}$. In view of 2.3.7 it is enough to calculate $\lim_{n\to\infty}\frac{a_{n+1}}{a_n}$. We have
$$\frac{\binom{(n+1)k}{n+1}}{\binom{nk}{n}} = \frac{(nk+1)(nk+2)\cdot\ldots\cdot(nk+k)}{(n+1)(nk-n+1)(nk-n+2)\cdot\ldots\cdot(nk-n+k-1)}.$$
Therefore the limit is equal to $\frac{k^k}{(k-1)^{k-1}}$.

2.3.18. Let $\{a_n\}$ be an arithmetic progression with the common difference $d > 0$. Set
$$c_n = \frac{n^n(a_1\cdot\ldots\cdot a_n)}{(a_1+\ldots+a_n)^n}.$$
Then
$$\frac{c_{n+1}}{c_n} = \frac{(n+1)a_{n+1}}{a_1+a_2+\ldots+a_{n+1}}\left(\frac{\frac{a_1+\ldots+a_n}{n}}{\frac{a_1+\ldots+a_{n+1}}{n+1}}\right)^n$$
$$= \frac{2a_{n+1}}{a_1+a_{n+1}}\left(\frac{2a_1+(n-1)d}{2a_1+nd}\right)^n \xrightarrow[n\to\infty]{} 2e^{-1}.$$
Hence by 2.3.7, the limit equals $2e^{-1}$. If $d=0$, the limit is 1.

2.3.19. Since $b_n = 2a_n + a_{n-1}$, $a_n = \frac{b_n - a_{n-1}}{2}$ and $a_{n-1} = \frac{b_{n-1}-a_{n-2}}{2}$. Thus $a_n = \frac{2b_n - b_{n-1}+a_{n-2}}{2^2}$. Repeated application of this procedure $n-1$ times gives
$$a_n = \frac{2^{n-1}b_n - 2^{n-2}b_{n-1}+\ldots+(-1)^{n-2}2^1 b_2 + (-1)^{n-1}a_1}{2^n}.$$

Thus by 2.3.16 (b), $\lim\limits_{n\to\infty} a_n = \frac{1}{3}b$.

2.3.20. Put $c_n = (a_1 \cdot \ldots \cdot a_n) n^{nx}$. Then

$$\frac{c_{n+1}}{c_n} = \left(1 + \frac{1}{n}\right)^{nx} (n+1)^x a_{n+1} \xrightarrow[n\to\infty]{} e^x a.$$

Therefore by 2.3.7, $\lim\limits_{n\to\infty} n^x (a_1 \cdot a_2 \cdot \ldots \cdot a_n)^{\frac{1}{n}} = e^x a$.

2.3.21.

(a) We apply the Stolz theorem with $x_n = 1 + \frac{1}{2} + \ldots + \frac{1}{n}$ and $y_n = \ln n$. This gives

$$\frac{x_n - x_{n-1}}{y_n - y_{n-1}} = \frac{1}{\ln\left(1 + \frac{1}{n-1}\right)^n} \xrightarrow[n\to\infty]{} 1,$$

because $\lim\limits_{n\to\infty} \ln\left(1 + \frac{1}{n}\right)^n = 1$, which follows from the inequalities $\left(1 + \frac{1}{n}\right)^n < e < \left(1 + \frac{1}{n}\right)^{n+1}$ (see 2.1.41).

(b) The limit is $\frac{1}{2}$ (see the solution of (a)).

2.3.22. We apply the Stolz theorem to

$$x_n = \frac{a_1}{1} + \frac{a_2}{2} + \ldots + \frac{a_n}{n} \quad \text{and} \quad y_n = \ln n.$$

Consequently,

$$\frac{x_n - x_{n-1}}{y_n - y_{n-1}} = \frac{a_n}{\ln\left(1 + \frac{1}{n-1}\right)^n} \xrightarrow[n\to\infty]{} a.$$

2.3.23. Use the result in 2.3.7.

(a) 1,

(b) e^{-2},

(c) e^{-2},

(d) e^3.

(e) We have

$$\lim_{n\to\infty} \frac{\sqrt[k]{n}}{\sqrt[n]{n!}} = \begin{cases} e & \text{for } k = 1, \\ 0 & \text{for } k > 1. \end{cases}$$

2.3. Toeplitz Transformation and Stolz Theorem 187

2.3.24. By the Stolz theorem (see 2.3.11),

$$\lim_{n\to\infty} \frac{\sum_{k=1}^{n} \frac{a_k}{k}}{\ln n} = \lim_{n\to\infty} \frac{a_{n+1}}{\ln\left(1+\frac{1}{n}\right)^{n+1}} = a.$$

2.3.25. One can easily verify that

$$a_1 = A_1, \quad a_2 = 2A_2 - A_1, \quad a_n = nA_n - (n-1)A_{n-1}, \quad n \geq 2.$$

Thus

$$\lim_{n\to\infty} \frac{\sum_{k=1}^{n} \frac{a_k}{k}}{\ln n} = \lim_{n\to\infty} \frac{\frac{1}{2}A_1 + \frac{1}{3}A_2 + \ldots + \frac{1}{n}A_{n-1} + A_n}{\ln n} = A,$$

where the last equality follows from the foregoing problem.

2.3.26. [O. Toeplitz, Prace Matematyczno-Fizyczne, 22(1911), 113-119] Let $\{a_n\}$ be the sequence all of whose terms are equal to 1. Then $\lim_{n\to\infty} a_n = 1$ and $b_n = \sum_{k=1}^{n} c_{n,k} a_k = \sum_{k=1}^{n} c_{n,k}$. Hence $1 = \lim_{n\to\infty} b_n = \lim_{n\to\infty} \sum_{k=1}^{n} c_{n,k}$. Thus (ii) holds. Now let $\{a_n^{(k)}\}$ be a sequence whose kth term is 1 and whose other terms are all 0. Then $\lim_{n\to\infty} a_n^{(k)} = 0$ and $0 = \lim_{n\to\infty} b_n = \lim_{n\to\infty} c_{n,k}$. Therefore (i) also holds. Suppose that (iii) is not satisfied. Then for any $C > 0$ there exists n_C such that $\sum_{k=1}^{n_C} |c_{n_C,k}| \geq C$. In fact, given $C > 0$, there are infinitely many such indices n_C. Now let n_1 be the least positive integer such that $\sum_{k=1}^{n_1} |c_{n_1,k}| > 10^2$. We define the first n_1 terms of $\{a_n\}$ by setting

$$\operatorname{sgn} c_{n_1,k} = \operatorname{sgn} a_k \quad \text{and} \quad |a_k| = \frac{1}{10}.$$

Then

$$b_{n_1} = \sum_{k=1}^{n_1} c_{n_1,k} a_k = \sum_{k=1}^{n_1} \frac{1}{10} |c_{n_1,k}| > 10.$$

By (i), there exists n_0 such that

$$\sum_{k=1}^{n_1} |c_{n,k}| < 1 \quad \text{for} \quad n \geq n_0.$$

Consequently,

$$\left| \sum_{k=1}^{n_1} c_{n,k} a_k \right| < \frac{1}{10} \quad \text{for} \quad n \geq n_0.$$

Now we take the least integer n_2 such that $n_2 \geq \max\{n_0, n_1\}$ and $\sum_{k=1}^{n_2} |c_{n_2,k}| > 10^4 + 1 + 10$. Define the consecutive terms of $\{a_n\}$ by setting

$$\operatorname{sgn} c_{n_2,k} = \operatorname{sgn} a_k \quad \text{and} \quad |a_k| = \frac{1}{10^2} \quad \text{for} \quad n_1 + 1 \leq k \leq n_2.$$

Then

$$b_{n_2} = \sum_{k=1}^{n_2} c_{n_2,k} a_k = \sum_{k=1}^{n_1} c_{n_2,k} a_k + \sum_{k=n_1+1}^{n_2} c_{n_2,k} a_k$$

$$= \sum_{k=1}^{n_1} c_{n_2,k} a_k + \frac{1}{10^2} \sum_{k=n_1+1}^{n_2} |c_{n_2,k}|.$$

It follows from the above that

$$b_{n_2} > -\frac{1}{10} + \frac{1}{10^2}\left(10^4 + 1 + 10 - 1\right) = 10^2.$$

We can construct inductively the sequence $\{a_n\}$ whose terms with indices from $n_{k-1}+1$ through n_k are equal either to $+\frac{1}{10^k}$ or $-\frac{1}{10^k}$; then the transformed sequence $\{b_n\}$ satisfies

$$b_{n_k} > 10^k \quad \text{for} \quad k = 1, 2, 3, \ldots.$$

Thus the sequence $\{a_n\}$ converges to zero whereas the transformed sequence $\{b_n\}$ has a divergent subsequence $\{b_{n_k}\}$. This is a contradiction, and so (iii) holds true.

2.4. Limit Points. Limit Superior and Limit Inferior

2.4.1.

(a) First, let us show that the given subsequences have a common limit. Suppose that $\lim_{k\to\infty} a_{2k} = a$, $\lim_{k\to\infty} a_{2k+1} = b$ and $\lim_{k\to\infty} a_{3k} = c$. Then $\lim_{k\to\infty} a_{6k} = a = c$ and $\lim_{k\to\infty} a_{6k+3} = b = c$. Therefore $a = b = c$. Now we prove that the sequence $\{a_n\}$ also converges to a. Given any $\varepsilon > 0$, there exist positive integers k_1 and k_2 such that

$$k > k_1 \quad \text{implies} \quad |a_{2k} - a| < \varepsilon,$$
$$k > k_2 \quad \text{implies} \quad |a_{2k+1} - a| < \varepsilon.$$

Hence

$$n > n_0 = \max\{2k_1, 2k_2 + 1\} \quad \text{implies} \quad |a_n - a| < \varepsilon.$$

(b) No. Consider the sequence $\{a_n\}$ defined by $a_n = (-1)^n$. Then $\lim_{k\to\infty} a_{2k} = 1$, $\lim_{k\to\infty} a_{2k+1} = -1$. But $\lim_{n\to\infty} a_n$ does not exist. Now take the sequence $\{a_n\}$ defined as follows

$$a_n = \begin{cases} 0 & \text{if } n = 2^k,\ k = 0, 1, 2, ..., \\ 1 & \text{otherwise.} \end{cases}$$

Then $\lim_{k\to\infty} a_{3k} = 1$ and $\lim_{k\to\infty} a_{2k+1} = 1$, but $\lim_{k\to\infty} a_{2k}$ does not exist. Of course, the sequence $\{a_n\}$ is divergent.
Finally, consider the third sequence

$$a_n = \begin{cases} 0 & \text{if } n \text{ is a prime number,} \\ 1 & \text{if } n \text{ is a composite number.} \end{cases}$$

For this sequence we have $\lim_{k\to\infty} a_{3k} = 1$ and $\lim_{k\to\infty} a_{2k} = 1$, but $\lim_{k\to\infty} a_{2k+1}$ does not exist, because the sequence $\{a_{2k+1}\}$ contains a subsequence with prime indexes and a subsequence with composite indexes.(Note that there are infinitely many prime numbers. Otherwise, if $p_1, p_2, ..., p_n$ are prime, $p_1 < p_2 < ... < p_n$, and no prime greater than p_n does exist, then $p_1 \cdot p_2 \cdot ... \cdot p_n + 1 >$

p_n is also prime, because it has no prime divisors except for itself and 1. This is a contradiction.)

2.4.2. No. Define the sequence $\{a_n\}$ by putting

$$a_n = \begin{cases} 0 & \text{if } n \text{ is a prime number,} \\ 1 & \text{if } n \text{ is a composite number.} \end{cases}$$

Then every subsequence $\{a_{s \cdot n}\}$, $s > 1$, $n \geq 2$, is a constant sequence and therefore it is convergent. The sequence $\{a_n\}$ is divergent (see the solution of Problem 2.4.1(b)).

2.4.3. Evidently, $\mathbf{S_p} \cup \mathbf{S_q} \cup ... \cup \mathbf{S_s} \subset \mathbf{S}$. To obtain the inclusion in the other direction, assume $x \notin \mathbf{S_p} \cup \mathbf{S_q} \cup ... \cup \mathbf{S_s}$. Then there exist positive numbers $\varepsilon_p, \varepsilon_q, ..., \varepsilon_s$ and positive integers $n_p, n_q, ..., n_s$ such that

$$n > n_p \quad \text{implies} \quad |x - a_{p_n}| > \varepsilon_p,$$
$$n > n_q \quad \text{implies} \quad |x - a_{q_n}| > \varepsilon_q,$$
$$...$$
$$n > n_s \quad \text{implies} \quad |x - a_{s_n}| > \varepsilon_s.$$

Setting $\varepsilon = \min\{\varepsilon_p, \varepsilon_q, ..., \varepsilon_s\}$ and $m = \max\{p_{n_p}, q_{n_q}, ..., s_{n_s}\}$, we obtain $|x - a_n| > \varepsilon$ for $n > m$, This implies that x cannot be a limit point of the sequence $\{a_n\}$. Therefore

$$\mathbf{S} \subset \mathbf{S_p} \cup \mathbf{S_q} \cup ... \cup \mathbf{S_s}.$$

It follows from the equality $\mathbf{S} = \mathbf{S_p} \cup \mathbf{S_q} \cup ... \cup \mathbf{S_s}$ just proved that, if every subsequence $\{a_{p_n}\}, \{a_{q_n}\}, ..., \{a_{s_n}\}$ converges to a, then the sequence $\{a_n\}$ also converges to a.

2.4.4. No. Define the sequence $\{a_n\}$ by the following formula:

$$a_n = \begin{cases} 0 & \text{if } n = 2^k, \ k = 0, 1, 2, ..., \\ 1 & \text{otherwise.} \end{cases}$$

Every subsequence

$$\{a_{2k-1}\}, \{a_{2(2k-1)}\}, \{a_{2^2(2k-1)}\}, ..., \{a_{2^m(2k-1)}\}, ...$$

converges to 1, whereas the sequence $\{a_n\}$ diverges.

2.4. Limit Points. Limit Superior and Limit Inferior

2.4.5. Assume that the sequence $\{a_n\}$ does not converge to a. Then there exists $\varepsilon > 0$ such that for any positive integer k there is $n_k > k$ satisfying $|a_{n_k} - a| \geq \varepsilon$. If we assume that n_k is the minimum of such numbers, then the sequence $\{n_k\}$ is monotonically increasing. Moreover, $\lim_{k \to \infty} n_k = +\infty$. Such a sequence $\{a_{n_k}\}$ does not contain any subsequence converging to a, which contradicts our hypothesis. Therefore $\{a_n\}$ converges to a.

2.4.6.

(a) It is obvious that 1 is the only limit point of the sequence. Hence **S** is a singleton, $\mathbf{S} = \{1\}$.

(b) We have $a_{3k} = 0$, $a_{3k+1} = 1$, $a_{3k+2} = 0$. Hence, by Problem 2.4.3, the set **S** of the limit points of this sequence has two members, $\mathbf{S} = \{0, 1\}$.

(c) We have
$$a_{2k} = \frac{1}{2^{2k} + 3} \quad \text{and} \quad a_{2k+1} = \frac{2^{2k+2} + 1}{2^{2k+1} + 3}.$$
Hence $\mathbf{S} = \{0, 2\}$.

(d) We have
$$a_{2k} = \frac{2\ln(6k) + \ln(2k)}{\ln(4k)} \quad \text{and} \quad a_{2k+1} = \frac{\ln(2k+1)}{\ln(2(2k+1))}.$$
Therefore $\mathbf{S} = \{1, 3\}$.

(e)
$$a_{6k} = 1, \quad a_{6k+1} = (0.5)^{6k+1}, \quad a_{6k+2} = (-0.5)^{6k+2},$$
$$a_{6k+3} = -1, \quad a_{6k+4} = (-0.5)^{6k+4}, \quad a_{6k+5} = (0.5)^{6k+5}.$$
Thus $\mathbf{S} = \{-1, 0, 1\}$.

(f) We have
$$a_{7k} = 0, \quad a_{7k+1} = \frac{2}{7}, \quad a_{7k+2} = \frac{1}{7}, \quad a_{7k+3} = \frac{4}{7},$$
$$a_{7k+4} = \frac{4}{7}, \quad a_{7k+5} = \frac{1}{7}, \quad a_{7k+6} = \frac{2}{7}.$$

Therefore $\mathbf{S} = \{0, \frac{1}{7}, \frac{2}{7}, \frac{4}{7}\}$.

2.4.7.

(a) Let $\alpha = \frac{p}{q}$, $p \in \mathbb{Z}, q \in \mathbb{N}$, where p and q are co-prime. Then

$$a_{kq} = 0 \quad \text{and} \quad a_{kq+l} = kp + \frac{lp}{q} - \left[kp + \left[\frac{lp}{q}\right] + r\right] = \frac{lp}{q} - \left[\frac{lp}{q}\right],$$

where $l = 1, 2, ..., q-1$ and $r = \frac{lp}{q} - \left[\frac{lp}{q}\right]$. Thus

$$\mathbf{S} = \left\{0, \frac{p}{q} - \left[\frac{p}{q}\right], \frac{2p}{q} - \left[\frac{2p}{q}\right], ..., \frac{(q-1)p}{q} - \left[\frac{(q-1)p}{q}\right]\right\}.$$

(b) We will show that every real number $x \in [0, 1]$ is a limit point of the sequence $\{n\alpha - [n\alpha]\}$. By Problem 1.1.20, there exist $p_n \in \mathbb{Z}$ and $q_n \in \mathbb{N}$ such that $0 < \alpha - \frac{p_n}{q_n} < \frac{1}{q_n^2}$. Since $\lim_{n \to \infty} q_n = +\infty$, $\lim_{n \to \infty} (\alpha q_n - p_n) = 0$. Let $x \in (0, 1)$ and let $\varepsilon > 0$ be so small that $0 < x - \varepsilon < x + \varepsilon < 1$. Now suppose that n_1 is so large that

$$0 < \alpha q_{n_1} - p_{n_1} < \frac{1}{q_{n_1}} < \varepsilon.$$

Then there is $n_0 \in \mathbb{N}$ satisfying

(1) $\qquad n_0(\alpha q_{n_1} - p_{n_1}) \in (x - \varepsilon, x + \varepsilon).$

(see the solution of Problem 1.1.21). It follows from (1) that $[n_0 \alpha q_{n_1} - n_0 p_{n_1}] = 0$, or equivalently, $n_0 p_{n_1} = [n_0 \alpha q_{n_1}]$. Therefore the term $n_0 \alpha q_{n_1} - [n_0 \alpha q_{n_1}]$ from the range of our sequence belongs to the interval $(x - \varepsilon, x + \varepsilon)$, which means that x is a limit point of the sequence under consideration. Similarly, one can show that 0 and 1 are also limit points.

(c) Assume first that α is a rational number of the interval $(0, 1)$. Let $\alpha = \frac{p}{q}$, where p and q are co-prime and $p < q$. Then $a_{2kq} = a_{2kq+q} = 0$, and

$$a_{2kq+l} = \sin \frac{lp\pi}{q} \quad \text{for} \quad l = 1, 2, ..., q-1, q+1, ..., 2q-1.$$

Hence

$$\mathbf{S} = \left\{0, \sin \frac{p\pi}{q}, \sin \frac{2p\pi}{q}, ..., \sin \frac{(q-1)p\pi}{q}\right\}.$$

2.4. Limit Points. Limit Superior and Limit Inferior 193

If $\alpha \in \mathbb{Z}$, then the sequence is constant. Taking $\alpha \in \mathbb{Q} \setminus \mathbb{Z}$, we can write

$$\alpha = [\alpha] + (\alpha - [\alpha]) \quad \text{and} \quad \alpha - [\alpha] \in (0, 1).$$

So, $\sin n\pi\alpha = (-1)^{[\alpha]} \sin(\alpha - [\alpha])n\pi$, and this case can be reduced to the foregoing special case.

(d) Let $t \in [-1, 1]$ be an arbitrarily chosen number. Then there exists $x \in \mathbb{R}_+$ such that $\sin x = -t$. We can restrict our consideration to the case $\alpha > 0$, because the sine is an odd function. Since α is irrational, there exist sequences of positive integers $\{p_n\}$ and $\{q_n\}$ such that

$$\frac{x}{2\pi} = \lim_{n \to \infty} \left(p_n - q_n \frac{\alpha}{2} \right).$$

(See the solution of 1.1.21.) Therefore $x = \lim_{n \to \infty} (2\pi p_n - \alpha \pi q_n)$. Hence, by continuity and periodicity of the sine function, we get

$$-t = \sin x = \lim_{n \to \infty} \sin(2\pi p_n - \alpha \pi q_n) = -\lim_{n \to \infty} \sin \alpha \pi q_n.$$

It follows from the above consideration that every number in the interval $[-1, 1]$ is a limit point of the sequence.

2.4.8. We will show that in any interval (a, b) there is at least one term of our sequence. Since $\lim_{n \to \infty} (\sqrt[3]{n+1} - \sqrt[3]{n}) = 0$, there exists $n_0 \in \mathbb{N}$ such that

$$\sqrt[3]{n+1} - \sqrt[3]{n} < b - a, \quad n > n_0.$$

Let m_0 be a positive integer satisfying $\sqrt[3]{m_0} > \sqrt[3]{n_0} - a$ and let $\mathbf{A} = \{n \in \mathbb{N} : \sqrt[3]{n} - \sqrt[3]{m_0} \leq a\}$. The set $\mathbf{A}$ is nonempty (e.g., $n_0 \in \mathbf{A}$) and bounded above. Putting $n_1 = \max \mathbf{A}$ and $n_2 = n_1 + 1$, we get $\sqrt[3]{n_2} - \sqrt[3]{m_0} > a$ and $\sqrt[3]{n_2} > a + \sqrt[3]{m_0} > \sqrt[3]{n_0}$. Therefore $n_2 > n_0$. Hence $\sqrt[3]{n_2} < \sqrt[3]{n_1} + b - a \leq \sqrt[3]{m_0} + a + b - a$, or equivalently, $a < \sqrt[3]{n_2} - \sqrt[3]{m_0} < b$.

2.4.9. Boundedness of the set of the limit points of a bounded sequence is evident. Let $\mathbf{S}$ denote the set of limit points of the sequence $\{a_n\}$. If $\mathbf{S}$ is finite, then it is closed. Assume that $\mathbf{S}$ is infinite and let s be its limit point. Define the sequence $\{s_k\}$ of members of $\mathbf{S}$

in the following way: for s_1 take any member of **S** different from s. For s_2 choose any member of **S** different from s and such that $|s_2 - s| < \frac{1}{2}|s_1 - s|$, and inductively, $|s_{k+1} - s| < \frac{1}{2}|s_k - s|$, $s_{k+1} \neq s$. Such a sequence $\{s_k\}$ satisfies the following condition:

$$|s_k - s| \leq \frac{1}{2^{k-1}}|s_1 - s|, \quad k \in \mathbb{N}.$$

Since s_k is a limit point of the sequence $\{a_n\}$, there exists a_{n_k} such that $|a_{n_k} - s_k| < \frac{1}{2^{k-1}}|s_1 - s|$. Hence

$$|a_{n_k} - s| \leq |a_{n_k} - s_k| + |s_k - s| < \frac{1}{2^{k-2}}|s_1 - s|,$$

which implies that s is a limit of the subsequence $\{a_{n_k}\}$. Therefore $s \in \mathbf{S}$.

2.4.10. Let **S** denote the set of limit points of $\{a_n\}$.

(a) The sequence $\{a_n\}$ is bounded. By 2.4.6, $\mathbf{S} = \{0, \frac{1}{7}, \frac{2}{7}, \frac{4}{7}\}$. Therefore $\varliminf\limits_{n\to\infty} a_n = 0$ and $\varlimsup\limits_{n\to\infty} a_n = \frac{4}{7}$.

(b) We have $\mathbf{S} = \{-1, -\frac{1}{2}, \frac{1}{2}, 1\}$, which together with the boundedness of the sequence gives $\varliminf\limits_{n\to\infty} a_n = -1$ and $\varlimsup\limits_{n\to\infty} = 1$.

(c) The sequence is unbounded and the set of its limit points is empty. Therefore

$$\varliminf_{n\to\infty} a_n = -\infty \quad \text{and} \quad \varlimsup_{n\to\infty} a_n = +\infty.$$

(d) The sequence is unbounded above because its subsequence $a_{2k} = (2k)^{2k}$ diverges to infinity. The subsequence with odd indexes tends to zero. This shows that

$$\varliminf_{n\to\infty} a_n = 0 \quad \text{and} \quad \varlimsup_{n\to\infty} a_n = +\infty.$$

(e) The sequence is unbounded because $a_{4k+1} = 4k + 2 \xrightarrow[k\to\infty]{} +\infty$ and $a_{4k+3} = -4k - 2 \xrightarrow[k\to\infty]{} -\infty$. Consequently, $\varliminf\limits_{n\to\infty} a_n = -\infty$ and $\varlimsup\limits_{n\to\infty} a_n = +\infty$.

2.4. Limit Points. Limit Superior and Limit Inferior

(f) It is clear that the sequence is bounded. Moreover,
$$S = \left\{-e - \frac{\sqrt{2}}{2}, -e + \frac{\sqrt{2}}{2}, e - 1, e, e + 1\right\}.$$
It then follows that
$$\varliminf_{n\to\infty} a_n = -e - \frac{\sqrt{2}}{2} \quad \text{and} \quad \varlimsup_{n\to\infty} a_n = e + 1.$$

(g) $\varliminf\limits_{n\to\infty} a_n = 1$ and $\varlimsup\limits_{n\to\infty} a_n = 2$.

(h) The sequence is not bounded above, as $a_{3k} = 2^{3k} \xrightarrow[k\to\infty]{} +\infty$. Moreover, $S = \{-1, 1\}$. So, $\varliminf\limits_{n\to\infty} a_n = -1$ and $\varlimsup\limits_{n\to\infty} a_n = +\infty$.

(i) We will show first that $\lim\limits_{n\to\infty} \frac{n}{\ln n} = +\infty$. Indeed, applying the Stolz theorem (see 2.3.11), we get
$$\lim_{n\to\infty} \frac{\ln n}{n} = \lim_{n\to\infty} \frac{\ln n - \ln(n-1)}{n - n + 1} = \lim_{n\to\infty} \ln\left(1 + \frac{1}{n-1}\right) = 0.$$
This shows that
$$\lim_{k\to\infty} a_{2k} = \lim_{k\to\infty} \frac{\ln(2k) - 4k}{\ln 2 + \ln(2k)} = -\infty.$$
So, the sequence $\{a_n\}$ is not bounded below. Moreover,
$$\lim_{k\to\infty} a_{2k+1} = \lim_{k\to\infty} \frac{\ln(2k+1)}{\ln 2 + \ln(2k+1)} = 1.$$
This gives $\varliminf\limits_{n\to\infty} a_n = -\infty$ and $\varlimsup\limits_{n\to\infty} a_n = 1$.

2.4.11. It is enough to apply Problem 2.4.7.

(a) $\varliminf\limits_{n\to\infty} a_n = \min S = 0$ and $\varlimsup\limits_{n\to\infty} a_n = \max S$, where
$$S = \left\{0, \frac{p}{q} - \left[\frac{p}{q}\right], \frac{2p}{q} - \left[\frac{2p}{q}\right], \ldots, \frac{(q-1)p}{q} - \left[\frac{(q-1)p}{q}\right]\right\}.$$

(b) $\varliminf\limits_{n\to\infty} a_n = 0$ and $\varlimsup\limits_{n\to\infty} a_n = 1$.

(c) $\varliminf\limits_{n\to\infty} a_n = \min S$ and $\varlimsup\limits_{n\to\infty} a_n = \max S$, where S is the set of all limit points of the sequence described in Problem 2.4.7(c).

(d) $\lim_{n\to\infty} a_n = -1$ and $\overline{\lim}_{n\to\infty} a_n = 1$.

2.4.12.

(a) If the set **S** of the limit points of $\{a_n\}$ is empty, then $\overline{\lim}_{n\to\infty} a_n = -\infty \leq A$. Now assume that **S** is nonempty. Since **S** is closed (see Problem 2.4.9), $\sup \mathbf{S} = \overline{\lim}_{n\to\infty} a_n = L \in \mathbf{S}$. It follows from the definition of a limit point that there exists a subsequence $\{a_{n_k}\}$ converging to L. Therefore, for any $\varepsilon > 0$, there is $k_0 \in \mathbb{N}$ such that
$$L - \varepsilon < a_{n_k} \leq A \quad \text{for} \quad k > k_0.$$
Since ε is arbitrary, we get $L \leq A$.

(b) If the sequence $\{a_n\}$ is not bounded below, then $\underline{\lim}_{n\to\infty} a_n = -\infty \leq A$. So, assume that the sequence $\{a_n\}$ is bounded below, that is, there exists $B \in \mathbb{R}$ such that $a_n \geq B$ for all $n \in \mathbb{N}$. Moreover, by assumption, there is a sequence n_k, $n_k > k$, such that $a_{n_k} \leq A$. Thus, by the Bolzano-Weierstrass theorem (see 2.4.30), the sequence $\{a_{n_k}\}$ contains a convergent subsequence. Let g denote its limit. Then $B \leq g \leq A$. Therefore the set **S** of the limit points of the sequence $\{a_n\}$ is nonempty and $\underline{\lim}_{n\to\infty} a_n = \inf \mathbf{S} \leq g \leq A$.

(c) It is enough to apply the argument presented in the proof of (a).

(d) It is enough to use analysis similar to that in the proof of (b).

2.4.13.

(a) Let $L = \overline{\lim}_{n\to\infty} a_n$. Suppose that (i) is not satisfied, contrary to what is to be proved. Then there is $\varepsilon > 0$ such that for any $k \in \mathbb{N}$ there is $n > k$ for which $a_n \geq L + \varepsilon$. Thus, by Problem 2.4.12 (d), $\overline{\lim}_{n\to\infty} a_n \geq L + \varepsilon$, which contradicts our hypothesis. Now suppose that (ii) is not satisfied. Then there are $\varepsilon > 0$ and $k \in \mathbb{N}$ such that $a_n \leq L - \varepsilon$ for all $n > k$. By 2.4.12(a), we get $\overline{\lim}_{n\to\infty} a_n \leq L - \varepsilon$, which again contradicts our hypothesis. Thus we have proved that $L = \overline{\lim}_{n\to\infty} a_n$ implies (i) and (ii).
Now we prove that conditions (i) and (ii) imply $\overline{\lim}_{n\to\infty} a_n = L$.

2.4. Limit Points. Limit Superior and Limit Inferior

It follows from (i) that the sequence $\{a_n\}$ is bounded above. On the other hand, it follows from (ii) that the sequence contains a subsequence which is bounded below. According to the Bolzano-Weierstrass theorem (see 2.4.30), the sequence contains at least one convergent subsequence. Therefore the set **S** of all limit points of $\{a_n\}$ is nonempty. We will show that $L = \sup \mathbf{S}$. Indeed, if s is an element of **S**, then by (i), $s \leq L + \varepsilon$. By the arbitrariness of ε we get $s \leq L$. Moreover, from condition (ii), we see that for any $\varepsilon > 0$ there is a subsequence of the sequence $\{a_n\}$ converging to $\tilde{s}$ which satisfies the inequality $L - \varepsilon \leq \tilde{s}$. Of course $\tilde{s} \in \mathbf{S}$. In this way the second implication is also proved. Thus the proof is complete.

(b) This follows by the same method as in (a).

Now we state necessary and sufficient conditions for infinite limit superior and inferior. The limit superior of $\{a_n\}$ is $+\infty$ if and only if the sequence is not bounded above. Therefore

(1) $\quad \overline{\lim}\limits_{n\to\infty} a_n = +\infty \quad$ if and only if for every $M \in \mathbb{R}$ and for every $k \in \mathbb{N}$ there exists $n_k > k$ such that $a_{n_k} > M$.

The limit superior of $\{a_n\}$ is $-\infty$ if and only if the sequence is bounded above, say by L, and the set of its limit points is empty. Therefore there is a finite number of terms of $\{a_n\}$ in every bounded interval $[M, L]$. Hence $a_n < M$ for all sufficiently large n. This implies that

(2) $\quad \overline{\lim}\limits_{n\to\infty} a_n = -\infty \quad$ if and only if for every $M \in \mathbb{R}$ there is $k \in \mathbb{N}$ such that for every $n > k$, $a_n < M$.

Similar arguments give

(3) $\quad \underline{\lim}\limits_{n\to\infty} a_n = -\infty \quad$ if and only if for every $M \in \mathbb{R}$ and for every $k \in \mathbb{N}$ there exists $n_k > k$ such that $a_{n_k} < M$,

(4) $\quad \underline{\lim}\limits_{n\to\infty} a_n = +\infty \quad$ if and only if for every $M \in \mathbb{R}$ there is $k \in \mathbb{N}$ such that for every $n > k$, $a_n > M$.

2.4.14. We prove only inequality (a), because the proof of (b) is analogous. Inequality (a) is obvious in the case of $\varliminf\limits_{n\to\infty} b_n = +\infty$ or $\varliminf\limits_{n\to\infty} a_n = -\infty$. If $\varliminf\limits_{n\to\infty} a_n = +\infty$, then, combining condition (4) given in the solution of Problem 2.4.13 with the inequality $a_n \leq b_n$, we get $\varliminf\limits_{n\to\infty} b_n = +\infty$. Similarly, if $\varliminf\limits_{n\to\infty} b_n = -\infty$, then combining condition (3) given in the solution of Problem 2.4.13 with the inequality $a_n \leq b_n$, we obtain $\varliminf\limits_{n\to\infty} a_n = -\infty$.

Assume now that both limits are finite and let

$$\varliminf_{n\to\infty} a_n = l_1 \quad \text{and} \quad \varliminf_{n\to\infty} b_n = l_2.$$

We wish to show that $l_1 \leq l_2$. Suppose, contrary to our aim, that $l_2 < l_1$. Let $\varepsilon > 0$ be so small that $l_2 + \varepsilon < l_1 - \varepsilon$. Then there is c such that $l_2 + \varepsilon < c < l_1 - \varepsilon$. By (ii) of Problem 2.4.13(b), we have $b_{n_k} < l_2 + \varepsilon < c$. On the other hand, by (i) we get $c < l_1 - \varepsilon < a_n$. Hence, in particular, $c < a_{n_k}$, and therefore the inequality $b_{n_k} < a_{n_k}$ holds for infinitely many n_k, contrary to our hypothesis.

2.4.15. Set

$$\varliminf_{n\to\infty} a_n = l_1, \quad \varliminf_{n\to\infty} b_n = l_2, \quad \varlimsup_{n\to\infty} a_n = L_1, \quad \varlimsup_{n\to\infty} b_n = L_2.$$

We show first that

(1) $$\varliminf_{n\to\infty} (a_n + b_n) \geq \varliminf_{n\to\infty} a_n + \varliminf_{n\to\infty} b_n.$$

Assume that l_1 and l_2 are finite. Then, by Problem 2.4.13 (b), for any $\varepsilon > 0$ there is k_1 such that $a_n > l_1 - \varepsilon$ for $n > k_1$, and there exists k_2 for which $b_n > l_2 - \varepsilon$ if $n > k_2$. As a result,

$$a_n + b_n > l_1 + l_2 - 2\varepsilon \quad \text{for} \quad n > \max\{k_1, k_2\}.$$

Combining this with Problem 2.4.12(c), we obtain $\varliminf\limits_{n\to\infty} (a_n + b_n) \geq l_1 + l_2 - 2\varepsilon$. Letting $\varepsilon \to 0^+$, we get (1).

If l_1 or l_2 is $-\infty$, then inequality (1) is obvious. Now we show that if one of the limits l_1 or l_2 is $+\infty$, then $\varliminf\limits_{n\to\infty} (a_n + b_n) = +\infty$.

2.4. Limit Points. Limit Superior and Limit Inferior

Assume, for example, that $l_1 = +\infty$. This is equivalent to condition (4) given in the solution of Problem 2.4.13:

(∗) for every $M \in \mathbb{R}$ there is $k \in \mathbb{N}$ such that $a_n > M$ if $n > k$

Since $l_2 \neq -\infty$, the sequence $\{b_n\}$ is bounded below. So, condition (∗) is satisfied by $\{a_n + b_n\}$. In other words, $\lim\limits_{n\to\infty} (a_n + b_n) = +\infty$. Thus inequality (1) is proved.

The proofs of remaining inequalities are similar, and we will give them only for finite limits. According to Problem 2.4.13, for any $\varepsilon > 0$ there exists a subsequence $\{a_{n_k}\}$ such that $a_{n_k} < l_1 + \varepsilon$ and there is n_0 for which $b_n < L_2 + \varepsilon$ when $n > n_0$. This implies that $a_{n_k} + b_{n_k} < l_1 + L_2 + 2\varepsilon$ for sufficiently large k. Therefore, by Problem 2.4.12(b), we get $\lim\limits_{n\to\infty}(a_n + b_n) \leq l_1 + L_2 + 2\varepsilon$. Since $\varepsilon > 0$ is arbitrary, we get

(2) $$\varliminf_{n\to\infty}(a_n + b_n) \leq \varliminf_{n\to\infty} a_n + \varlimsup_{n\to\infty} b_n.$$

Similarly, for any $\varepsilon > 0$ there is a subsequence $\{b_{n_k}\}$ such that $b_{n_k} > L_2 - \varepsilon$ and there exists n_0 for which $a_n > l_1 - \varepsilon$, for $n > n_0$. Hence $a_{n_k} + b_{n_k} > l_1 + L_2 - 2\varepsilon$ for sufficiently large k. Therefore, by Problem 2.4.12(c), we have $\varlimsup\limits_{n\to\infty}(a_n + b_n) \geq l_1 + L_2 - 2\varepsilon$. Since ε can be made arbitrarily small, we conclude that

(3) $$\varlimsup_{n\to\infty}(a_n + b_n) \geq \varliminf_{n\to\infty} a_n + \varlimsup_{n\to\infty} b_n.$$

Moreover, for any $\varepsilon > 0$ there is k_1 for which $a_n < L_1 + \varepsilon$ when $n > k_1$, and there is k_2 for which $b_n < L_2 + \varepsilon$, for $n > k_2$. Thus

$$a_n + b_n < L_1 + L_2 + 2\varepsilon \quad \text{for} \quad n > \max\{k_1, k_2\}.$$

Combining this with Problem 2.4.12(a), we obtain $\varlimsup\limits_{n\to\infty}(a_n + b_n) \leq L_1 + L_2 + 2\varepsilon$. Since ε can be made arbitrarily small, we get

(4) $$\varlimsup_{n\to\infty}(a_n + b_n) \leq \varlimsup_{n\to\infty} a_n + \varlimsup_{n\to\infty} b_n.$$

Now we give examples of sequences $\{a_n\}$ and $\{b_n\}$ for which the inequalities (1)-(4) are strict. Let

$$a_n = \begin{cases} 0 & \text{if } n = 4k, \\ 1 & \text{if } n = 4k+1, \\ 2 & \text{if } n = 4k+2, \\ 1 & \text{if } n = 4k+3, \end{cases}$$

$$b_n = \begin{cases} 2 & \text{if } n = 4k, \\ 1 & \text{if } n = 4k+1, \\ 1 & \text{if } n = 4k+2, \\ 0 & \text{if } n = 4k+3. \end{cases}$$

In this case the inequalities given in the problem are of the form $0 < 1 < 2 < 3 < 4$.

2.4.16. No. It is enough to consider sequences $\{a_n^m\}$, $m = 1, 2, 3, \ldots$, defined by setting

$$a_n^m = \begin{cases} 1 & \text{for } n = m, \\ 0 & \text{for } n \neq m. \end{cases}$$

Then

$$\varlimsup_{n \to \infty} (a_n^1 + a_n^2 + \ldots) = 1 > 0 = \varlimsup_{n \to \infty} a_n^1 + \varlimsup_{n \to \infty} a_n^2 + \ldots.$$

Now let

$$a_n^m = \begin{cases} -1 & \text{for } n = m, \\ 0 & \text{for } n \neq m. \end{cases}$$

In this case

$$\varliminf_{n \to \infty} (a_n^1 + a_n^2 + \ldots) = -1 < 0 = \varliminf_{n \to \infty} a_n^1 + \varliminf_{n \to \infty} a_n^2 + \ldots.$$

2.4.17. Let

$$\varliminf_{n\to\infty} a_n = l_1, \quad \varliminf_{n\to\infty} b_n = l_2, \quad \varlimsup_{n\to\infty} a_n = L_1, \quad \varlimsup_{n\to\infty} b_n = L_2.$$

We will only show the inequality

(1) $$l_1 l_2 \leq \varliminf_{n\to\infty} (a_n b_n) \leq l_1 L_2.$$

The same reasoning applies to the other cases.

2.4. Limit Points. Limit Superior and Limit Inferior

Assume first that l_1 and l_2 are positive. Then, by Problem 2.4.13(b), for any $\varepsilon > 0$, there exists n_0 such that

$$a_n > l_1 - \varepsilon, \quad b_n > l_2 - \varepsilon \quad \text{for} \quad n > n_0.$$

Consequently, $a_n b_n > l_1 l_2 - \varepsilon(l_1 + l_2) + \varepsilon^2$ for ε so small that $l_1 - \varepsilon > 0$ and $l_2 - \varepsilon > 0$. Therefore, on account of Problem 2.4.12(c), $\varliminf_{n \to \infty}(a_n b_n) \geq l_1 l_2 - \varepsilon(l_1 + l_2) + \varepsilon^2$. Letting $\varepsilon \to 0^+$, we get

(i) $$l_1 l_2 \leq \varliminf_{n \to \infty}(a_n b_n).$$

If $l_1 = 0$ or $l_2 = 0$, then inequality (i) is obvious. If $l_1 = +\infty$ and $l_2 = +\infty$, then (by condition (4) in the solution of Problem 2.4.13), for any preassigned positive number M, we can find n_0 such that

$$a_n > \sqrt{M}, \quad b_n > \sqrt{M} \quad \text{for} \quad n > n_0.$$

Therefore $a_n b_n > M$, which means that $\varliminf_{n \to \infty}(a_n b_n) = +\infty$.

Assume now that one of the limits, say l_1, is infinite and the second one is finite and positive. Then for any $0 < \varepsilon < l_2$ and any $M > 0$ there exists a positive integer n_0 such that for $n > n_0$ we have

$$b_n > l_2 - \varepsilon, \quad a_n > \frac{M}{l_2 - \varepsilon}.$$

Hence $a_n b_n > M$ for $n > n_0$. Therefore $\varliminf_{n \to \infty}(a_n b_n) = +\infty$, and inequality (i) is proved.

Now our task is to prove that

(ii) $$\varlimsup_{n \to \infty}(a_n b_n) \leq l_1 L_2.$$

If l_1 and L_2 are finite, then on account of Problem 2.4.13, one can find a subsequence $\{n_k\}$ such that $a_{n_k} < l_1 + \varepsilon$ and $b_{n_k} < L_2 + \varepsilon$. This gives

$$a_{n_k} b_{n_k} < l_1 L_2 + \varepsilon(l_1 + L_2) + \varepsilon^2.$$

Therefore $\varlimsup_{n \to \infty}(a_n b_n) \leq l_1 L_2 + \varepsilon(l_1 + L_2) + \varepsilon^2$. Letting $\varepsilon \to 0^+$ yields (ii). If $l_1 = +\infty$ or $L_2 = +\infty$, then inequality (ii) is apparent.

Now we give examples of sequences $\{a_n\}$ and $\{b_n\}$ for which all the inequalities are strict. Let

$$a_n = \begin{cases} 1 & \text{for } n = 4k, \\ 2 & \text{for } n = 4k+1, \\ 3 & \text{for } n = 4k+2, \\ 2 & \text{for } n = 4k+3, \end{cases}$$

$$b_n = \begin{cases} 3 & \text{for } n = 4k, \\ 2 & \text{for } n = 4k+1, \\ 2 & \text{for } n = 4k+2, \\ 1 & \text{for } n = 4k+3. \end{cases}$$

In this case our inequalities are of the form $1 < 2 < 3 < 6 < 9$.

2.4.18. Assume that $\varliminf_{n\to\infty} a_n = \varlimsup_{n\to\infty} a_n = g$. Then, by 2.4.13,

(i) for any $\varepsilon > 0$ there is $k \in \mathbb{N}$ such that $a_n < g+\varepsilon$ if $n > k$; and
(i') for any $\varepsilon > 0$ there is $k \in \mathbb{N}$ such that $g - \varepsilon < a_n$ if $n > k$.

Thereby g is a limit of the sequence $\{a_n\}$.

On the other hand, if $\lim_{n\to\infty} a_n = g$, then (i) and (ii) in Problem 2.4.13(a) and (b) are satisfied with $L = g$ and $l = g$. Consequently, $\varliminf_{n\to\infty} a_n = \varlimsup_{n\to\infty} a_n = g$.

Assume now that $\lim_{n\to\infty} a_n = +\infty$. Then statements (1) and (4) in the solution of Problem 2.4.13 are obvious. If $\varliminf_{n\to\infty} a_n = \varlimsup_{n\to\infty} a_n = +\infty$, then condition (4) means that $\lim_{n\to\infty} a_n = +\infty$. Similar arguments apply to the case $\lim_{n\to\infty} a_n = -\infty$.

2.4.19. By Problem 2.4.15,

$$\varliminf_{n\to\infty} a_n + \varliminf_{n\to\infty} b_n \leq \varliminf_{n\to\infty} (a_n + b_n) \leq \varlimsup_{n\to\infty} a_n + \varliminf_{n\to\infty} b_n.$$

On the other hand, on account of the last problem, $a = \lim_{n\to\infty} a_n = \varliminf_{n\to\infty} a_n$. Therefore $\varliminf_{n\to\infty} (a_n + b_n) = a + \varliminf_{n\to\infty} b_n$. The proof of the second equality runs as before.

2.4. Limit Points. Limit Superior and Limit Inferior   203

2.4.20. Using the inequalities given in Problem 2.4.17, we can apply the same method as in the solution of the preceding exercise.

2.4.21. We will apply Problem 2.4.13. Let $\varlimsup_{n\to\infty} a_n = L$. Then conditions (i) and (ii) in 2.4.13 (a) are fulfilled. Multiplying both sides of the inequalities in (i) and (ii) by -1, we get:
  (i) for every $\varepsilon > 0$ there is $k \in \mathbb{N}$ such that for every $n > k$ we have $-L - \varepsilon < -a_n$; and
  (ii) for every $\varepsilon > 0$ and for every $k \in \mathbb{N}$ there exists $n_k > k$ for which $-a_{n_k} < -L + \varepsilon$.

By 2.4.13(b), we obtain
$$\lim_{n\to\infty} (-a_n) = -L = -\varlimsup_{n\to\infty} a_n.$$

The proof of the second equality runs as before. In the case of infinite limits it is enough to apply statements (1)-(4) given in the solution of Problem 2.4.13.

2.4.22. We will apply Problem 2.4.13. Let $\varlimsup_{n\to\infty} a_n = L$. Then by conditions (i) and (ii) in 2.4.13(a), we have
  (i) for every $\varepsilon > 0$, there exists $k \in \mathbb{N}$ such that for every $n > k$ we have $a_n < L + \varepsilon L^2$; and
  (ii) for every $\varepsilon > 0$ and for every $k \in \mathbb{N}$ there exists $n_k > k$ for which $L - \varepsilon \frac{L^2}{2} < a_{n_k}$.

Assume first that $L \neq 0$. Then by (i),
$$\frac{1}{a_n} > \frac{1}{L + \varepsilon L^2} = \frac{1}{L} - \frac{\varepsilon L^2}{L(L + \varepsilon L^2)} > \frac{1}{L} - \varepsilon.$$

Assume now that $0 < \varepsilon < \frac{1}{L}$. Then by (ii),
$$\frac{1}{a_{n_k}} < \frac{1}{L - \frac{\varepsilon L^2}{2}} = \frac{1}{L} + \frac{\varepsilon \frac{L^2}{2}}{L(L - \varepsilon \frac{L^2}{2})} < \frac{1}{L} + \varepsilon.$$

The above conditions imply (by 2.4.13(b))
$$\lim_{n\to\infty} \frac{1}{a_n} = \frac{1}{L} = \frac{1}{\varlimsup_{n\to\infty} a_n}.$$

Now suppose that $\varliminf\limits_{n\to\infty} a_n = 0$. Given $M > 0$, by (i) in Problem 2.4.13(a), there exists an integer k such that $a_n < \frac{1}{M}$ for $n > k$. Therefore $\frac{1}{a_n} > M$ for $n > k$, which in turn, by statement (4) given in the solution of Problem 2.4.13, means that $\varlimsup\limits_{n\to\infty} \frac{1}{a_n} = +\infty$. Finally, suppose $\varlimsup\limits_{n\to\infty} a_n = +\infty$. Then for any $\varepsilon > 0$ and for any $k \in \mathbb{N}$ there exists $n_k > k$ such that $a_{n_k} > \frac{1}{\varepsilon}$ (see statement (1) in the solution of Problem 2.4.13(a)). The above inequality is equivalent to $\frac{1}{a_{n_k}} < \varepsilon$. Of course $-\varepsilon < \frac{1}{a_n}$. Thus both conditions given in 2.4.13(b) are fulfilled for the sequence $\{\frac{1}{a_n}\}$ with $l = 0$, which means that $\varliminf\limits_{n\to\infty} \frac{1}{a_n} = 0$. The proof of the first equality is finished. The proof of the second equality is similar.

2.4.23. It follows from our hypothesis that $0 < \varlimsup\limits_{n\to\infty} a_n < +\infty$. The equality $\varlimsup\limits_{n\to\infty} a_n \cdot \varliminf\limits_{n\to\infty} \frac{1}{a_n} = 1$ combined with the preceding problem yields

$$\varlimsup_{n\to\infty} a_n = \frac{1}{\varliminf\limits_{n\to\infty} \frac{1}{a_n}} = \varliminf_{n\to\infty} a_n.$$

Therefore, by Problem 2.4.18, the sequence $\{a_n\}$ is convergent.

2.4.24. Assume that $\{a_n\}$ is a sequence such that for any sequence $\{b_n\}$ the first equality holds. Take $b_n = -a_n$. From Problem 2.4.21 it follows that

$$0 = \lim_{n\to\infty}(a_n + (-a_n)) = \varliminf_{n\to\infty} a_n + \varlimsup_{n\to\infty}(-a_n) = \varliminf_{n\to\infty} a_n - \varlimsup_{n\to\infty} a_n.$$

From this, by Problem 2.4.18, we conclude that the sequence $\{a_n\}$ is convergent.

2.4.25. Assume that $\{a_n\}$ is a positive valued sequence such that for any positive valued sequence $\{b_n\}$ the first equality holds. Take $b_n = \frac{1}{a_n}$. Hence, by Problem 2.4.22, we get

$$1 = \lim_{n\to\infty}\left(a_n \cdot \frac{1}{a_n}\right) = \varliminf_{n\to\infty} a_n \cdot \varlimsup_{n\to\infty}\left(\frac{1}{a_n}\right) = \varliminf_{n\to\infty} a_n \cdot \frac{1}{\varliminf\limits_{n\to\infty} a_n}.$$

2.4. Limit Points. Limit Superior and Limit Inferior

From this it follows that both the limits superior and inferior of $\{a_n\}$ are positive and $\varlimsup\limits_{n\to\infty} a_n = \varliminf\limits_{n\to\infty} a_n$. Therefore the sequence $\{a_n\}$ is convergent (see 2.4.18).

2.4.26. Evidently, $\varliminf\limits_{n\to\infty} \sqrt[n]{a_n} \leq \varlimsup\limits_{n\to\infty} \sqrt[n]{a_n}$. Now we will show that $\varlimsup\limits_{n\to\infty} \sqrt[n]{a_n} \leq \varlimsup\limits_{n\to\infty} \frac{a_{n+1}}{a_n}$. If $\varlimsup\limits_{n\to\infty} \frac{a_{n+1}}{a_n} = +\infty$, then the inequality is obvious. So, assume that $\varlimsup\limits_{n\to\infty} \frac{a_{n+1}}{a_n} = L < +\infty$. Then for any $\varepsilon > 0$ there exists k such that

$$\frac{a_{n+1}}{a_n} < L + \varepsilon \quad \text{for} \quad n \geq k.$$

Hence
$$\frac{a_n}{a_k} = \frac{a_n}{a_{n-1}} \cdot \frac{a_{n-1}}{a_{n-2}} \cdot \ldots \cdot \frac{a_{k+1}}{a_k} < (L+\varepsilon)^{n-k}.$$

Consequently,
$$\sqrt[n]{a_n} < \sqrt[n]{a_k}(L+\varepsilon)^{\frac{-k}{n}}(L+\varepsilon).$$

Since $\sqrt[n]{a_k}(L+\varepsilon)^{\frac{-k}{n}} \xrightarrow[n\to\infty]{} 1$,

$$\sqrt[n]{a_k}(L+\varepsilon)^{\frac{-k}{n}} < 1 + \varepsilon$$

for sufficiently large n. From what has already been proved, we have

$$\sqrt[n]{a_n} < (1+\varepsilon)(L+\varepsilon) = L + (L+1)\varepsilon + \varepsilon^2$$

for sufficiently large n. Combining this with Problem 2.4.12(a), we obtain $\varlimsup\limits_{n\to\infty} \sqrt[n]{a_n} \leq L + (L+1)\varepsilon + \varepsilon^2$. Since ε can be made arbitrarily small, $\varlimsup\limits_{n\to\infty} \sqrt[n]{a_n} \leq L = \varlimsup\limits_{n\to\infty} \frac{a_{n+1}}{a_n}$. To prove $\varliminf\limits_{n\to\infty} \frac{a_{n+1}}{a_n} \leq \varliminf\limits_{n\to\infty} \sqrt[n]{a_n}$ it is enough to apply Problem 2.4.22 and the inequality just proved to the sequence $\{\frac{1}{a_n}\}$.

2.4.27. We first prove that $\varlimsup\limits_{n\to\infty} b_n \leq \varlimsup\limits_{n\to\infty} a_n$. To this end, assume that $\varlimsup\limits_{n\to\infty} a_n = L < +\infty$. (For $L = +\infty$ the above inequality is clear). Then, given $\varepsilon > 0$, there exists $k \in \mathbb{N}$ such that $a_n < L + \varepsilon$ for $n > k$. Hence

$$b_n = \frac{a_1 + a_2 + \ldots + a_k + a_{k+1} + \ldots + a_n}{n}$$
$$< \frac{a_1 + a_2 + \ldots + a_k}{n} - \frac{k(L+\varepsilon)}{n} + L + \varepsilon.$$

Since $\frac{a_1+a_2+\ldots+a_k}{n} - \frac{k(L+\varepsilon)}{n} \xrightarrow[n\to\infty]{} 0$, $\frac{a_1+a_2+\ldots+a_k}{n} - \frac{k(L+\varepsilon)}{n} < \varepsilon$ for sufficiently large n. It follows from the above consideration that $b_n < \varepsilon + L + \varepsilon$ for sufficiently large n. According to Problem 2.4.12(a), as ε can be made arbitrarily small, we get $\varlimsup_{n\to\infty} b_n \leq L = \varlimsup_{n\to\infty} a_n$. The proof of the inequality $\varliminf_{n\to\infty} a_n \leq \varliminf_{n\to\infty} b_n$ is analogous.

2.4.28.

(a) (b) It is enough to apply Problem 2.4.13.

(c) The equality is not true. To see this, it is enough to consider the sequences defined by setting

$$a_n = \begin{cases} 0 & \text{for } n = 2k, \\ 1 & \text{for } n = 2k+1, \end{cases}$$

$$b_n = \begin{cases} 1 & \text{for } n = 2k, \\ 0 & \text{for } n = 2k+1. \end{cases}$$

Then

$$0 = \varlimsup_{n\to\infty} \min\{a_n, b_n\} \neq \min\{\varlimsup_{n\to\infty} a_n, \varlimsup_{n\to\infty} b_n\} = 1.$$

(d) This equality is likewise not true, as can be seen by considering the sequences defined in (c).

2.4.29. Assume that the sequence $\{a_n\}$ has the property that there are infinitely many n such that

(1) $\quad\quad\quad\quad$ for every $k \geq n$, $a_k \leq a_n$.

Let n_1 be the first such n, n_2 the second, etc. Then the sequence $\{a_{n_k}\}$ is a monotonically decreasing subsequence of $\{a_n\}$. On the other hand, if the sequence $\{a_n\}$ fails to have the above property, that is, there are only finitely many n satisfying (1), choose an integer m_1 such that the sequence $\{a_{m_1+n}\}$ does not satisfy (1). Let m_2 be the first integer greater than m_1 for which $a_{m_2} > a_{m_1}$. Continuing the process, we obtain a subsequence $\{a_{m_n}\}$ of $\{a_n\}$ which is monotonically increasing.

2.4.30. By the preceding problem such a sequence contains a monotonic subsequence which is bounded and therefore convergent.

2.4. Limit Points. Limit Superior and Limit Inferior   207

2.4.31. Assume first that $\varlimsup\limits_{n\to\infty} \frac{a_{n+1}}{a_n} = +\infty$. Then, by 2.4.14(b),

$$\varlimsup_{n\to\infty} \frac{a_1 + \ldots + a_n + a_{n+1}}{a_n} = +\infty.$$

Now let

$$\varlimsup_{n\to\infty} \frac{a_{n+1}}{a_n} = \alpha < +\infty.$$

Then, given $\varepsilon > 0$, there exists k such that

(1) $$\frac{a_{n+1}}{a_n} < \alpha + \varepsilon \quad \text{for} \quad n \geq k.$$

In other words,

(2) $$\frac{a_n}{a_{n+1}} > \frac{1}{\alpha + \varepsilon} \quad \text{for} \quad n \geq k.$$

Hence, for sufficiently large n, we have

$$b_n = \frac{a_1 + \ldots + a_n + a_{n+1}}{a_n} \geq \frac{a_k + \ldots + a_n + a_{n+1}}{a_n}$$
$$= \frac{a_k}{a_{k+1}} \cdot \ldots \cdot \frac{a_{n-2}}{a_{n-1}} \cdot \frac{a_{n-1}}{a_n} + \frac{a_{k+1}}{a_{k+2}} \cdot \ldots \cdot \frac{a_{n-2}}{a_{n-1}} \cdot \frac{a_{n-1}}{a_n}$$
$$+ \ldots + \frac{a_{n-2}}{a_{n-1}} \cdot \frac{a_{n-1}}{a_n} + \frac{a_{n-1}}{a_n} + 1 + \frac{a_{n+1}}{a_n}$$
$$\geq \left(\frac{1}{\alpha + \varepsilon}\right)^{n-k} + \left(\frac{1}{\alpha + \varepsilon}\right)^{n-k-1} + \ldots + \frac{1}{\alpha + \varepsilon} + 1 + \frac{a_{n+1}}{a_n}.$$

If $0 < \alpha < 1$, then the above inequality and Problem 2.4.14(b) yield $\varlimsup\limits_{n\to\infty} b_n = +\infty$.

On the other hand, if $\alpha \geq 1$, then by Problems 2.4.14 (b) and 2.4.19 we conclude that

(3) $$\varlimsup_{n\to\infty} b_n \geq \alpha + \lim_{n\to\infty} \frac{1 - \left(\frac{1}{\alpha+\varepsilon}\right)^{n-k+1}}{1 - \frac{1}{\alpha+\varepsilon}} = \alpha + \frac{\alpha + \varepsilon}{\alpha + \varepsilon - 1}.$$

In case $\alpha = 1$ ($\varepsilon > 0$ can be arbitrary) we get $\varlimsup\limits_{n\to\infty} b_n = +\infty$. If $\alpha > 1$, then (3) implies

$$\varlimsup_{n\to\infty} b_n \geq 1 + \alpha + \frac{1}{\alpha - 1} = 2 + (\alpha - 1) + \frac{1}{\alpha - 1} \geq 4.$$

4 is an optimal estimate because it is attained for the sequence $a_n = 2^n$, $n \in \mathbb{N}$.

2.5. Miscellaneous Problems

2.5.1. Assume first that $\lim_{n\to\infty} a_n = +\infty$. Put $b_n = [a_n]$. Then $b_n \le a_n < b_n + 1$. Hence

$$\left(1 + \frac{1}{b_n+1}\right)^{b_n} < \left(1 + \frac{1}{a_n}\right)^{a_n} < \left(1 + \frac{1}{b_n}\right)^{b_n+1}.$$

Thus the result in 2.1.38 and the squeeze principle imply

$$\lim_{n\to\infty}\left(1 + \frac{1}{a_n}\right)^{a_n} = e.$$

Moreover,

$$\lim_{n\to\infty}\left(1 - \frac{1}{a_n}\right)^{a_n} = e^{-1},$$

because

$$\lim_{n\to\infty}\left(1 - \frac{1}{a_n}\right)^{a_n} = \lim_{n\to\infty} \frac{1}{\left(1 + \frac{1}{a_n-1}\right)^{a_n}} = e^{-1}.$$

This implies that

$$\lim_{n\to\infty}\left(1 + \frac{1}{a_n}\right)^{a_n} = e, \quad \text{if } \{a_n\} \text{ diverges to } -\infty.$$

2.5.2. One can apply the foregoing problem with $a_n = \frac{n}{x}$, $x \ne 0$.

2.5.3. By 2.1.39, 2.1.40 and 2.5.2, $\left(1 + \frac{x}{n}\right)^n < e^x < \left(1 + \frac{x}{n}\right)^{l+n}$ for $l > x > 0$, $l \in \mathbb{N}$. Hence for any positive x and for any positive integer n, $\frac{x}{n+l} < \ln\left(1 + \frac{x}{n}\right) < \frac{x}{n}$ if $l > x$. Taking $n = 1$ we get $\ln(1+x) < x$ for $x > 0$. Now, set $l = [x] + 1$. Then we obtain

$$\ln\left(1 + \frac{x}{n}\right) > \frac{\frac{x}{n}}{2 + \frac{x}{n}}.$$

Therefore $\ln(1+x) > \frac{x}{2+x}$ for $x > 0$.

Consider now the function $f(x) = \ln(1+x) - \frac{2x}{2+x}$, $x > 0$. We have

$$f'(x) = \frac{x^2}{(x+1)(x+2)^2} > 0 \quad \text{for} \quad x > 0.$$

2.5. Miscellaneous Problems

Hence
$$f(x) = \ln(1+x) - \frac{2x}{2+x} > f(0) = 0 \quad \text{for} \quad x > 0.$$

2.5.4.

(a) Assume first that $a > 1$. Set $a_n = \sqrt[n]{a} - 1$. By the inequality in 2.5.3,
$$\frac{2a_n}{a_n+2} < \frac{1}{n}\ln a = \ln(a_n+1) < a_n.$$
Therefore the squeeze principle implies that $\lim\limits_{n\to\infty} n(\sqrt[n]{a}-1) = \ln a$ for $a > 1$. We see at once that the claim holds if $a = 1$. To prove it for $0 < a < 1$, it is enough to apply the above with $\frac{1}{a} > 1$.

(b) Put $a_n = \sqrt[n]{n} - 1$. Then $(a_n+1)^n = n$. Hence by 2.5.3, $\ln n = n\ln(a_n+1) < na_n$. Consequently, $\lim\limits_{n\to\infty} na_n = +\infty$.

2.5.5. Using differentiation we can show that, for $x > -1$, $\frac{x}{1+x} \leq \ln(1+x) \leq x$. Since $\lim\limits_{n\to\infty} a_n = 1$, $a_n > 0$ beginning with some value of the index n. It follows that $\frac{a_n-1}{1+a_n-1} \leq \ln a_n = \ln(1+(a_n-1)) \leq a_n-1$. Dividing the inequalities by $a_n - 1$ and using the squeeze principle yields the desired result.

2.5.6. By the definition (see 2.1.38), $e = \lim\limits_{n\to\infty}\left(1+\frac{1}{n}\right)^n$. Moreover,

$$\left(1+\frac{1}{n}\right)^n = 1 + \binom{n}{1}\frac{1}{n} + \binom{n}{2}\frac{1}{n^2} + \ldots + \binom{n}{k}\frac{1}{n^k} + \ldots + \binom{n}{n}\frac{1}{n^n}$$
$$= 1 + 1 + \frac{1}{2!}\left(1-\frac{1}{n}\right) + \ldots + \frac{1}{k!}\left(1-\frac{1}{n}\right)\cdots\left(1-\frac{k-1}{n}\right)$$
$$+ \ldots + \frac{1}{n!}\left(1-\frac{1}{n}\right)\cdots\left(1-\frac{n-1}{n}\right).$$

Hence

(i) $$\left(1+\frac{1}{n}\right)^n < a_n.$$

On the other hand,

$$\left(1+\frac{1}{n}\right)^n > 2 + \frac{1}{2!}\left(1-\frac{1}{n}\right) + \frac{1}{3!}\left(1-\frac{1}{n}\right)\left(1-\frac{2}{n}\right)$$
$$+ ... + \frac{1}{k!}\left(1-\frac{1}{n}\right)\cdot ... \cdot\left(1-\frac{k-1}{n}\right).$$

Passage to the limit as $n \to \infty$ gives

(ii) $\qquad e \geq a_k.$

By (i) and (ii), the limit of the sequence $\{a_n\}$ is e. Moreover,

$$a_{n+m} - a_n = \frac{1}{(n+1)!} + \frac{1}{(n+2)!} + ... + \frac{1}{(n+m)!}$$
$$< \frac{1}{(n+1)!}\left\{1 + \frac{1}{n+2} + \frac{1}{(n+2)^2} + ... + \frac{1}{(n+2)^{m-1}}\right\}$$
$$< \frac{1}{(n+1)!}\frac{n+2}{n+1}.$$

Keeping n fixed and letting $m \to \infty$, we get

$$e - a_n \leq \frac{1}{(n+1)!}\frac{n+2}{n+1}.$$

This and (ii) imply that $0 < e - a_n < \frac{1}{nn!}$.

2.5.7. We know (see 2.5.2) that $e^x = \lim\limits_{n\to\infty}\left(1+\frac{x}{n}\right)^n$, $x \in \mathbb{R}$. For a fixed $x \in \mathbb{R}$, put $a_n = \left(1 + \frac{x}{1!} + \frac{x^2}{2!} + ... + \frac{x^n}{n!}\right)$. We get

$$\left|a_n - \left(1+\frac{x}{n}\right)^n\right| = \left|\sum_{k=2}^n\left(1 - \left(1-\frac{1}{n}\right)\cdot ... \cdot\left(1-\frac{k-1}{n}\right)\right)\frac{x^k}{k!}\right|$$
$$\leq \sum_{k=2}^n\left(1 - \left(1-\frac{1}{n}\right)\cdot ... \cdot\left(1-\frac{k-1}{n}\right)\right)\frac{|x|^k}{k!}.$$

By 1.2.1,

$$\left(1-\frac{1}{n}\right)\cdot ... \cdot\left(1-\frac{k-1}{n}\right) \geq 1 - \sum_{j=1}^{k-1}\frac{j}{n} = 1 - \frac{k(k-1)}{2n} \text{ for } 2 \leq k \leq n.$$

2.5. Miscellaneous Problems 211

Therefore

$$\left|a_n - \left(1 + \frac{x}{n}\right)^n\right| \leq \sum_{k=2}^{n} \frac{k(k-1)}{2n} \frac{|x|^k}{k!} = \frac{1}{2n} \sum_{k=2}^{n} \frac{|x|^k}{(k-2)!}.$$

Since $\lim_{n\to\infty} \frac{1}{2n} \sum_{k=2}^{n} \frac{|x|^k}{(k-2)!} = 0$ (which follows easily from the Stolz theorem, see 2.3.11), we get $\lim_{n\to\infty} a_n = \lim_{n\to\infty} \left(1 + \frac{x}{n}\right)^n = e^x$.

2.5.8.

(a) By 2.1.38, $\frac{1}{n+1} < \ln\left(1 + \frac{1}{n}\right) < \frac{1}{n}$. So for $n > 1$ we get

$$\ln \frac{2n+1}{n} < \frac{1}{n} + \frac{1}{n+1} + \ldots + \frac{1}{2n} < \ln \frac{2n}{n-1}.$$

Thus the desired result follows from the continuity of the logarithm function and the squeeze principle.

(b) We have

$$\frac{1}{n+1} + \ldots + \frac{1}{2n+1} < \frac{1}{\sqrt{n(n+1)}} + \ldots + \frac{1}{\sqrt{2n(2n+1)}}$$

$$< \frac{1}{n} + \frac{1}{n+1} + \ldots + \frac{1}{2n}.$$

Therefore the claim follows from (a).

2.5.9. Analysis similar to that in the proof of 2.5.3 gives

$$(*) \qquad x - \frac{x^2}{2} < \ln(1+x) < x \quad \text{for} \quad x > 0.$$

Set $b_n = \ln a_n = \sum_{k=1}^{n} \ln\left(1 + \frac{k}{n^2}\right)$. By $(*)$,

$$\frac{k}{n^2} - \frac{k^2}{2n^4} < \ln\left(1 + \frac{k}{n^2}\right) < \frac{k}{n^2}.$$

This and the equalities

$$\sum_{k=1}^{n} k = \frac{n(n+1)}{2}, \qquad \sum_{k=1}^{n} k^2 = \frac{n(n+1)(2n+1)}{6}$$

imply that $\lim_{n\to\infty} b_n = \frac{1}{2}$. Finally, the continuity of the logarithm function yields $\lim_{n\to\infty} a_n = \sqrt{e}$.

2.5.10. One can show by induction that

$$a_n = n + n(n-1) + \ldots + n(n-1)\cdot\ldots\cdot 2 + n(n-1)\cdot\ldots\cdot 2\cdot 1$$
$$= \frac{n!}{(n-1)!} + \frac{n!}{(n-2)!} + \ldots + \frac{n!}{1!} + \frac{n!}{0!}.$$

Hence

$$\lim_{n\to\infty}\prod_{k=1}^{n}\left(1+\frac{1}{a_k}\right) = \lim_{n\to\infty}\left(\frac{a_1+1}{a_1}\cdot\ldots\cdot\frac{a_n+1}{a_n}\right)$$
$$= \lim_{n\to\infty}\frac{a_n+1}{n!} = \lim_{n\to\infty}\left(1+\frac{1}{1!}+\ldots+\frac{1}{n!}\right) = e,$$

where the last equality follows from 2.5.6.

2.5.11. By 2.5.6,

$$e = 1 + \frac{1}{1!} + \ldots + \frac{1}{n!} + \frac{\theta_n}{nn!}, \quad \text{where} \quad 0 < \theta_n < 1.$$

Therefore $0 < n!e - [n!e] = \frac{\theta_n}{n} < \frac{1}{n}$, which proves our claim.

2.5.12. By the arithmetic-geometric mean inequality, the monotonicity of the logarithm function and the inequality proved in 2.5.3, we get

$$\frac{1}{n}\ln\sqrt{ab} \leq \ln\frac{1}{2}(\sqrt[n]{a}+\sqrt[n]{b}) = \ln\left(\frac{1}{2}(\sqrt[n]{a}-1)+\frac{1}{2}(\sqrt[n]{b}-1)+1\right)$$
$$< \frac{1}{2}\left((\sqrt[n]{a}-1)+(\sqrt[n]{b}-1)\right).$$

To get the desired result it is enough to multiply these inequalities by n and use the result in 2.5.4 (a).

2.5.13. Note first that if $\lim\limits_{n\to\infty} a_n^n = a > 0$, then $\lim\limits_{n\to\infty} a_n = 1$. Assume now that $\{a_n\}$ and $\{b_n\}$ are sequences whose terms are different from 1. By Problem 2.5.5,

$$(*) \qquad \lim_{n\to\infty}\frac{n\ln a_n}{n(a_n-1)} = 1.$$

2.5. Miscellaneous Problems

The assumption $\lim\limits_{n\to\infty} a_n^n = a > 0$ and the continuity of the logarithm function imply that $\lim\limits_{n\to\infty} n \ln a_n = \ln a$. Thus, by (*),

$$\lim_{n\to\infty} n(a_n - 1) = \lim_{n\to\infty} n \ln a_n = \ln a.$$

Observe that these equalities remain valid if $a_n = 1$. Finally,

$$\lim_{n\to\infty} n \ln(pa_n + qb_n) = \lim_{n\to\infty} n(p(a_n - 1) + q(b_n - 1)) = \ln a^p b^q.$$

2.5.14. We have $a_{n+1} - a_n = -\frac{1}{n}(a_n - a_{n-1})$. Consequently,

$$a_n = a + (b-a) + \ldots + (a_n - a_{n-1})$$
$$= a + (b-a)\left(1 - \frac{1}{2!} + \frac{1}{3!} - \ldots + (-1)^{n-2}\frac{1}{(n-1)!}\right).$$

Finally, by 2.5.7, $\lim\limits_{n\to\infty} a_n = b - (b-a)e^{-1}$.

2.5.15. Consider the sequence $\{b_n\}$, where $b_n = \frac{a_n}{n!}$, and apply the same method as in the solution of the foregoing problem to conclude that $a_n = n!$.

2.5.16. As in the solution of 2.5.14, $a_{n+1} - a_n = -\frac{1}{2n}(a_n - a_{n-1})$. Thus $\lim\limits_{n\to\infty} a_n = 2b - a - 2(b-a)e^{\frac{-1}{2}}$.

2.5.17. (a) We have

$$a_n = 3 - \sum_{k=1}^{n}\left(\frac{1}{k} - \frac{1}{k+1}\right)\frac{1}{(k+1)!}$$
$$= 3 - \sum_{k=1}^{n}\frac{k+1-k}{k(k+1)!} + \sum_{k=1}^{n}\frac{1}{(k+1)(k+1)!}$$
$$= 3 - \sum_{k=1}^{n}\frac{1}{kk!} + \sum_{k=1}^{n}\frac{1}{(k+1)!} + \sum_{k=1}^{n}\frac{1}{(k+1)(k+1)!}$$
$$= \sum_{k=0}^{n+1}\frac{1}{k!} + \frac{1}{(n+1)(n+1)!}.$$

Therefore, by 2.5.6, $\lim\limits_{n\to\infty} a_n = e$.

(b) By (a) and 2.5.6,
$$0 < a_n - e < \frac{1}{(n+1)(n+1)!}.$$

Remark. It is interesting to note that this sequence tends to e faster than the sequence considered in Problem 2.5.6.

2.5.18. It follows from 2.5.6 that $e = 1 + \frac{1}{1!} + ... + \frac{1}{n!} + r_n$, where $\lim\limits_{n\to\infty} n!r_n = 0$. Moreover,

(*) $$\frac{1}{n+1} < n!r_n < \frac{1}{n}.$$

So,
$$\lim_{n\to\infty} n\sin(2\pi n!e) = \lim_{n\to\infty} n\sin(2\pi n!r_n)$$
$$= \lim_{n\to\infty} n2\pi n!r_n \frac{\sin(2\pi n!r_n)}{2\pi n!r_n} = \lim_{n\to\infty} n2\pi n!r_n = 2\pi.$$

The last equality follows from (*).

2.5.19. We will show that $\lim\limits_{n\to\infty}\left(1 - \frac{a_n}{n}\right)^n = 0$. By assumption, for an arbitrarily chosen $M > 0$ we have $a_n > M$ if n is large enough. Hence
$$0 < 1 - \frac{a_n}{n} < 1 - \frac{M}{n}.$$

Consequently,
$$0 < \left(1 - \frac{a_n}{n}\right)^n < \left(1 - \frac{M}{n}\right)^n.$$

So, by 2.4.12, 2.4.14 and 2.5.2, we get
$$0 \le \varliminf_{n\to\infty}\left(1 - \frac{a_n}{n}\right)^n \le \varlimsup_{n\to\infty}\left(1 - \frac{a_n}{n}\right)^n \le e^{-M}.$$

Letting $M \to \infty$ yields
$$0 \le \varliminf_{n\to\infty}\left(1 - \frac{a_n}{n}\right)^n \le \varlimsup_{n\to\infty}\left(1 - \frac{a_n}{n}\right)^n \le 0.$$

Therefore $\lim\limits_{n\to\infty}\left(1 - \frac{a_n}{n}\right)^n = 0$, as claimed.

2.5. Miscellaneous Problems

2.5.20. We will show that $\lim\limits_{n\to\infty}\left(1+\frac{b_n}{n}\right)^n = +\infty$. Given $M > 0$, we have $b_n > M$ for sufficiently large n. So, as in the solution of the foregoing problem, we get

$$\lim_{n\to\infty}\left(1+\frac{b_n}{n}\right)^n \geq \lim_{n\to\infty}\left(1+\frac{M}{n}\right)^n = e^M.$$

Since M can be arbitrarily large, we see that $\lim\limits_{n\to\infty}\left(1+\frac{b_n}{n}\right)^n = +\infty$.

2.5.21. It is easy to see that the sequence $\{a_n\}$ is monotonically decreasing to zero.

(a) We have $\frac{1}{a_{n+1}} - \frac{1}{a_n} = \frac{1}{1-a_n} \xrightarrow[n\to\infty]{} 1$. So, by 2.3.14, $\lim\limits_{n\to\infty}\frac{1}{na_n} = 1$.

(b) By (a),

$$\lim_{n\to\infty}\frac{n(1-na_n)}{\ln n} = \lim_{n\to\infty}\frac{n\left(\frac{1}{na_n}-1\right)na_n}{\ln n} = \lim_{n\to\infty}\frac{\frac{1}{a_n}-n}{\ln n}.$$

Using the Stolz theorem (see 2.3.11), we obtain

$$\lim_{n\to\infty}\frac{n(1-na_n)}{\ln n} = \lim_{n\to\infty}\frac{n\left(\frac{1}{a_{n+1}}-\frac{1}{a_n}-1\right)}{n\ln\left(1+\frac{1}{n}\right)}$$

$$= \lim_{n\to\infty}\frac{n\left(\frac{1}{1-a_n}-1\right)}{\ln\left(1+\frac{1}{n}\right)^n} = \lim_{n\to\infty}\frac{na_n}{1-a_n} = 1.$$

2.5.22. It is easy to see that the sequence $\{a_n\}$ is monotonically decreasing to zero. Moreover, an application of l'Hospital's rule gives

$$\lim_{x\to 0}\frac{x^2-\sin^2 x}{x^2\sin^2 x} = \frac{1}{3}.$$

Therefore

$$\lim_{n\to\infty}\left(\frac{1}{a_{n+1}^2}-\frac{1}{a_n^2}\right) = \frac{1}{3}.$$

Now, by the result in Problem 2.3.14, $\lim\limits_{n\to\infty}na_n^2 = 3$.

2.5.23. Clearly, the sequence is monotonically increasing. We will show that it diverges to $+\infty$. We have

$$a_{n+1}^2 = \left(a_n + \frac{1}{a_1 + \ldots + a_n}\right)^2 > \left(a_n + \frac{1}{na_n}\right)^2 > a_n^2 + \frac{2}{n}$$

and

$$a_{2n}^2 - a_n^2 > \frac{2}{2n-1} + \frac{2}{2n-2} + \ldots + \frac{2}{n} > \frac{2n}{2n-1} > 1.$$

Thus $\{a_n^2\}$ is not a Cauchy sequence. Since it increases, it must diverge to $+\infty$. Moreover,

$$(*) \qquad 1 \leq \frac{a_{n+1}}{a_n} \leq 1 + \frac{1}{na_n}.$$

By the Stolz theorem

$$\lim_{n\to\infty} \frac{a_n^2}{2\ln n} = \lim_{n\to\infty} \frac{n(a_{n+1}^2 - a_n^2)}{2\ln\left(1 + \frac{1}{n}\right)^n} = \lim_{n\to\infty} \frac{n}{2}(a_{n+1}^2 - a_n^2)$$

$$= \lim_{n\to\infty} \frac{n}{2}\left(\frac{2a_n}{a_1 + a_2 + \ldots + a_n} + \frac{1}{(a_1 + a_2 + \ldots + a_n)^2}\right) = 1,$$

because

$$0 < \frac{n}{(a_1 + a_2 + \ldots + a_n)^2} < \frac{1}{n};$$

and again, by Stolz's theorem,

$$\lim_{n\to\infty} \frac{na_n}{a_1 + a_2 + \ldots + a_n} = \lim_{n\to\infty} \frac{(n+1)a_{n+1} - na_n}{a_{n+1}}$$

$$= \lim_{n\to\infty} \left(n + 1 - n\frac{a_n}{a_{n+1}}\right) = 1.$$

The last equality follows from $(*)$. Indeed, we have

$$1 \leq n + 1 - n\frac{a_n}{a_{n+1}} \leq \frac{1 + \frac{n+1}{na_n}}{1 + \frac{1}{na_n}}.$$

Since $\lim\limits_{n\to\infty} a_n = +\infty$,

$$\lim_{n\to\infty} \frac{1 + \frac{n+1}{na_n}}{1 + \frac{1}{na_n}} = 1.$$

2.5. Miscellaneous Problems

2.5.24. By the inequality $\arctan x < x$ for $x > 0$, the sequence is monotonically decreasing. Moreover, it is bounded below by zero. Hence it converges, say, to g, which has to satisfy the equation $g = \arctan g$. Thus $g = 0$.

2.5.25. Note that all the terms of the sequence $\{a_n\}$ belong to the interval $(0,1)$. Denote by x_0 the unique root of the equation $\cos x = x$. If $x > x_0$, then $\cos(\cos x) < x$. The function $f(x) = \cos(\cos x) - x$ is monotonically decreasing, because $f'(x) = \sin x \sin(\cos x) - 1 < 0$ for $x \in \mathbb{R}$. Thus, for $x > x_0$, $\cos(\cos x) - x < f(x_0) = 0$. Analogously, if $x < x_0$, then $\cos(\cos x) > x$.

Assume first that $a_1 > x_0$. It follows from the above that $a_3 = \cos(\cos a_1) < a_1$. Since the function $y = \cos(\cos x)$ is monotonically increasing in $(0, \frac{\pi}{2})$, we get $a_5 < a_3$. It can be shown by induction that the sequence $\{a_{2n-1}\}$ is monotonically decreasing. On the other hand, $a_2 = \cos a_1 < \cos x_0 = x_0$, which implies that $a_4 = \cos(\cos a_2) > a_2$, and consequently, $\{a_{2n}\}$ is monotonically increasing.

Similar arguments can be applied to the case where $0 < a_1 < x_0$. If $a_1 = x_0$, then all the terms of the sequence $\{a_n\}$ are equal to x_0. In all these cases the sequences $\{a_{2n-1}\}$ and $\{a_{2n}\}$ both tend to the unique root of the equation $\cos(\cos x) = x$. It is easily seen that x_0 is such a root.

2.5.26. We get, inductively,

$$a_n = 1 - (-1)^{n-1} \underbrace{\sin(\sin(\ldots \sin 1)\ldots)}_{(n-1) \text{ times}}, \qquad n > 1.$$

Hence

$$\frac{1}{n}\sum_{k=1}^{n} a_k = \frac{n - 1 - \sum_{k=1}^{n-1}(-1)^{k-1}\underbrace{\sin(\sin(\ldots \sin 1)\ldots)}_{(k-1) \text{ times}}}{n}.$$

We will show now that

(∗) $$\lim_{n\to\infty} \frac{\sum_{k=1}^{n-1}(-1)^{k-1}\underbrace{\sin(\sin(\ldots \sin 1)\ldots)}_{(k-1) \text{ times}}}{n} = 0.$$

If $n-1$ is even, then

$$\frac{-\sin 1 + \underbrace{\sin(\sin(\ldots \sin 1)\ldots)}_{(n-1) \text{ times}}}{n} < \frac{\sum_{k=1}^{n-1}(-1)^{k-1}\underbrace{\sin(\sin(\ldots \sin 1)\ldots)}_{(k-1) \text{ times}}}{n} < 0.$$

Obviously, for odd $n-1$, (∗) also holds. Finally, $\lim_{n\to\infty} \frac{1}{n}\sum_{k=1}^{n} a_k = 1$.

2.5.27. Clearly, $a_n \in (n\pi, n\pi + \frac{\pi}{2})$, $n = 1, 2, \ldots$, and thus $\lim_{n\to\infty} a_n = +\infty$. Moreover,

$$\lim_{n\to\infty} \tan(\frac{\pi}{2} + n\pi - a_n) = \lim_{n\to\infty} \frac{1}{\tan a_n} = \lim_{n\to\infty} \frac{1}{a_n} = 0.$$

By the continuity of the arctan function we get $\lim_{n\to\infty}(\frac{\pi}{2}+n\pi-a_n) = 0$. Therefore

$$\lim_{n\to\infty} (a_{n+1} - a_n - \pi)$$
$$= \lim_{n\to\infty} \left(\frac{\pi}{2} + n\pi - a_n - \left(\frac{\pi}{2} + (n+1)\pi - a_{n+1}\right)\right) = 0.$$

Consequently, $\lim_{n\to\infty} (a_{n+1} - a_n) = \pi$.

2.5.28. Observe that without loss of generality we can assume that $|a_1| \le \frac{\pi}{2}$. Indeed, if not, then by assumption $|a_2| \le \frac{\pi}{2}$. Consider first the case where $0 < a \le 1$ and $0 < a_1 \le \frac{\pi}{2}$. Then $a_{n+1} = a \sin a_n < a_n$. This means that $\{a_n\}$ is monotonically decreasing and, since it is bounded, it converges. Its limit is equal to zero, which is the unique root of the equation $x = a\sin x$, $0 < a \le 1$. Assume now that $1 < a \le \frac{\pi}{2}$ and $0 < a_1 \le \frac{\pi}{2}$. Then the equation $x = a \sin x$ has two nonnegative solutions 0 and $x_0 > 0$. If $a_1 < x_0$, then $\{a_n\}$ is monotonically increasing and bounded above by x_0. Indeed, $a_2 = a \sin a_1 > a_1$. Moreover, $a_2 = a \sin a_1 < a \sin x_0 = x_0$ and, inductively, $a_n < a_{n+1} < x_0$. Similarly, $x_0 < a_1 \le \frac{\pi}{2}$ implies that $a_n > a_{n+1} > x_0$. Hence $\lim_{n\to\infty} a_n = x_0$ for $1 < a \le \frac{\pi}{2}$. If $-\frac{\pi}{2} \le a < 0$, $a_1 > 0$, then we consider the sequence $\{b_n\}$ defined by setting $b_1 = a_1$, $b_{n+1} = -a \sin b_n$. Obviously, $b_n = (-1)^{n-1}a_n$. It

2.5. Miscellaneous Problems

follows from the above that in the case where $0 < a_1 \leq \frac{\pi}{2}$ we have

$$\lim_{n\to\infty} a_n = 0 \qquad \text{if} \quad |a| \leq 1,$$

$$\lim_{n\to\infty} a_n = x_0 \qquad \text{if} \quad 1 < a \leq \frac{\pi}{2},$$

$$\lim_{n\to\infty} a_n \quad \text{does not exist} \quad \text{if} \quad -\frac{\pi}{2} \leq a < -1.$$

If $-\frac{\pi}{2} \leq a_1 < 0$, then one can consider the sequence given by $b_1 = -a_1$, $b_{n+1} = a \sin b_n$, and apply the foregoing. If $a_1 = 0$, then all the terms of the sequence are also equal to 0.

2.5.29.

(a) Note that $a_n > 0$ and $a_{n+1} = \ln(1 + a_n) < a_n$. Therefore the sequence converges to a g which satisfies $g = \ln(1+g)$, so $g = 0$. We now show that $\lim_{n\to\infty} na_n = 2$. Using differentiation, one can prove that (see also 2.5.3)

$$\frac{2x}{2+x} < \ln(1+x) < x - \frac{x^2}{2} + \frac{x^3}{3} \quad \text{for} \quad x > 0.$$

This implies that

$$(*) \qquad \frac{1}{a_n} - \frac{1}{a_n} + \frac{1}{a_n\left(1 - \frac{1}{2}a_n + \frac{1}{3}a_n^2\right)} < \frac{1}{a_{n+1}} < \frac{1}{a_n} + \frac{1}{2}.$$

Putting

$$b_n = -\frac{1}{a_n} + \frac{1}{a_n\left(1 - \frac{1}{2}a_n + \frac{1}{3}a_n^2\right)}$$

we see that $\lim_{n\to\infty} b_n = \frac{1}{2}$. Summing both sides of $(*)$, we get

$$\frac{1}{a_1} + \frac{1}{a_2} + \cdots + \frac{1}{a_n} + b_1 + b_2 + \cdots + b_n < \frac{1}{a_2} + \cdots + \frac{1}{a_n} + \frac{1}{a_{n+1}}$$

$$< \frac{1}{a_1} + \frac{1}{a_2} + \cdots + \frac{1}{a_n} + \frac{n}{2}.$$

Consequently,

$$\frac{1}{(n+1)a_1} + \frac{b_1 + b_2 + \cdots + b_n}{n+1} < \frac{1}{(n+1)a_{n+1}}$$

$$< \frac{1}{(n+1)a_1} + \frac{n}{2(n+1)}.$$

Hence $\lim\limits_{n\to\infty} \frac{1}{(n+1)a_{n+1}} = \frac{1}{2}$.

(b) We have

(1) $$\lim_{n\to\infty} (na_n - 2)\frac{n}{\ln n} = \lim_{n\to\infty} na_n \frac{n - \frac{2}{a_n}}{\ln n}.$$

To prove that $\lim\limits_{n\to\infty} \frac{n - \frac{2}{a_n}}{\ln n}$ exists we will use the Stolz theorem (see 2.3.11). We get

$$\lim_{n\to\infty} \frac{n - \frac{2}{a_n}}{\ln n} = \lim_{n\to\infty} \frac{1 - \frac{2}{a_{n+1}} + \frac{2}{a_n}}{\ln\left(1 + \frac{1}{n}\right)}.$$

Since $\lim\limits_{n\to\infty} \frac{a_{n+1}}{a_n} = \lim\limits_{n\to\infty} \frac{\ln(1+a_n)}{a_n} = 1$ and $\lim\limits_{n\to\infty} n\ln\left(1+\frac{1}{n}\right) = 1$, we see that

(2) $$\lim_{n\to\infty} \frac{n - \frac{2}{a_n}}{\ln n} = \lim_{n\to\infty} \frac{n(2a_{n+1} - 2a_n + a_n a_{n+1})}{a_n^2}.$$

Now it is enough to show that $\lim\limits_{n\to\infty} \frac{2a_{n+1} - 2a_n + a_n a_{n+1}}{a_n^3}$ exists. To this end, we use the inequality (which can be proved by differentiation)

$$x - \frac{x^2}{2} + \frac{x^3}{3} - \frac{x^4}{4} < \ln(1+x) < x - \frac{x^2}{2} + \frac{x^3}{3}, \quad x > 0.$$

Thus

$$\frac{1}{6}a_n^3 - \frac{1}{6}a_n^4 - \frac{1}{4}a_n^5 < 2a_{n+1} - 2a_n + a_n a_{n+1} < \frac{1}{6}a_n^3 + \frac{1}{3}a_n^4.$$

This gives

$$\lim_{n\to\infty} \frac{2a_{n+1} - 2a_n + a_n a_{n+1}}{a_n^3} = \frac{1}{6}.$$

Combining this with (1) and (2), we see that $\lim\limits_{n\to\infty} \frac{n(na_n - 2)}{\ln n} = \frac{2}{3}$.

2.5.30. Set $f(x) = \left(\frac{1}{4}\right)^x$ and $F(x) = f(f(x)) - x$. We first show that $F'(x) < 0$ for positive x. We have

$$F'(x) = \left(\frac{1}{4}\right)^{\left(\frac{1}{4}\right)^x + x} \ln^2 4 - 1.$$

2.5. Miscellaneous Problems

Hence

$$F'(x) < 0 \quad \text{if and only if} \quad \left(\frac{1}{4}\right)^{\left(\frac{1}{4}\right)^x + x} < \frac{1}{\ln^2 4}.$$

It is a simple matter to verify that the function on the left-hand side of the last inequality attains its maximum value of $\frac{1}{e \ln 4}$ at $x = \frac{\ln \ln 4}{\ln 4}$. This implies that $F'(x) < 0$, which means that F strictly decreases on $(0, +\infty)$. Moreover, $F(\frac{1}{2}) = 0$. Therefore $F(x) > 0$ for $0 < x < \frac{1}{2}$ and $F(x) < 0$ for $x > \frac{1}{2}$. Consequently,

$$f(f(x)) < x \quad \text{for} \quad x > \frac{1}{2}.$$

Since $a_2 = 1 > \frac{1}{2}$, it follows that $a_4 = f(f(a_2)) < a_2$, and inductively, we find that $\{a_{2n}\}$ strictly decreases. Thus it tends to a g_1 such that $f(f(g_1)) = g_1$. The convergence of $\{a_{2n-1}\}$ to a g_2 satisfying $f(f(g_2)) = g_2$ can be established in much the same way. Clearly, $g_1 = g_2 = \frac{1}{2}$.

2.5.31. Observe first that $0 < a_n < 2$ for $n \geq 2$. If $a_n > 1$, then $a_{n+1} < 1$. Set $f(x) = 2^{1-x}$ and $F(x) = f(f(x)) - x$. One can show that $F'(x) < 0$ for $0 < x < 2$. Therefore

$$F(x) < F(1) = 0 \quad \text{for} \quad 1 < x < 2,$$
$$F(x) > F(1) = 0 \quad \text{for} \quad 0 < x < 1.$$

Next, as in the proof of the foregoing problem, we show that if $a_1 < 1$, then the sequence $\{a_{2n}\}$ is monotonically decreasing and the sequence $\{a_{2n-1}\}$ is monotonically increasing, and both tend to the same limit 1. Similar considerations apply to the case $a_1 > 1$.

2.5.32. Observe first the all the terms of the sequence are in the interval $(1, 2)$. Since the function $F(x) = 2^{\frac{x}{2}} - x$ is monotonically decreasing on this interval, $F(x) > F(2) = 0$ for $x \in (1, 2)$. Therefore the sequence is monotonically increasing and its limit g satisfies $g = 2^{\frac{g}{2}}$, so $g = 2$.

2.5.33. We apply 2.3.14 to the sequence $\{a_n + a_{n-1}\}$ and obtain $\lim\limits_{n\to\infty} \frac{a_n+a_{n-1}}{n} = 0$. Next we consider the sequence $b_n = (-1)^n a_n$. Since $\lim\limits_{n\to\infty}(b_n - b_{n-2}) = 0$, we see that

$$0 = \lim_{n\to\infty} \frac{b_n + b_{n-1}}{n} = \lim_{n\to\infty} \frac{a_n - a_{n-1}}{n}.$$

2.5.34. By the Stolz theorem (see 2.3.11),

$$\lim_{n\to\infty} \frac{\ln \frac{1}{a_n}}{\ln n} = \lim_{n\to\infty} \frac{\ln \frac{1}{a_{n+1}} - \ln \frac{1}{a_n}}{\ln(n+1) - \ln n} = \lim_{n\to\infty} \frac{-n\ln \frac{a_{n+1}}{a_n}}{\ln\left(1+\frac{1}{n}\right)^n}$$
$$= -\lim_{n\to\infty} n\ln \frac{a_{n+1}}{a_n}.$$

If $\lim\limits_{n\to\infty} n\left(1 - \frac{a_{n+1}}{a_n}\right) = g$ is finite, then $\lim\limits_{n\to\infty} \left(\frac{a_{n+1}}{a_n} - 1\right) = 0$. Now the desired result follows from the inequalities

$$n\frac{\frac{a_{n+1}}{a_n} - 1}{1 + \frac{a_{n+1}}{a_n} - 1} \le n\ln\left(1 + \left(\frac{a_{n+1}}{a_n} - 1\right)\right) \le n\left(\frac{a_{n+1}}{a_n} - 1\right).$$

If $g = +\infty$, then the right inequality shows that $\lim\limits_{n\to\infty} n\ln \frac{a_{n+1}}{a_n} = -\infty$, and consequently, $\lim\limits_{n\to\infty} \frac{\ln \frac{1}{a_n}}{\ln n} = +\infty$. Finally, if $g = -\infty$, then for any $M > 0$ there is n_0 such that $\frac{a_{n+1}}{a_n} > \frac{M}{n} + 1$ for $n > n_0$. Hence

$$n\ln \frac{a_{n+1}}{a_n} > \ln\left(1 + \frac{M}{n}\right)^n \xrightarrow[n\to\infty]{} M.$$

Since M can be arbitrarily large, we see that $\lim\limits_{n\to\infty} \frac{\ln \frac{1}{a_n}}{\ln n} = -\infty$.

2.5.35. By the definition of the sequences,

$$a_{n+1} + b_{n+1} = (a_1 + b_1)(1 - (a_n + b_n)) + (a_n + b_n).$$

Set $d_n = a_n + b_n$. Then $d_{n+1} = d_1(1 - d_n) + d_n$ and by induction we show that $d_n = 1 - (1 - d_1)^n$. Likewise,

$$a_n = \frac{a_1}{d_1}(1 - (1-d_1)^n) \quad \text{and} \quad b_n = \frac{b_1}{d_1}(1 - (1-d_1)^n).$$

2.5. Miscellaneous Problems

Since $|1 - d_1| < 1$, we get

$$\lim_{n \to \infty} a_n = \frac{a_1}{a_1 + b_1} \quad \text{and} \quad \lim_{n \to \infty} b_n = \frac{b_1}{a_1 + b_1}.$$

2.5.36. Define the sequence $\{b_n\}$ by setting $b_n = aa_n$. Then $b_{n+1} = b_n(2 - b_n) = -(b_n - 1)^2 + 1$. Hence $b_{n+1} - 1 = -(b_n - 1)^2$. Obviously, the sequence $\{a_n\}$ converges if and only if $\{b_n - 1\}$ does or, in other words, when $|b_1 - 1| = |aa_1 - 1| \leq 1$. Moreover, if $a_1 = \frac{2}{a}$, then $\lim_{n \to \infty} a_n = 0$, and if $0 < aa_1 < 2$, then $\lim_{n \to \infty} a_n = \frac{1}{a}$.

2.5.37. This result is contained as a special case in Problem 2.5.38.

2.5.38. One can show that the function f is continuous at $(a, a, ..., a)$ and $f(a, a, ..., a) = a$. Define the sequence $\{b_n\}$ by setting

$$b_1 = b_2 = ... = b_k = \min\{a_1, a_2, ..., a_k\},$$
$$b_n = f(b_{n-1}, b_{n-2}, ..., b_{n-k}) \quad \text{for} \quad n > k.$$

Note that if $\min\{a_1, a_2, ..., a_k\} < a$, then $\{b_n\}$ is strictly increasing and bounded above by a. On the other hand, if $\min\{a_1, a_2, ..., a_k\} > a$, then $\{b_n\}$ is strictly decreasing and bounded below by a. Hence, in both cases, the sequence $\{b_n\}$ converges and $\lim_{n \to \infty} b_n = a$. Moreover, the monotonicity of f with respect to every variable implies $a_n \leq b_n$ for $n \in \mathbb{N}$. Now define the sequence $\{c_n\}$ by setting

$$c_1 = c_2 = ... = c_k = \max\{a_1, a_2, ..., a_k\},$$
$$c_n = f(c_{n-1}, c_{n-2}, ..., c_{n-k}) \quad \text{for} \quad n > k.$$

As above, we show that $\lim_{n \to \infty} c_n = a$ and $c_n \leq a_n$ for $n \in \mathbb{N}$. Finally, by the squeeze principle, $\lim_{n \to \infty} a_n = a$.

2.5.39. We have $a_3 = a_2 e^{a_2 - a_1}$, $a_4 = a_3 e^{a_3 - a_2} = a_2 e^{a_3 - a_1}$ and, inductively, $a_{n+1} = a_2 e^{a_n - a_1}$ for $n \geq 2$. Suppose g is the limit of the sequence. Then

(*) $$\frac{e^{a_1}}{a_2} g = e^g.$$

Note that if $\frac{e^{a_1}}{a_2} = e$, then the equation (*) has only one solution $g = 1$. If $\frac{e^{a_1}}{a_2} > e$ this equation has two solutions, and if $0 < \frac{e^{a_1}}{a_2} < e$

it has no solutions. Consider first the case where $0 < \frac{e^{a_1}}{a_2} < e$. Then the sequence $\{a_n\}$ diverges because in this case (∗) does not have any solutions. Moreover, one can show that the sequence $\{a_n\}$ is monotonically increasing and therefore diverges to $+\infty$.

Consider now the case where $\frac{e^{a_1}}{a_2} = e$. Then $a_2 = e^{a_1-1} \geq a_1$ and, inductively, $a_{n+1} \geq a_n$. Moreover, if $a_1 \leq 1$, then one can show by induction that also $a_n \leq 1$. Hence, in such a case, $\lim\limits_{n\to\infty} a_n = 1$. If $a_1 > 1$, then $\{a_n\}$ is monotonically increasing and diverges to $+\infty$. Next, consider the case where $\frac{e^{a_1}}{a_2} > e$. Then (∗) has two solutions, say g_1, g_2, where, e.g. $g_1 < g_2$. Assume that $a_1 < g_1$. Then

$$e^{a_1} - \frac{e^{a_1}}{a_2} a_1 > 0$$

or, in other words, $a_2 > a_1$. It follows by induction that $\{a_n\}$ is monotonically increasing and bounded above by g_1, which is its limit. If $g_1 < a_1 < g_2$, then $\{a_n\}$ is monotonically decreasing and bounded below by g_1, which is also its limit. If $a_1 = g_1$ or $a_1 = g_2$, then the sequence is constant. Finally, if $a_1 > g_2$, then the sequence increases to $+\infty$.

2.5.40. (This problem and its solution is due to Euler in a more general case. See also [13]). Applying differentiation, we show that $\ln x \leq \frac{x}{e}$ for $x > 0$. Hence $\frac{a_n}{e} \geq \ln a_n = a_{n-1} \ln a$, $n > 1$, and consequently, $a_n \geq a_{n-1} \ln a^e$. Therefore, if $a > e^{\frac{1}{e}}$, the sequence $\{a_n\}$ is monotonically increasing. We will show that in this case, $\lim\limits_{n\to\infty} a_n = +\infty$. We have $a_{n+1} - a_n = a^{a_n} - a_n$. So for $a > e^{\frac{1}{e}}$, we consider the function $g(x) = a^x - x$. This function attains its minimum at $x_0 = -\frac{\ln(\ln a)}{\ln a} < e$. It follows that $a^x - x > \frac{1+\ln(\ln a)}{\ln a} > 0$, and consequently, $a_{n+1} - a_n > \frac{1+\ln(\ln a)}{\ln a} > 0$. Since the difference between two consecutive terms is greater than a positive number, the sequence diverges to infinity.

Now we will consider the case where $1 < a < e^{\frac{1}{e}}$. We first show that in this case the equation $a^x - x = 0$ has two positive roots. The derivative of the function $g(x) = a^x - x$ vanishes at the point $x_0 > 0$ such that $a^{x_0} = \frac{1}{\ln a}$. The function g attains its minimum value at x_0, and $g(x_0) = a^{x_0} - x_0 = \frac{1}{\ln a} - x_0 = \frac{1-x_0 \ln a}{\ln a} < 0$, because

2.5. Miscellaneous Problems

if $1 < a < e^{\frac{1}{e}}$, then $\frac{1}{\ln a} > e$. Since g is a continuous function on $\mathbb{R}$, it possesses the intermediate value property. Thus the equation $a^x = x$ has one root in the interval $(0, x_0)$ and the other in $(x_0, +\infty)$. Denote these roots by α and β, respectively. Note that since $g(e) = a^e - e < \left(e^{\frac{1}{e}}\right)^e - e = 0$, the number e is between α and β.

If $x > \beta$, then $a^x > a^\beta = \beta$ and $g(x) > 0$. This means that in such a case the sequence $\{a_n\}$ is monotonically increasing and bounded below by β. Hence $\lim\limits_{n \to \infty} a_n = +\infty$.

If $\alpha < x < \beta$, then $\alpha < a^x < \beta$ and $g(x) < 0$. Consequently, the sequence $\{a_n\}$ is bounded and monotonically decreasing. Thus it converges to α.

For either $x = \alpha$ or $x = \beta$, we get a constant sequence.

Now if $0 < x < \alpha$, then $1 < a^x < \alpha$ and $g(x) > 0$. Therefore the sequence $\{a_n\}$ increases to α.

Finally, if $a = e^{\frac{1}{e}}$, then the number e is the only solution of the equation $a^x = x$, and the function g attains its minimum value of 0 at e. Thus for $0 < x \leq e$, we get $0 < a^x \leq e$ and $g(x) \geq g(e) = 0$. This implies that the sequence $\{a_n\}$ is monotonically increasing and its limit is equal to e. On the other hand, if $x > e$, the sequence increases to infinity.

We can summarize the results as follows:

$$\lim_{n \to \infty} a_n = \begin{cases} +\infty & \text{if } a > e^{\frac{1}{e}} \text{ and } x > 0, \\ +\infty & \text{if } 1 < a < e^{\frac{1}{e}} \text{ and } x > \beta, \\ \beta & \text{if } 1 < a < e^{\frac{1}{e}} \text{ and } x = \beta, \\ \alpha & \text{if } 1 < a < e^{\frac{1}{e}} \text{ and } 0 < x < \beta, \\ +\infty & \text{if } a = e^{\frac{1}{e}} \text{ and } x > e, \\ e & \text{if } a = e^{\frac{1}{e}} \text{ and } 0 < x \leq e. \end{cases}$$

2.5.41. The equality can be proved by induction. We have $\lim\limits_{n \to \infty} a_n = 2$ (compare with the solution of 2.1.16).

2.5.42. [20] Note first that
$$a_n \leq \underbrace{\sqrt{2 + \sqrt{2 + \ldots + \sqrt{2}}}}_{n \text{ roots}} < 2.$$
Observe that if $\varepsilon_1 = 0$, then all the terms of the sequence $\{a_n\}$ are equal to zero. Assume now that $\varepsilon_1 \neq 0$. We will show by induction that the given equality holds. It is evident for $n = 1$. So suppose that
$$a_n = \varepsilon_1 \sqrt{2 + \varepsilon_2 \sqrt{2 + \ldots + \varepsilon_n \sqrt{2}}} = 2\sin\left(\frac{\pi}{4} \sum_{k=1}^{n} \frac{\varepsilon_1 \varepsilon_2 \ldots \varepsilon_k}{2^{k-1}}\right).$$
Then
$$a_{n+1}^2 - 2 = 2\sin\left(\frac{\pi}{4} \sum_{k=2}^{n+1} \frac{\varepsilon_2 \ldots \varepsilon_k}{2^{k-2}}\right) = -2\cos\left(\frac{\pi}{2} + \frac{\pi}{2} \sum_{k=2}^{n+1} \frac{\varepsilon_2 \ldots \varepsilon_k}{2^{k-1}}\right)$$
$$= -2\cos\left(\frac{\pi}{2} \sum_{k=1}^{n+1} \frac{\varepsilon_1 \ldots \varepsilon_k}{2^{k-1}}\right) = 4\sin^2\left(\frac{\pi}{4} \sum_{k=1}^{n+1} \frac{\varepsilon_1 \ldots \varepsilon_k}{2^{k-1}}\right) - 2,$$
which completes the proof of equality. Now, by the continuity of the sine function,
$$\lim_{n \to \infty} a_n = 2\sin\left(\frac{\pi}{4} \sum_{k=1}^{\infty} \frac{\varepsilon_1 \ldots \varepsilon_k}{2^{k-1}}\right).$$

2.5.43. One can show by induction that
$$\arctan \frac{1}{2} + \ldots + \arctan \frac{1}{2n^2} = \arctan \frac{n}{n+1}.$$
Therefore $\lim_{n \to \infty} \left(\arctan \frac{1}{2} + \ldots + \arctan \frac{1}{2n^2}\right) = \frac{\pi}{4}$.

2.5.44. We have
$$\sin^2(\pi\sqrt{n^2 + n}) = \sin^2(\pi\sqrt{n^2 + n} - \pi n) = \sin^2 \frac{\pi}{1 + \sqrt{1 + \frac{1}{n}}} \xrightarrow[n \to \infty]{} 1.$$

2.5.45. One can show by induction that the sequence is monotonically increasing and bounded above, e.g. by 3. Hence it is a convergent sequence whose limit g satisfies $g = \sqrt{2 + \sqrt{3 + g}}$ and $g \in (2, 3)$.

2.5. Miscellaneous Problems

2.5.46. [13] We have

$$3 = \sqrt{1+2\cdot 4} = \sqrt{1+2\sqrt{16}} = \sqrt{1+2\sqrt{1+3\sqrt{25}}}$$
$$= \sqrt{1+2+\sqrt{1+3\sqrt{1+4\sqrt{36}}}},$$

and, inductively,

(1) $$\sqrt{1+2\sqrt{1+3\sqrt{1+...\sqrt{1+n\sqrt{(n+2)^2}}}}} = 3.$$

Therefore

(2) $$3 \geq \sqrt{1+2\sqrt{1+3\sqrt{1+...\sqrt{1+(n-1)\sqrt{n+1}}}}}.$$

Now we will use the following (easily verifiable) inequality:

(3) $$\sqrt{1+x\alpha} \leq \sqrt{\alpha}\sqrt{1+x}, \quad x \geq 0, \ \alpha > 1.$$

By (3) with $x = n$ and $\alpha = n+2$,

$$\sqrt{1+n\sqrt{(n+2)^2}} < \sqrt{n+2}\sqrt{1+n}.$$

Hence

$$\sqrt{1+(n-1)\sqrt{1+n\sqrt{(n+2)^2}}} < \sqrt{1+\sqrt{n+2}(n-1)\sqrt{1+n}}$$
$$< (n+2)^{\frac{1}{4}}\sqrt{1+(n-1)\sqrt{n+1}},$$

where the last inequality follows from (3) with $\alpha = \sqrt{n+2}$. In view of (1), repeating this argument n times gives

(4) $$3 \leq (n+2)^{2^{-n}}\sqrt{1+2\sqrt{1+3\sqrt{1+...\sqrt{1+(n-1)\sqrt{n+1}}}}}.$$

Combining (2) with (4) yields

$$\lim_{n\to\infty} \sqrt{1+2\sqrt{1+3\sqrt{1+...\sqrt{1+(n-1)\sqrt{n+1}}}}} = 3.$$

2.5.47. The equation $x^2 + x - a = 0$, $a > 0$, has two roots α and β such that $\alpha > 0 > \beta$. Furthermore, we have

$$a_{n+1} - \alpha = \frac{a}{a_n} - 1 - \alpha = \frac{a - a_n - \alpha a_n}{a_n}$$
$$= \frac{a - (1+\alpha)(a_n - \alpha) - \alpha(1+\alpha)}{a_n} = \frac{-(1+\alpha)(a_n - \alpha)}{a_n}.$$

Since $\alpha + \beta = -1$, we see that $a_{n+1} - \alpha = \beta \frac{a_n - \alpha}{a_n}$. Likewise, $a_{n+1} - \beta = \alpha \frac{a_n - \beta}{a_n}$. Thus

$$\frac{a_{n+1} - \beta}{a_{n+1} - \alpha} = \frac{\alpha}{\beta} \frac{a_n - \beta}{a_n - \alpha},$$

and inductively,

$$\frac{a_n - \beta}{a_n - \alpha} = \left(\frac{\alpha}{\beta}\right)^{n-1} \frac{a_1 - \beta}{a_1 - \alpha}.$$

Since $\left|\frac{\alpha}{\beta}\right| = \frac{\alpha}{1+\alpha} < 1$, we get $\lim\limits_{n\to\infty} \left(\frac{\alpha}{\beta}\right)^{n-1} = 0$, and consequently, $\lim\limits_{n\to\infty} a_n = \beta$.

2.5.48. Let α and β be the roots of $x^2 + x - a = 0$, $a > 0$. Then $\alpha > 0 > \beta$. In much the same manner as in the solution of the foregoing problem we get

$$\frac{a_n - \alpha}{a_n - \beta} = \left(\frac{\alpha}{\beta}\right)^{n-1} \frac{a_1 - \alpha}{a_1 - \beta}.$$

Thus $\lim\limits_{n\to\infty} a_n = \alpha$.

2.5.49. For any positive integer k, we have

$$|a_{n+1+k} - a_{n+1}| = \left|\frac{1}{1+a_{n+k}} - \frac{1}{1+a_n}\right| = \frac{|a_{n+k} - a_n|}{(1+a_{n+k})(1+a_n)}$$
$$\leq \frac{1}{4}|a_{n+k} - a_n|.$$

Now by induction we get

$$|a_{n+1+k} - a_{n+1}| \leq \left(\frac{1}{4}\right)^n |a_{k+1} - a_1|.$$

2.5. Miscellaneous Problems

Moreover,

$$|a_{k+1} - a_1| \le |a_{k+1} - a_k| + |a_k - a_{k-1}| + \ldots + |a_2 - a_1|$$
$$\le \frac{1}{1-\frac{1}{4}}|a_2 - a_1| = \frac{4}{3}|a_2 - a_1|.$$

Thus $\{a_n\}$ is a Cauchy sequence. Its limit is $\sqrt{2}$.

2.5.50. One can proceed as in the solution of the foregoing problem and show that $\lim\limits_{n\to\infty} a_n = 1 + \sqrt{2}$.

2.5.51. Let $f(x) = \frac{a}{2+x}$, $x > 0$, and $F(x) = f(f(x))$. Then $F'(x) > 0$ for $x > 0$. It is easily verified that $a_1 < a_3$ and $a_4 < a_2$. Moreover, since F is strictly increasing, we see that the sequence $\{a_{2n}\}$ is strictly decreasing and the sequence $\{a_{2n+1}\}$ is strictly increasing. The sequence $\{a_n\}$ is bounded. Thus both its subsequences $\{a_{2n}\}$ and $\{a_{2n+1}\}$ converge. One can check that they have the same limit $\sqrt{1+a} - 1$.

2.5.52. If $a_1 \le 0$, then $a_2 = 1 - a_1 > 1$ and $a_3 = a_2 - \frac{1}{2} > \frac{1}{2}$. By induction, $a_{n+1} = a_n - \frac{1}{2^{n-1}}$ for $n \ge 2$. Consequently,

$$a_{n+1} = -\left(\frac{1}{2^{n-1}} + \frac{1}{2^{n-2}} + \ldots + \frac{1}{2}\right) + a_2,$$

and therefore $\lim\limits_{n\to\infty} a_n = -a_1$ if $a_1 \le 0$. Now if $a_1 \in (0,2)$, then $a_2 \in [0,1)$ and inductively, we see that $a_{n+1} \in [0, \frac{1}{2^{n-1}}]$, which implies that in this case $\lim\limits_{n\to\infty} a_n = 0$. Finally, if $a_1 \ge 2$, then $a_2 = a_1 - 1 \ge 1$. By induction, we get $a_{n+1} = a_n - \frac{1}{2^{n-1}}$, and consequently, as in the first case, we show that $\lim\limits_{n\to\infty} a_n = a_1 - 2$.

2.5.53.
(a) We have

$$\sum_{j=1}^{n-1} \frac{ja^j}{n-j} = \frac{a}{n-1} + \frac{2a^2}{n-2} + \frac{3a^3}{n-3} + \ldots + \frac{(n-1)a^{n-1}}{1}$$
$$= \frac{1}{n-1}\left(\frac{a}{1} + \frac{2(n-1)a^2}{n-2} + \frac{3(n-1)a^3}{n-3} + \ldots + \frac{(n-1)^2 a^{n-1}}{1}\right).$$

Since
$$j\frac{n-1}{n-j} = j\left(\frac{n-j}{n-j} + \frac{j-1}{n-j}\right) \leq j(1+j-1) = j^2,$$
we obtain
$$\sum_{j=1}^{n-1} \frac{ja^j}{n-j} \leq \frac{a + 2^2a^2 + 3^2a^3 + \ldots + (n-1)^2a^{n-1}}{n-1}.$$

Now it is enough to observe that, by the result in Problem 2.3.2,
$$\lim_{n\to\infty} \frac{a + 2^2a^2 + 3^2a^3 + \ldots + (n-1)^2a^{n-1}}{n-1} = 0.$$

(b) Observe that
$$na^n \sum_{j=1}^n \frac{1}{ja^j} = \sum_{j=1}^n \frac{n}{j} a^{n-j} = \sum_{k=0}^{n-1} \frac{n}{n-k} a^k = \sum_{k=0}^{n-1} a^k + \sum_{k=0}^{n-1} \frac{ka^k}{n-k}$$
and apply (a).

(c) Apply (b) with $a = \frac{1}{6}$.

2.5.54. Since for positive x, $x - \frac{x^3}{6} < \sin x < x$, we see that
$$\sum_{k=1}^n \frac{\pi}{n+k} - \sum_{k=1}^n \frac{\pi^3}{6(n+k)^3} < \sum_{k=1}^n \sin \frac{\pi}{n+k} < \sum_{k=1}^n \frac{\pi}{n+k}.$$

It is easy to check that $\lim_{n\to\infty} \sum_{k=1}^n \frac{\pi^3}{6(n+k)^3} = 0$. Moreover, by 2.5.8 (a), $\lim_{n\to\infty} \sum_{k=1}^n \frac{\pi}{n+k} = \pi \ln 2$. Therefore the limit is $\pi \ln 2$.

2.5.55.

(a) Let $a_n = \prod_{k=1}^n \left(1 + \frac{k^2}{cn^3}\right)$. In view of the inequality (see 2.5.3) $\frac{x}{x+1} < \ln(1+x) < x$ for $x > 0$, we get
$$\sum_{k=1}^n \frac{k^2}{cn^3 + k^2} < \ln a_n < \sum_{k=1}^n \frac{k^2}{cn^3}.$$

2.5. Miscellaneous Problems

Hence, by the equality $\sum\limits_{k=1}^{n} k^2 = \frac{n(n+1)(2n+1)}{6}$,

$$\frac{n(n+1)(2n+1)}{6(cn^3 + n^2)} < \ln a_n < \frac{n(n+1)(2n+1)}{6cn^3}.$$

Therefore $\lim\limits_{n \to \infty} a_n = e^{\frac{1}{3c}}$.

(b) One can show that the inequality $\frac{x}{x+1} < \ln(1+x) < x$ also holds if $-1 < x < 0$. Therefore, as in the proof of (a), we get

$$\lim_{n \to \infty} \prod_{k=1}^{n} \left(1 - \frac{k^2}{cn^3}\right) = e^{-\frac{1}{3c}}.$$

2.5.56. Since for positive x, $x - \frac{x^3}{6} < \sin x < x - \frac{x^3}{6} + \frac{x^5}{5!}$, we see that

(1) $\quad \dfrac{\sqrt{n^{3n}}}{n!} \dfrac{n!}{(n\sqrt{n})^n} \prod\limits_{k=1}^{n} \left(1 - \dfrac{k^2}{6n^3}\right) < \dfrac{\sqrt{n^{3n}}}{n!} \prod\limits_{k=1}^{n} \sin \dfrac{k}{n\sqrt{n}}$

and

(2) $\quad \dfrac{\sqrt{n^{3n}}}{n!} \prod\limits_{k=1}^{n} \sin \dfrac{k}{n\sqrt{n}} < \dfrac{\sqrt{n^{3n}}}{n!} \dfrac{n!}{(n\sqrt{n})^n} \prod\limits_{k=1}^{n} \left(1 - \dfrac{k^2}{6n^3} + \dfrac{k^4}{5!n^6}\right).$

It follows from (1) and from the result in the foregoing problem that the limit is greater than or equal to $e^{-\frac{1}{18}}$. We will now show that

$$\lim_{n \to \infty} \prod_{k=1}^{n} \left(1 - \frac{k^2}{6n^3} + \frac{k^4}{5!n^6}\right) \leq e^{-\frac{1}{18}}.$$

Indeed,

$$\ln \prod_{k=1}^{n} \left(1 - \frac{k^2}{6n^3} + \frac{k^4}{5!n^6}\right) < \sum_{k=1}^{n} \left(-\frac{k^2}{6n^3} + \frac{k^4}{5!n^6}\right)$$
$$= -\frac{n(n+1)(2n+1)}{36n^3} + \frac{n(n+1)(2n+1)(3n^2 + 3n - 1)}{30 \cdot 5!n^6}.$$

Finally, by (2) and the squeeze principle,

$$\lim_{n \to \infty} \frac{\sqrt{n^{3n}}}{n!} \prod_{k=1}^{n} \sin \frac{k}{n\sqrt{n}} = e^{-\frac{1}{18}}.$$

2.5.57. We will first show that $a_n = \frac{n+1}{2n}a_{n-1} + 1$, $n \geq 2$. We have

$$a_n = \sum_{k=0}^{n} \frac{1}{\binom{n}{k}} = \sum_{k=0}^{n-1} \frac{k!(n-1-k)!(n-k)}{(n-1)!n} + 1$$

(1)
$$= \sum_{k=0}^{n-1} \frac{1}{\binom{n-1}{k}} - \sum_{k=0}^{n-1} \frac{k}{n} \frac{(n-1-k)!k!}{(n-1)!} + 1$$

$$= a_{n-1} - \sum_{k=0}^{n-1} \frac{k}{n} \frac{1}{\binom{n-1}{k}} + 1.$$

Moreover,

$$\sum_{k=0}^{n-1} \frac{k}{n} \frac{1}{\binom{n-1}{k}} = \sum_{k=0}^{n-1} \frac{n-1-k}{n} \frac{1}{\binom{n-1}{k}} = \frac{n-1}{n} a_{n-1} - \sum_{k=0}^{n-1} \frac{k}{n} \frac{1}{\binom{n-1}{k}}.$$

Therefore

$$\sum_{k=0}^{n-1} \frac{k}{n} \frac{1}{\binom{n-1}{k}} = \frac{n-1}{2n} a_{n-1}.$$

Finally, by (1), $a_n = a_{n-1} - \frac{n-1}{2n}a_{n-1} + 1 = \frac{n+1}{2n}a_{n-1} + 1$, which establishes our first assertion. From this we deduce that $\lim_{n\to\infty} a_n = 2$.

2.5.58. If $\alpha = 0$, then obviously $\lim_{n\to\infty} a_n = 0$. If $\alpha > 0$, then $0 < a_n < 1 - \frac{(n-1)^\alpha}{n^\alpha}$, and consequently, $\lim_{n\to\infty} a_n = 0$. Assume now that $\alpha < 0$. Then

$$a_n = (-1)^{n-1}\left(n^{-\alpha} - 1\right)\left(\left(\frac{n}{2}\right)^{-\alpha} - 1\right) \cdot \ldots \cdot \left(\left(\frac{n}{n-1}\right)^{-\alpha} - 1\right).$$

Therefore, if we take $\alpha = -1$, we obtain the divergent sequence $a_n = (-1)^{n-1}$. If we take $\alpha < -1$, we get

$$\left(\left(\frac{n}{p}\right)^{-\alpha} - 1\right)\left(\left(\frac{n}{n-p}\right)^{-\alpha} - 1\right) > \left(\frac{n}{p} - 1\right)\left(\frac{n}{n-p} - 1\right) = 1$$

for $1 \leq p < n$. Moreover, $\left(\frac{n}{\frac{n}{2}}\right)^{-\alpha} - 1 > 2 - 1 = 1$. Therefore

$$|a_n| \geq (n^{-\alpha} - 1)\left(\left(\frac{n}{n-1}\right)^{-\alpha} - 1\right) \xrightarrow[n\to\infty]{} +\infty.$$

2.5. Miscellaneous Problems 233

Likewise, we can see that if $-1 < \alpha < 0$, then

$$|a_n| \leq (n^{-\alpha} - 1)\left(\left(\frac{n}{n-1}\right)^{-\alpha} - 1\right) \xrightarrow[n \to \infty]{} 0.$$

2.5.59. We have $(2+\sqrt{3})^n = \sum_{k=0}^{n} \binom{n}{k}(\sqrt{3})^k 2^{n-k}$. If we group the terms with odd and even indices, respectively, we can write

$$(2+\sqrt{3})^n = A_n + \sqrt{3}B_n \quad \text{and} \quad (2-\sqrt{3})^n = A_n - \sqrt{3}B_n.$$

Hence $\lim_{n \to \infty}(A_n + \sqrt{3}B_n) = +\infty$ and $\lim_{n \to \infty}(A_n - \sqrt{3}B_n) = 0$. Also,

$$\lim_{n \to \infty} \frac{\sqrt{3}B_n}{A_n} = 1.$$

Since the A_n are integers and $\frac{\sqrt{3}B_n}{A_n} < 1$, it follows that $[\sqrt{3}B_n] = A_n - 1$ for sufficiently large n. Consequently,

$$\{\sqrt{3}B_n\} \xrightarrow[n \to \infty]{} 1 \quad \text{or} \quad \{A_n + \sqrt{3}B_n\} = \{\sqrt{3}B_n\} \xrightarrow[n \to \infty]{} 1.$$

2.5.60. The sequence $\{S_n\}$ is monotonically increasing. If it were bounded above it would converge, and then

$$\lim_{n \to \infty} a_n = \lim_{n \to \infty}(S_n - S_{n-1}) = 0.$$

Suppose now that $\lim_{n \to \infty} S_n = +\infty$. By our assumption, $S_{n+1}a_{n+1} + a_n \leq S_n a_n + a_{n-1}$, and consequently, $S_n a_n + a_{n-1} \leq S_2 a_2 + a_1$. Hence

$$a_n \leq a_n + \frac{a_{n-1}}{S_n} \leq \frac{S_2 a_2 + a_1}{S_n}.$$

Finally, $\lim_{n \to \infty} a_n = 0$.

2.5.61. By assumption, for any $\varepsilon > 0$ there is a positive integer n_0 such that $a_n < \varepsilon n$ if $n > n_0$. Hence

$$\frac{a_1^2 + a_2^2 + \ldots + a_n^2}{n^2} = \frac{a_1^2 + a_2^2 + \ldots + a_{n_0}^2}{n^2} + \frac{a_{n_0+1}^2 + \ldots + a_n^2}{n^2}$$

$$\leq \frac{a_1^2 + a_2^2 + \ldots + a_{n_0}^2}{n^2} + \frac{n\varepsilon(a_{n_0+1} + \ldots + a_n)}{n^2}.$$

By 2.4.14 and 2.4.19,
$$\overline{\lim_{n\to\infty}} \frac{a_1^2 + a_2^2 + \ldots + a_n^2}{n^2} \leq \varepsilon \overline{\lim_{n\to\infty}} \frac{a_1 + \ldots + a_n}{n}.$$

This obviously implies that $\lim\limits_{n\to\infty} \frac{a_1^2+a_2^2+\ldots+a_n^2}{n^2} = 0$.

2.5.62. We will apply the Toeplitz theorem (see 2.3.1). Set
$A_n = a_1 + a_2 + \ldots + a_n$, $B_n = b_1 + b_2 + \ldots + b_n$, $C_n = c_1 + c_2 + \ldots + c_n$
and
$$d_{n,k} = \frac{a_{n-k+1}B_k}{a_1B_n + a_2B_{n-1} + \ldots + a_nB_1}.$$

Now we will show that the positive numbers $d_{n,k}$ satisfy conditions (i) and (ii) in 2.3.1 (see also 2.3.3 (a)). For fixed k,
$$d_{n,k} \leq \frac{a_{n-k+1}}{a_1 + a_2 + \ldots + a_{n-k+1}} \xrightarrow[n\to\infty]{} 0.$$

Clearly, $\sum\limits_{k=1}^{n} d_{n,k} = 1$. Observe also that
$$\frac{c_n}{C_n} = \frac{a_1b_n + a_2b_{n-1} + \ldots + a_nb_1}{a_1B_n + a_2B_{n-1} + \ldots + a_nB_1} = d_{n,1}\frac{b_1}{B_1} + d_{n,2}\frac{b_2}{B_2} + \ldots + d_{n,n}\frac{b_n}{B_n}.$$

Finally, by the Toeplitz theorem, $\lim\limits_{n\to\infty} \frac{c_n}{C_n} = \lim\limits_{n\to\infty} \frac{b_n}{B_n} = 0$.

2.5.63. We know that $x - \frac{x^2}{2} < \ln(1+x) < x - \frac{x^2}{2} + \frac{x^3}{3}$ for $x > 0$. Put $a_n = \left(1 + \frac{1}{n}\right)^{n^2} e^{-n}$. Then $-\frac{1}{2} < \ln a_n < -\frac{1}{2} + \frac{1}{3n}$, which implies $\lim\limits_{n\to\infty} \ln a_n = -\frac{1}{2}$. Therefore the limit is equal to $e^{-\frac{1}{2}}$.

2.5.64. We have $a_{n+1} - a_n > -\frac{1}{n^2} > -\frac{1}{n(n-1)} = -\frac{1}{n-1} + \frac{1}{n}$ for $n > 1$. Let $b_n = a_n - \frac{1}{n-1}$. Then the sequence $\{b_n\}$ is monotonically increasing and bounded above; hence it converges. Therefore $\{a_n\}$ also converges.

2.5.65. By the assumption $a_{n+1} \sqrt[2^n]{2} \geq a_n$, we see that
$$a_{n+1}2^{-\frac{1}{2^n}} \geq a_n 2^{-\frac{1}{2^{n-1}}}.$$

Hence the sequence $b_n = a_n 2^{-\frac{1}{2^{n-1}}}$ is monotonically increasing and bounded. So, it converges. Obviously, $\lim\limits_{n\to\infty} b_n = \lim\limits_{n\to\infty} a_n$.

2.5. Miscellaneous Problems

2.5.66. Let $a \in (l, L)$. Suppose, contrary to our claim, that a is not a limit point of $\{a_n\}$. Then there is a neighborhood of a which contains only finitely many terms of the sequence. Let $\varepsilon > 0$ be so small that

(∗) $\quad l < a - \varepsilon < a < a + \varepsilon < L \quad$ and $\quad a_n \notin [a - \varepsilon, a + \varepsilon] \quad$ for $\quad n > n_1$.

By the assumption, $|a_{n+1} - a_n| < \varepsilon$ for $n > n_2$. By 2.4.13 (b), we know that there exists a_{n_k} such that $a_{n_k} < l + \varepsilon < a$ for $n_k > \max\{n_1, n_2\}$. Hence $a_{n_k+1} \leq a_{n_k} + |a_{n_k+1} - a_{n_k}| < a + \varepsilon$. Thus by (∗), $a_{n_k+1} < a - \varepsilon$. Therefore, by 2.4.12 (a), $L \leq a - \varepsilon < L$, a contradiction.

2.5.67. Let $a \in (l, L)$. Suppose, contrary to our claim, that a is not a limit point of $\{a_n\}$. Then there is a neighborhood of a which contains only finitely many terms of the sequence. Let $\varepsilon > 0$ be so small that

(∗) $\quad l < a - \varepsilon < a < a + \varepsilon < L \quad$ and $\quad a_n \notin [a - \varepsilon, a + \varepsilon] \quad$ for $\quad n > n_1$.

By the assumption,

(∗∗) $$a_n - a_{n+1} < \alpha_n < \varepsilon \quad \text{for} \quad n > n_2.$$

It follows from 2.4.13 (a) that there is a_{n_k} such that $a_{n_k} > L - \varepsilon > a$. Hence, by (∗∗), we get $a_{n_k+1} = a_{n_k} + (a_{n_k+1} - a_{n_k}) > a - \varepsilon$. Now, by (∗), $a_{n_k+1} > a + \varepsilon$ for $n_k > \max\{n_1, n_2\}$. Thus by 2.4.12 (c), $l \geq a + \varepsilon > a > l$, a contradiction.

2.5.68. We will use the result proved in the solution of the last problem. By the monotonicity of $\{a_n\}$,

$$\frac{a_{n+1}}{n+1+a_{n+1}} - \frac{a_n}{n+a_n} \geq \frac{-a_n}{(n+1+a_{n+1})(n+a_n)} \geq \frac{-1}{n}.$$

Thus, by the result in the preceding problem, the set of all limit points of the given sequence is the interval $[l, L]$, where

$$l = \lim_{n \to \infty} \frac{a_n}{n + a_n} \quad \text{and} \quad L = \varlimsup_{n \to \infty} \frac{a_n}{n + a_n}.$$

2.5.69. Note that
$$\left|a_{2n+1}-\frac{1}{3}\right|=\frac{1}{2}\left|a_{2n}-\frac{2}{3}\right|=\frac{1}{2}\left|\frac{1+a_{2n-1}}{2}-\frac{2}{3}\right|=\frac{1}{4}\left|a_{2n-1}-\frac{1}{3}\right|.$$
This implies that the sequence has the two limit points: $\frac{1}{3}$ and $\frac{2}{3}$.

2.5.70. We know, by 1.1.14, that for any positive integer n there exist a positive integer q_n and an integer p_n such that
$$\left|2\pi-\frac{p_n}{q_n}\right|<\frac{1}{q_n^2}.$$
Thus $|p_n|<(2\pi+1)q_n$, and consequently,
$$|\sqrt{|p_n|}\sin p_n|=|\sqrt{|p_n|}\sin(2\pi q_n-p_n)|\leq\left|\sqrt{|p_n|}\sin\frac{1}{q_n}\right|\leq\frac{\sqrt{2\pi+1}}{\sqrt{q_n}}.$$
Since the sequence $\{q_n\}$ is unbounded, it contains a subsequence divergent to infinity. Therefore zero is a limit point of $\{a_n\}$.

2.5.71. It is enough to show that there is a subsequence $\{a_{n_k}\}$ for which
$$\left(\frac{n_k(a_1+a_{n_k+1})}{(n_k+1)a_{n_k}}\right)^{n_k}\geq 1.$$
Suppose that the above condition does not hold. Then there exists n_0 such that
$$\frac{n(a_1+a_{n+1})}{(n+1)a_n}<1 \quad \text{if} \quad n\geq n_0.$$
Hence $\frac{a_1}{n+1}+\frac{a_{n+1}}{n+1}<\frac{a_n}{n}$ for $n\geq n_0$. Thus
$$\frac{a_n}{n}-\frac{a_{n-1}}{n-1}<-\frac{a_1}{n},$$
$$\frac{a_{n-1}}{n-1}-\frac{a_{n-2}}{n-2}<-\frac{a_1}{n-1},$$
$$\vdots$$
$$\frac{a_{n_0+1}}{n_0+1}-\frac{a_{n_0}}{n_0}<-\frac{a_1}{n_0+1}.$$
Summing the above inequalities, we get
$$\frac{a_n}{n}<\frac{a_{n_0}}{n_0}-a_1\left(\frac{1}{n_0+1}+...+\frac{1}{n}\right).$$

2.5. Miscellaneous Problems 237

Therefore, by 2.2.50 (c), $\lim\limits_{n\to\infty} \frac{a_n}{n} = -\infty$, which is impossible because $a_n > 0$.

2.5.72. In much the same way as in the proof of the foregoing problem we show that there is a subsequence $\{a_{n_k}\}$ for which

$$\left(\frac{n_k(a_1 + a_{n_k+p})}{(n_k+p)a_{n_k}}\right)^{n_k} \geq 1.$$

2.5.73. Suppose that our claim does not hold. Then there exists an n_0 such that for $n \geq n_0$, $n\left(\frac{1+a_{n+1}}{a_n} - 1\right) < 1$. The last inequality can be rewritten as $\frac{1}{n+1} < \frac{a_n}{n} - \frac{a_{n+1}}{n+1}$. This, in turn, implies (see the solution of 2.5.71) that

$$\frac{1}{n_0+1} + \ldots + \frac{1}{n} < \frac{a_{n_0}}{n_0} - \frac{a_n}{n}.$$

Hence $\lim\limits_{n\to\infty} \frac{a_n}{n} = -\infty$, which contradicts the fact that $\{a_n\}$ is a positive sequence.

To show that 1 is the best possible constant, take $a_n = n\ln n$. Then

$$\lim_{n\to\infty} \left(n\frac{1+(n+1)\ln(n+1)}{n\ln n} - n\right)$$
$$= \lim_{n\to\infty} \frac{1+(n+1)\ln(n+1) - n\ln n}{\ln n}$$
$$= \lim_{n\to\infty} \frac{1+\ln(n+1) + \ln\left(1+\frac{1}{n}\right)^n}{\ln n} = 1.$$

2.5.74. Note that $a_n^2 = 1 + a_{n-1}$ and $a_1 = 1$. Clearly, the sequence strictly increases. We will show by induction that it is bounded above by $\frac{1}{2}(1+\sqrt{5})$. Indeed, if $a_{n-1} < \frac{1}{2}(1+\sqrt{5})$, then $a_n^2 = 1 + a_{n-1} < \frac{3}{2} + \frac{1}{2}\sqrt{5}$. Therefore $a_n < \sqrt{\frac{3}{2} + \frac{1}{2}\sqrt{5}} = \frac{1}{2} + \frac{1}{2}\sqrt{5}$, and $\{a_n\}$ converges to $\frac{1}{2}(1+\sqrt{5})$.

2.5.75. [20] Obviously, the sequence $\{b_n\}$ is strictly increasing. Assume first that $\alpha < \ln 2$. Then, by assumption, there is $n_0 \in \mathbb{N}$ such that $\ln(\ln a_n) < n \ln 2$ if $n \geq n_0$; or, equivalently, $a_n < e^{2^n}$ if

$n \geq n_0$. We have

$$b_n = \sqrt{a_1 + \ldots + \sqrt{a_{n_0} + \ldots + \sqrt{a_n}}}$$

$$\leq \sqrt{a_1 + \ldots + \sqrt{a_{n_0-1} + \sqrt{e^{2^{n_0}} + \ldots + \sqrt{e^{2^n}}}}}$$

$$\leq \sqrt{a_1 + \ldots + \sqrt{a_{n_0-1} + e^{2^{n_0}}\sqrt{1 + \ldots + \sqrt{1}}}}$$

By the foregoing problem,

$$b_n \leq \sqrt{a_1 + \ldots + \sqrt{a_{n_0-1} + e^{2^{n_0}}\frac{1+\sqrt{5}}{2}}}.$$

This means that $\{b_n\}$ is bounded above and convergent. Assume now that $\alpha > \ln 2$. By assumption, given $\varepsilon > 0$, there is an n_0 such that $\ln(\ln a_n) > n(\alpha + \varepsilon)$ for $n \geq n_0$. Setting $\alpha + \varepsilon = \ln \beta$, we get $a_n > e^{\beta^n}$ for $n \geq n_0$, where $\beta > 2$. Thus

$$b_n = \sqrt{a_1 + \sqrt{a_2 + \ldots + \sqrt{a_{n_0} + \ldots + \sqrt{a_n}}}}$$

$$> \sqrt{a_1 + \ldots + \sqrt{a_{n_0-1} + e^{\frac{\beta^n}{2^{n-n_0+1}}}}} > e^{\left(\frac{\beta}{2}\right)^n}.$$

In this case, the sequence $\{b_n\}$ diverges to infinity.

Note, additionally, that if $0 < a_n \leq 1$, then, although $\ln \ln a_n$ is not defined, the sequence $\{b_n\}$ is monotonically increasing and bounded above by $\frac{1+\sqrt{5}}{2}$, and is convergent.

2.5.76. [20] It follows from the assumption that $0 \leq a_n \leq na_1$. Thus the sequence $\left\{\frac{a_n}{n}\right\}$ is bounded. Denote by L its limit superior. Then there is a sequence $\{m_k\}$ of positive integers such that $\lim\limits_{k\to\infty} \frac{a_{m_k}}{m_k} = L$. For an arbitrarily fixed $n \in \mathbb{N}$, we can write $m_k = nl_k + r_k$, where $r_k \in \{0, 1, \ldots, n-1\}$. Thus, by the assumption, $a_{m_k} \leq l_k a_n + a_{r_k}$. Hence

$$\frac{a_{m_k}}{m_k} \leq \frac{l_k}{nl_k + r_k}a_n + \frac{a_{r_k}}{m_k}.$$

2.5. Miscellaneous Problems

Letting $k \to \infty$, we get

$(*)$
$$L \le \frac{a_n}{n},$$

which implies

$$\varlimsup_{n\to\infty} \frac{a_n}{n} \le \varliminf_{n\to\infty} \frac{a_n}{n}.$$

As a result the sequence $\left\{\frac{a_n}{n}\right\}$ converges.

2.5.77. Analysis similar to that in the solution of the preceding problem can be applied.

2.5.78. [20] The sequences $\{a_n + 1\}$ and $\{1 - a_n\}$ satisfy the assumption of Problem 2.5.76. Hence $\lim\limits_{n\to\infty} \frac{a_n+1}{n}$ and $\lim\limits_{n\to\infty} \frac{1-a_n}{n}$ exist and are finite.

(a) By the above, since $\lim\limits_{n\to\infty} \frac{a_n+1}{n} = g$, $\lim\limits_{n\to\infty} \frac{a_n}{n} = g$.

(b) The inequalities follow immediately from $(*)$ in the solution of 2.5.76.

2.5.79. We will show that the sequence $\left\{\frac{a_n}{n}\right\}$ converges to $A = \sup\left\{\frac{a_n}{n} : n \in \mathbb{N}\right\}$. Let p be an arbitrarily fixed positive integer. Then

$$\frac{a_n}{n} = \frac{a_{pl_n + r_n}}{pl_n + r_n} \ge \frac{a_p \cdot l_n}{pl_n + r_n},$$

where $r_n \in \{0, 1, ..., p-1\}$. So, by our assumption, $\varliminf_{n\to\infty} \frac{a_n}{n} \ge \frac{a_p}{p}$. This, in turn, implies that $\varliminf_{n\to\infty} \frac{a_n}{n} \ge \varlimsup_{p\to\infty} \frac{a_p}{p}$. Thus the convergence of the sequence $\left\{\frac{a_n}{n}\right\}$ has been established. Moreover, $\frac{a_{m \cdot n}}{mn} \ge \frac{a_n}{n}$ implies that

$$A \ge \varliminf_{n\to\infty} \frac{a_n}{n} \ge \varlimsup_{p\to\infty} \frac{a_p}{p} = \inf_{p} \sup_{l \ge p} \frac{a_l}{l}$$
$$\ge \inf_{p} \sup_{m \in \mathbb{N}} \frac{a_{p \cdot m}}{pm} \ge \inf_{p} \sup_{m} \frac{a_m}{m} = A.$$

2.5.80. We first show the boundedness of the given sequence. Indeed, if $\frac{1}{a} \le a_n, a_{n+1} \le a$, then $\frac{1}{a} \le a_{n+2} = \frac{2}{a_n + a_{n+1}} \le a$. Thus, by the principle of induction stated in the solution of Problem 2.1.10, the sequence $\{a_n\}$ is bounded. Put

$$l = \varliminf_{n\to\infty} a_n, \qquad L = \varlimsup_{n\to\infty} a_n.$$

Then for an arbitrarily fixed $\varepsilon > 0$ there exist $n_1, n_2 \in \mathbb{N}$ such that

(i) $\qquad\qquad\qquad a_n < L + \varepsilon \quad \text{for} \quad n > n_1,$

(ii) $\qquad\qquad\qquad a_n > l - \varepsilon \quad \text{for} \quad n > n_2.$

By (i), $a_{n+2} = \frac{2}{a_n + a_{n+1}} > \frac{1}{L+\varepsilon}$, $n > n_1$. Since the positive ε can be arbitrarily small, we get $l \geq \frac{1}{L}$. In much the same way (ii) implies that $L \leq \frac{1}{l}$. Thus $l = \frac{1}{L}$. Let $\{n_k\}$ be a sequence of positive integers such that $\lim\limits_{k\to\infty} a_{n_k+2} = L$. We can assume that the sequences $\{a_{n_k+1}\}, \{a_{n_k}\}$ and $\{a_{n_k-1}\}$ converge to $l_1, l_2,$ and l_3, respectively. In fact, if this is not the case, we can choose subsequences which do. By the definition of $\{a_n\}$,

$$l_1 + l_2 = \frac{2}{L} = 2l \quad \text{and} \quad l_2 + l_3 = \frac{2}{l_1},$$

and since $l \leq l_1, l_2, l_3 \leq L$, we get $l_1 = l_2 = l$ and $l_2 = l_3 = L$. Hence $l = L$. This and the equality $l = \frac{1}{L}$ imply that the sequence $\{a_n\}$ converges to 1.

2.5.81. Since $0 < a_1 \leq b_1$, there exists $\varphi \in [0, \frac{\pi}{2})$ such that $a_1 = b_1 \cos\varphi$. Now, one can show by induction that, for $\varphi \neq 0$,

$$a_{n+1} = \frac{b_1 \sin\varphi}{2^n \tan\frac{\varphi}{2^n}} \quad \text{and} \quad b_{n+1} = \frac{b_1 \sin\varphi}{2^n \sin\frac{\varphi}{2^n}}, \quad n \in \mathbb{N}.$$

Therefore $\lim\limits_{n\to\infty} a_n = \lim\limits_{n\to\infty} b_n = \frac{b_1 \sin\varphi}{\varphi}$. If $\varphi = 0$, i.e. $a_1 = b_1$, then the given sequences $\{a_n\}$ and $\{b_n\}$ are constant.

2.5.82. [18] By assumption, $\frac{a_{k,n}}{b_{k,n}} = 1 + \varepsilon_{k,n}$, where $\varepsilon_{k,n}$ tends to zero, uniformly with respect to k. Thus

(*) $$\sum_{k=1}^{n} a_{k,n} = \sum_{k=1}^{n} b_{k,n} + \sum_{k=1}^{n} \varepsilon_{k,n} b_{k,n}.$$

2.5. Miscellaneous Problems

Since $\lim\limits_{n\to\infty}\sum\limits_{k=1}^{n} b_{k,n}$ exists, there is an $M > 0$ such that $\left|\sum\limits_{k=1}^{n} b_{k,n}\right| \leq M$, $n \in \mathbb{N}$. Moreover, for any $\varepsilon > 0$, $|\varepsilon_{k,n}| < \frac{\varepsilon}{M}$ for $k = 1, 2, \ldots, n$, provided n is sufficiently large. Hence $\left|\sum\limits_{k=1}^{n} \varepsilon_{k,n} b_{k,n}\right| < \varepsilon$, which means that $\lim\limits_{n\to\infty}\sum\limits_{k=1}^{n} \varepsilon_{k,n} b_{k,n} = 0$. Thus, by $(*)$,

$$\lim_{n\to\infty}\sum_{k=1}^{n} a_{k,n} = \lim_{n\to\infty}\sum_{k=1}^{n} b_{k,n}.$$

2.5.83. We have

$$\frac{\sin\frac{(2k-1)a}{n^2}}{\frac{(2k-1)a}{n^2}} \xrightarrow[n\to\infty]{} 1 \quad \text{uniformly with respect to } k.$$

Thus, by the foregoing problem,

$$\lim_{n\to\infty}\sum_{k=1}^{n} \sin\frac{(2k-1)a}{n^2} = \lim_{n\to\infty}\sum_{k=1}^{n} \frac{(2k-1)a}{n^2} = a.$$

2.5.84. It follows from 2.5.5 that, if the sequence $\{x_n\}$ converges to zero, then $\frac{a^{x_n}-1}{x_n \ln a} \xrightarrow[n\to\infty]{} 1$. This implies that

$$\frac{a^{\frac{k}{n^2}}-1}{\frac{k}{n^2}\ln a} \xrightarrow[n\to\infty]{} 1$$

uniformly with respect to k. Now applying Problem 2.5.82 we get

$$\lim_{n\to\infty}\sum_{k=1}^{n}\left(a^{\frac{k}{n^2}}-1\right) = \lim_{n\to\infty} \ln a \sum_{k=1}^{n}\frac{k}{n^2} = \frac{1}{2}\ln a.$$

2.5.85. If $\{x_n\}$ is a positive sequence that converges to zero, then by Problem 2.5.3, $\frac{\ln(1+x_n)}{x_n} \xrightarrow[n\to\infty]{} 1$. Applying 2.5.82, we see that

$$\lim_{n\to\infty}\sum_{k=1}^{n}\ln\left(1+\frac{k}{n^2}\right) = \lim_{n\to\infty}\sum_{k=1}^{n}\frac{k}{n^2} = \frac{1}{2}.$$

Thus $\lim\limits_{n\to\infty}\prod\limits_{k=1}^{n}\left(1+\frac{k}{n^2}\right) = e^{\frac{1}{2}}$.

2.5.86. One can show that if $\{x_n\}$ is a positive sequence that converges to zero, then

(*) $$\frac{(1+x_n)^{\frac{1}{p}}-1}{\frac{1}{p}x_n} \xrightarrow[n\to\infty]{} 1$$

Set
$$c_{k,n} = \frac{k^{q-1}}{n^q}, \quad k=1,2,...,n.$$

Then $c_{k,n} \leq \max\{\frac{1}{n}, \frac{1}{n^q}\}$, and consequently, $\{c_{k,n}\}$ converges to zero, uniformly with respect to k. Putting $a_{k,n} = (1+c_{k,n})^{\frac{1}{p}} - 1$ and $b_{k,n} = \frac{1}{p}c_{k,n}$, and using 2.5.82, we obtain

$$\lim_{n\to\infty} \sum_{k=1}^{n} \left(\left(1+\frac{k^{q-1}}{n^q}\right)^{\frac{1}{p}} - 1\right) = \frac{1}{p}\lim_{n\to\infty} \sum_{k=1}^{n} \frac{k^{q-1}}{n^q}.$$

By the Stolz theorem (see 2.3.11),

$$\lim_{n\to\infty} \sum_{k=1}^{n} \frac{k^{q-1}}{n^q} = \lim_{n\to\infty} \frac{n^{q-1}}{n^q - (n-1)^q} = \lim_{n\to\infty} \frac{n^{q-1}}{n^q - n^q\left(1-\frac{1}{n}\right)^q}$$

$$= \lim_{n\to\infty} \frac{n^{q-1}}{n^q - n^q\left(1 - q\frac{1}{n} + \frac{q(q-1)}{2}\frac{1}{n^2} - ...\right)} = \frac{1}{q}.$$

2.5.87. Set $a_n = \frac{a(a+d)...(a+nd)}{b(b+d)...(b+nd)}$. Then

$$a_n = \frac{a}{b}\frac{\left(1+\frac{d}{a}\right)...\left(1+n\frac{d}{a}\right)}{\left(1+\frac{d}{a}+\left(\frac{b}{a}-1\right)\right)\left(1+2\frac{d}{a}+\left(\frac{b}{a}-1\right)\right)...\left(1+n\frac{d}{a}+\left(\frac{b}{a}-1\right)\right)}.$$

Now let $x = \frac{b}{a} - 1$. Then $x > 0$ and

$$a_n = \frac{a}{b}\frac{1}{\left(1+\frac{x}{1+\frac{d}{a}}\right)\left(1+\frac{x}{1+2\frac{d}{a}}\right)...\left(1+\frac{x}{1+n\frac{d}{a}}\right)}.$$

Since

$$\frac{x}{1+\frac{d}{a}} + ... + \frac{x}{1+n\frac{d}{a}} < \left(1+\frac{x}{1+\frac{d}{a}}\right)...\left(1+\frac{x}{1+n\frac{d}{a}}\right),$$

2.5. Miscellaneous Problems

we get
$$a_n < \frac{a}{bx\left(\frac{1}{1+\frac{d}{a}} + \frac{1}{1+2\frac{d}{a}} + ... + \frac{1}{1+n\frac{d}{a}}\right)}.$$

Hence $\lim_{n\to\infty} a_n = 0$, because

$$\lim_{n\to\infty}\left(\frac{1}{1+\frac{d}{a}} + \frac{1}{1+2\frac{d}{a}} + ... + \frac{1}{1+n\frac{d}{a}}\right) = +\infty.$$

Chapter 3

Series of Real Numbers

3.1. Summation of Series

3.1.1.

(a) We have $a_1 = S_1 = 2$ and $a_n = S_n - S_{n-1} = -\frac{1}{n(n-1)}$, $n > 1$. So we get the series $2 - \sum_{n=2}^{\infty} \frac{1}{n(n-1)}$, whose sum is $S = \lim_{n \to \infty} S_n = 1$.

(b) As in the solution of (a), we get $a_n = \frac{1}{2^n}$, $\sum_{n=1}^{\infty} \frac{1}{2^{n+1}} = 1$.

(c) By a similar argument, $a_n = \arctan n - \arctan(n-1)$, and consequently, $\tan a_n = \frac{1}{n^2-n+1}$. Therefore $a_n = \arctan \frac{1}{n^2-n+1}$ and $\sum_{n=1}^{\infty} \arctan \frac{1}{n^2-n+1} = \frac{\pi}{2}$.

(d) $a_1 = -1$, $a_n = (-1)^n \frac{2n-1}{n(n-1)}$ for $n > 1$. Moreover,

$$-1 + \sum_{n=2}^{\infty} (-1)^n \frac{2n-1}{n(n-1)} = 0.$$

3.1.2.

(a) We have $a_n = \frac{1}{n^2} - \frac{1}{(n+1)^2}$. Thus $S_n = 1 - \frac{1}{(n+1)^2}$ and $S = \lim_{n \to \infty} S_n = 1$.

(b) Similarly, $a_n = \frac{1}{8}\left(\frac{1}{(2n-1)^2} - \frac{1}{(2n+1)^2}\right)$. It then follows that $S_n = \frac{1}{8}\left(1 - \frac{1}{(2n+1)^2}\right)$ and $S = \lim_{n\to\infty} S_n = \frac{1}{8}$.

(c) $a_n = \frac{\sqrt{n}}{\sqrt{n+1}} - \frac{\sqrt{n-1}}{\sqrt{n}}$. Therefore $S_n = \frac{\sqrt{n}}{\sqrt{n+1}}$ and $S = \lim_{n\to\infty} S_n = 1$.

(d) $a_n = \frac{1}{2}\left(\frac{1}{2n-1} - \frac{1}{2n+1}\right)$. Thus $S = \lim_{n\to\infty} S_n = \frac{1}{2}$.

(e) $a_n = \frac{\sqrt{n+1}-\sqrt{n}}{\sqrt{n(n+1)}} = \frac{1}{\sqrt{n}} - \frac{1}{\sqrt{n+1}}$. Consequently, $S = \lim_{n\to\infty} S_n = 1$.

3.1.3.

(a) $S_n = \ln 1 - \ln 4 + \ln 2 + \ln 4 - \ln 1 - \ln 7 + \ln 3 + \ln 7 - \ln 2$
$- \ln 10 + \ldots + \ln n + \ln(3n-2) - \ln(n-1) - \ln(3n+1)$
$+ \ln(n+1) + \ln(3n+1) - \ln n - \ln(3n+4) = \ln \frac{n+1}{3n+4}$.

Hence $S = \ln \frac{1}{3}$.

(b) $S = \ln 2$.

3.1.4. (a) We have
$$a_n = \frac{1}{n(n+1)\ldots(n+m)}$$
$$= \frac{1}{m}\left(\frac{1}{n(n+1)\ldots(n+m-1)} - \frac{1}{(n+1)(n+2)\ldots(n+m)}\right).$$

Hence $S_n = \frac{1}{m}\left(\frac{1}{1\cdot 2\cdot\ldots\cdot m} - \frac{1}{(n+1)(n+2)\ldots(n+m)}\right)$. So, $S = \frac{1}{mm!}$.

(b) Since $a_n = \frac{1}{m}\left(\frac{1}{n} - \frac{1}{n+m}\right)$, $S = \frac{1}{m}\left(1 + \frac{1}{2} + \ldots + \frac{1}{m}\right)$.

(c) We have
$$\frac{n^2}{(n+1)(n+2)(n+3)(n+4)} = \frac{2}{(n+1)(n+2)} - \frac{1}{(n+3)(n+4)}$$
$$- \frac{11}{2}\left(\frac{1}{(n+1)(n+3)} - \frac{1}{(n+2)(n+4)}\right)$$
$$+ \frac{11}{4}\left(\frac{1}{(n+1)(n+4)} - \frac{1}{(n+2)(n+3)}\right).$$

3.1. Summation of Series

Now, as in (b), after simple calculations we get $S = \frac{5}{36}$.

3.1.5.
(a) For $n \geq 5$, we have
$$S_n = \sin\frac{\pi}{720} + \sin\frac{\pi}{360} + \sin\frac{\pi}{120} + \sin\frac{\pi}{30} + \sin\frac{\pi}{6}.$$

(b) Observe that $0 \leq \frac{\ln n}{n - \ln n} < 1$, $n \in \mathbb{N}$. Thus $S = 0$.

3.1.6. Since $a_n = \sin\frac{1}{2^{n+1}}\cos\frac{3}{2^{n+1}} = \frac{1}{2}\left(\sin\frac{1}{2^{n-1}} - \sin\frac{1}{2^n}\right)$, we see that $S = \frac{1}{2}\sin 1$.

3.1.7. Note that
$$\frac{1}{n!(n^4 + n^2 + 1)}$$
$$= \frac{1}{2}\left(\frac{n}{(n+1)!((n+1)n+1)} - \frac{n-1}{n!(n(n-1)+1)} + \frac{1}{(n+1)!}\right).$$

Hence
$$S_n = \frac{1}{2}\left(\frac{n}{(n+1)!((n+1)n+1)} + 1 + \sum_{k=0}^{n}\frac{1}{(k+1)!}\right),$$

and by 2.5.6, we get $S = \lim_{n\to\infty} S_n = \frac{1}{2}e$.

3.1.8. Note that for $n > 1$,
$$a_n = \frac{1}{2} \cdot \frac{(2n+1) - 1}{3 \cdot 5 \cdot \ldots \cdot (2n+1)}$$
$$= \frac{1}{2}\left(\frac{1}{3 \cdot 5 \cdot \ldots \cdot (2n-1)} - \frac{1}{3 \cdot 5 \cdot \ldots \cdot (2n+1)}\right)$$

and $a_1 = \frac{1}{3}$. It follows that
$$S_n = \frac{1}{3} + \frac{1}{2}\left(\frac{1}{3} - \frac{1}{3 \cdot 5 \cdot \ldots \cdot (2n+1)}\right),$$

which implies the desired result.

3.1.9. As in the solution of the last problem, we have

$$\frac{a_n}{(a_1+1)(a_2+1)\ldots(a_n+1)} = \frac{a_n+1-1}{(a_1+1)(a_2+1)\ldots(a_n+1)}$$
$$= \frac{1}{(a_1+1)(a_2+1)\ldots(a_{n-1}+1)} - \frac{1}{(a_1+1)(a_2+1)\ldots(a_n+1)}$$

for $n > 1$. Hence $S_n = 1 - \frac{1}{(a_1+1)(a_2+1)\ldots(a_n+1)}$.

3.1.10.

(a) If we take $a_n = n - 1$ in the last problem, we have $g = +\infty$, and so the sum of the given series is equal to 1.

(b) Here we take $a_n = 2n - 1$, and, as in (a), the sum of the given series is equal to 1.

(c) We take $a_n = -\frac{1}{n^2}$. Then

$$\lim_{n\to\infty} ((a_2+1)(a_3+1)\ldots(a_n+1))$$
$$= \lim_{n\to\infty} \frac{(2-1)(2+1)}{2^2} \frac{(3-1)(3+1)}{3^2} \cdots \frac{(n-1)(n+1)}{n^2} = \frac{1}{2}.$$

Thus, by the result in 3.1.9, the sum of the given series is equal to 1.

3.1.11. By definition, the sequence $\{a_n\}$ increases to infinity. Moreover, we have $a_n^2 - 4 = a_{n-1}^2(a_{n-1}^2 - 4)$, and it can be shown by induction that $a_n^2 - 4 = a_1^2 \cdot a_2^2 \cdot \ldots \cdot a_{n-1}^2(a_1^2 - 4)$. Hence

(1) $$\lim_{n\to\infty} \frac{a_n}{a_1 \cdot a_2 \cdot \ldots \cdot a_{n-1}} = \sqrt{a_1^2 - 4}.$$

Note also that for $n > 1$,

$$\frac{1}{a_1 \cdot a_2 \cdot \ldots \cdot a_n} = \frac{1}{2}\left(\frac{a_n}{a_1 \cdot a_2 \cdot \ldots \cdot a_{n-1}} - \frac{a_{n+1}}{a_1 \cdot a_2 \cdot \ldots \cdot a_n}\right).$$

Thus by (1) the sum of the given series is equal to

$$\frac{1}{a_1} + \frac{1}{2}\frac{a_2}{a_1} - \frac{1}{2}\sqrt{a_1^2 - 4} = \frac{a_1 - \sqrt{a_1^2 - 4}}{2}.$$

3.1. Summation of Series

3.1.12. Observe that

$$\frac{1}{b} = \frac{1!}{b-2} - \frac{2!}{(b-2)b},$$

$$\frac{2!}{b(b+1)} = \frac{2!}{(b-2)b} - \frac{3!}{(b-2)b(b+1)},$$

$$\vdots$$

$$\frac{n!}{b(b+1)...(b+n-1)} = \frac{n!}{(b-2)b(b+1)...(b+n-2)} - \frac{(n+1)!}{(b-2)b(b+1)...(b+n-1)}.$$

Summing the above equalities, we obtain

$$S_n = \frac{1}{b-2} - \frac{(n+1)!}{(b-2)b(b+1)...(b+n-1)}.$$

Hence, by 2.5.87, $\lim\limits_{n\to\infty} S_n = \frac{1}{b-2}$.

3.1.13. For $n = 0, 1, ...$, put $a_n = \frac{a(a+1)...(a+n)}{b(b+1)...(b+n)}$, and define $A_n = a_n(a+n+1)$. Then $A_{n-1} - A_n = a_n(b-a-1)$, $n = 0, 1, ...$, where we additionally set $A_{-1} = a$ and $a_{-1} = 1$. Summing both sides of the above equalities from $n = 0$ to $n = N$, we get

$$a - A_N = A_{-1} - A_N = (b-a-1)\sum_{n=0}^{N} a_n = (b-a-1)S_{N+1},$$

or equivalently, $a - a_N(a+N+1) = (b-a-1)S_{N+1}$. Therefore

$$a\left(1 - \frac{(a+1)...(a+N+1)}{b(b+1)...(b+N)}\right) = (b-a-1)S_{N+1}$$

and, by 2.5.87, we see that $\lim\limits_{N\to\infty} S_{N+1} = \frac{a}{b-a-1}$.

3.1.14. By the foregoing problem,

$$(*) \quad 1 + \sum_{n=1}^{\infty} \frac{a(a+1)...(a+n-1)}{b(b+1)...(b+n-1)} = 1 + \frac{a}{b-a-1} = \frac{b-1}{b-a-1}.$$

Replacing a by $a+1$, we see that

(**) $$1 + \sum_{n=1}^{\infty} \frac{(a+1)...(a+n)}{b(b+1)...(b+n-1)} = \frac{b-1}{b-a-2}.$$

Subtracting (*) from (**), we obtain

$$\sum_{n=1}^{\infty} n \frac{(a+1)...(a+n-1)}{b(b+1)...(b+n-1)}$$
$$= \frac{b-1}{b-a-2} - \frac{b-1}{b-a-1} = \frac{(b-1)}{(b-a-1)(b-a-2)}.$$

3.1.15. Set $A_n = \frac{a_1 \cdot a_2 \cdot ... \cdot a_n}{(a_2+b)(a_3+b)...(a_{n+1}+b)}$ and $S_n = \sum_{k=1}^{n} A_k$. Then $\frac{A_k}{A_{k-1}} = \frac{a_k}{a_{k+1}+b}$, or equivalently, $A_k a_{k+1} + A_k b = A_{k-1} a_k$. Summing both sides of these equalities from $k=2$ to $k=n$, we get

(*) $$A_n a_{n+1} + S_n b - A_1 b = A_1 a_2.$$

Now note that

$$0 < A_n a_{n+1} = a_1 \frac{a_2 \cdot a_3 \cdot ... \cdot a_{n+1}}{(a_2+b)(a_3+b)...(a_{n+1}+b)}$$
$$= a_1 \frac{1}{\left(1+\frac{b}{a_2}\right)\left(1+\frac{b}{a_3}\right)...\left(1+\frac{b}{a_{n+1}}\right)}.$$

Hence, by 1.2.1,

$$0 < A_n a_{n+1} < \frac{a_1}{b \sum_{k=2}^{n+1} \frac{1}{a_k}}.$$

Therefore, by assumption, $\lim_{n \to \infty} A_n a_{n+1} = 0$. It then follows from (*) that $\lim_{n \to \infty} S_n = \frac{A_1(b+a_2)}{b} = \frac{a_1}{b}$.

3.1.16. By the trigonometric identity $4\cos^3 x = \cos 3x + 3\cos x$, we get

3.1. Summation of Series

$$4\cos^3 x = \cos 3x + 3\cos x,$$
$$4\cos^3 3x = \cos 3^2 x + 3\cos 3x,$$
$$4\cos^3 3^2 x = \cos 3^3 x + 3\cos 3^2 x,$$
$$\vdots$$
$$4\cos^3 3^n x = \cos 3^{n+1} x + 3\cos 3^n x.$$

Multiplying both sides of these equalities by $1, -\frac{1}{3}, \frac{1}{3^2}, ..., (-1)^n \frac{1}{3^n}$, respectively, and then summing them we obtain $4S_n = 3\cos x + (-1)^n \frac{1}{3^n} \cos 3^{n+1} x$. Thus $S = \frac{3}{4}\cos x$.

3.1.17.

(a) By assumption,

$$f(x) = af(bx) + cg(x),$$
$$af(bx) = a^2 f(b^2 x) + acg(bx),$$
$$a^2 f(b^2 x) = a^3 f(b^3 x) + ca^2 g(b^2 x),$$
$$\vdots$$
$$a^{n-1} f(b^{n-1} x) = a^n f(b^n x) + a^{n-1} cg(b^{n-1} x).$$

Hence $f(x) = a^n f(b^n x) + c\left(g(x) + ag(bx) + ... + a^{n-1} g(b^{n-1} x)\right)$.

Since $\lim\limits_{n\to\infty} a^n f(b^n x) = L(x)$, $\sum\limits_{n=0}^{\infty} a^n g(b^n x) = \frac{f(x) - L(x)}{c}$.

(b) As in (a),

$$f(x) = af(bx) + cg(x),$$
$$a^{-1} f(b^{-1} x) = f(x) + a^{-1} cg(b^{-1} x),$$
$$a^{-2} f(b^{-2} x) = a^{-1} f(b^{-1} x) + ca^{-2} g(b^{-2} x),$$
$$\vdots$$
$$a^{-n} f(b^{-n} x) = a^{1-n} f(b^{1-n} x) + a^{-n} cg(b^{-n} x).$$

Thus

$$af(bx) = a^{-n} f(b^{-n} x) - c\left(g(x) + a^{-1} g(b^{-1} x) + ... + a^{-n} g(b^{-n} x)\right)$$

and consequently,
$$\sum_{n=0}^{\infty} \frac{1}{a^n} g\left(\frac{x}{b^n}\right) = \frac{M(x) - af(bx)}{c}.$$

3.1.18. We may apply the foregoing problem to the functions $f(x) = \sin x$, $g(x) = \sin^3 \frac{x}{3}$ with $a = 3$, $b = \frac{1}{3}$ and $c = -4$. The desired results follow from the equalities $\lim_{n\to\infty} 3^n \sin \frac{x}{3^n} = x = L(x)$ and $\lim_{n\to\infty} 3^{-n} \sin 3^n x = 0 = M(x)$.

3.1.19. One can apply 3.1.17 to $f(x) = \cot x$, $g(x) = \tan x$, $a = 2$, $b = 2$ and $c = 1$, and then use the equality
$$\lim_{n\to\infty} \frac{1}{2^n} \cot \frac{x}{2^n} = \frac{1}{x}.$$

3.1.20. We apply 3.1.17 to
$$f(x) = \arctan x, \quad g(x) = \arctan \frac{(1-b)x}{1+bx^2}, \quad a = c = 1,$$
and use the following:
$$\lim_{n\to\infty} \arctan(b^n x) = \begin{cases} 0 & \text{for } 0 < b < 1, \\ \frac{\pi}{2} \operatorname{sign} x & \text{for } b > 1. \end{cases}$$

3.1.21. Since $a_{n+1} = a_n + a_{n-1}$, we have $a_{n+1} a_n = a_n^2 + a_{n-1} a_n$ for $n \geq 1$. Summing these equalities, we get

(∗) $\qquad S_n = a_n a_{n+1}, \quad n \geq 0.$

One can prove, by induction, that

(i) $\quad a_n = \frac{1}{\sqrt{5}}\left(\left(\frac{1+\sqrt{5}}{2}\right)^{n+1} - \left(\frac{1-\sqrt{5}}{2}\right)^{n+1}\right), \quad n \geq 0,$

(ii) $\quad a_{n-1} a_{n+1} - a_n^2 = (-1)^{n+1}, \quad n \geq 1.$

3.1. Summation of Series

Combining (*) with (ii) yields

$$\widetilde{S_n} = \sum_{k=0}^{n} \frac{(-1)^k}{S_k} = \sum_{k=0}^{n} \frac{(-1)^k}{a_k a_{k+1}} = 1 - \sum_{k=1}^{n} \frac{a_{k-1} a_{k+1} - a_k^2}{a_k a_{k+1}}$$

$$= 1 - \sum_{k=1}^{n} \left(\frac{a_{k-1}}{a_k} - \frac{a_k}{a_{k+1}} \right) = \frac{a_n}{a_{n+1}}.$$

By (i),

(iii)
$$\lim_{n \to \infty} \frac{a_n}{a_{n+1}} = \lim_{n \to \infty} \frac{\frac{1}{\sqrt{5}}\left(\left(\frac{1+\sqrt{5}}{2}\right)^{n+1} - \left(\frac{1-\sqrt{5}}{2}\right)^{n+1}\right)}{\frac{1}{\sqrt{5}}\left(\left(\frac{1+\sqrt{5}}{2}\right)^{n+2} - \left(\frac{1-\sqrt{5}}{2}\right)^{n+2}\right)}$$

$$= \frac{2}{1+\sqrt{5}}.$$

Hence $\sum_{n=0}^{\infty} \frac{(-1)^n}{S_n} = \frac{2}{1+\sqrt{5}}.$

3.1.22. It is easy to check that

(*) $\qquad (-1)^{n+1} = a_{n+1} a_{n+2} - a_n a_{n+3}, \quad n \geq 0.$

Thus

$$S_n = \sum_{k=0}^{n} \frac{(-1)^k}{a_k a_{k+2}} = -\sum_{k=0}^{n} \frac{a_{k+1} a_{k+2} - a_k a_{k+3}}{a_k a_{k+2}}$$

$$= -\sum_{k=0}^{n} \left(\frac{a_{k+1}}{a_k} - \frac{a_{k+3}}{a_{k+2}} \right) = -3 + \frac{a_{n+2}}{a_{n+1}} + \frac{a_{n+3}}{a_{n+2}}.$$

Now, by (iii) in the solution of the foregoing problem, $\lim_{n \to \infty} S_n = \sqrt{5} - 2$.

3.1.23. By (*) in the solution of the preceding problem,

$$\arctan \frac{1}{a_{2n+1}} - \arctan \frac{1}{a_{2n+2}} = \arctan \frac{a_{2n+2} - a_{2n+1}}{a_{2n+1} a_{2n+2} + 1}$$

$$= \arctan \frac{a_{2n}}{a_{2n} a_{2n+3}} = \arctan \frac{1}{a_{2n+3}}.$$

Summing up these equalities gives

$$\arctan \frac{1}{a_1} = \sum_{k=1}^{n+1} \arctan \frac{1}{a_{2k}} + \arctan \frac{1}{a_{2n+3}}.$$

So, $\sum_{n=1}^{\infty} \operatorname{arctg} \frac{1}{a_{2n}} = \frac{\pi}{4}$.

3.1.24.

(a) Note that $\arctan \frac{2}{n^2} = \arctan \frac{1}{n-1} - \arctan \frac{1}{n+1}$, $n > 1$. Hence $\sum_{n=1}^{\infty} \arctan \frac{2}{n^2} = \arctan 2 + \arctan 1 + \arctan \frac{1}{2} = \frac{3}{4}\pi$, where the last equality follows from the fact that $\arctan a + \arctan \frac{1}{a} = \frac{\pi}{2}$ for $a > 0$.

(b) For $n \in \mathbb{N}$, $\arctan \frac{1}{n^2+n+1} = \arctan \frac{1}{n} - \arctan \frac{1}{n+1}$. Hence we see that $\sum_{n=1}^{\infty} \arctan \frac{1}{n^2+n+1} = \arctan 1 = \frac{1}{4}\pi$.

(c) Since for $n > 1$, $\arctan \frac{8n}{n^4-2n^2+5} = \arctan \frac{2}{(n-1)^2} - \arctan \frac{2}{(n+1)^2}$, we get, as in (a), $\sum_{n=1}^{\infty} \arctan \frac{8n}{n^4-2n^2+5} = \arctan 2 + \arctan 2 + \arctan \frac{1}{2} = \frac{1}{2}\pi + \arctan 2$.

3.1.25. To obtain the desired result one can apply the trigonometric identity

$$\arctan x - \arctan y = \arctan \frac{x-y}{1+xy}.$$

It is worth noting here that the results in the preceding problem are contained as special cases of this one.

3.1.26. Let $\sum_{n=1}^{\infty} b_n$ be a rearrangement of $\sum_{n=1}^{\infty} a_n$. Moreover, let $S_n = a_1 + a_2 + ... + a_n$, $S'_n = b_1 + b_2 + ... + b_n$ and $S = \lim_{n \to \infty} S_n$. Clearly, $S'_n \leq S$. So $\{S'_n\}$ converges, say to S', and $S' \leq S$. By similar arguments, $S \leq S'$.

3.1.27. Since

$$S_{2n} = \sum_{k=1}^{2n} \frac{1}{k^2} = \sum_{k=1}^{n} \frac{1}{(2k)^2} + \sum_{k=1}^{n} \frac{1}{(2k-1)^2}$$

3.1. Summation of Series

and $\lim_{n\to\infty} S_{2n+1} = \lim_{n\to\infty} S_{2n}$, we get

$$\sum_{n=1}^{\infty} \frac{1}{n^2} = \sum_{n=1}^{\infty} \frac{1}{(2n-1)^2} + \frac{1}{2^2}\sum_{n=1}^{\infty} \frac{1}{n^2}.$$

3.1.28. Following A. M. Yaglom and I. M. Yaglom, Uspehi Matem. Nauk (N.S.) 8 (1953) no. 5(57), pp. 181-187 (in Russian), we present elementary proofs of these well known identities.

(a) For $0 < x < \frac{\pi}{2}$, the inequalities $\sin x < x < \tan x$ hold. Thus $\cot^2 x < \frac{1}{x^2} < 1 + \cot^2 x$. Putting $x = \frac{k\pi}{2m+1}$ with $k = 1, 2, ..., m$, and summing from $k = 1$ to $k = m$ we obtain

(i) $\quad \displaystyle\sum_{k=1}^{m} \cot^2 \frac{k\pi}{2m+1} < \frac{(2m+1)^2}{\pi^2} \sum_{k=1}^{m} \frac{1}{k^2} < m + \sum_{k=1}^{m} \cot^2 \frac{k\pi}{2m+1}.$

We now show that

(ii) $\quad \displaystyle\sum_{k=1}^{m} \cot^2 \frac{k\pi}{2m+1} = \frac{m(2m-1)}{3}.$

Let $0 < t < \frac{\pi}{2}$. By DeMoivre's law

$$\cos nt + i \sin nt = (\cos t + i \sin t)^n = \sin^n t (\cot t + i)^n$$

$$= \sin^n t \sum_{k=0}^{n} \binom{n}{k} i^k \cot^{n-k} t.$$

Taking $n = 2m+1$, and equating imaginary parts, we get

(iii) $\quad \sin(2m+1)t = \sin^{2m+1} t \, P_m(\cot^2 t),$

where

(iv) $\quad P_m(x) = \binom{2m+1}{1} x^m - \binom{2m+1}{3} x^{m-1} + ... \pm 1.$

Substituting $t = \frac{k\pi}{2m+1}$ in (iii) gives $P_m(\cot^2 \frac{k\pi}{2m+1}) = 0$. So, $x_k = \cot^2 \frac{k\pi}{2m+1}$, $k = 1, 2, ..., m$, are the zeros of P_m and their sum is

(v) $\quad \displaystyle\sum_{k=1}^{m} \cot^2 \frac{k\pi}{2m+1} = \frac{\binom{2m+1}{3}}{\binom{2m+1}{1}} = \frac{m(2m-1)}{3}.$

This, (i) and (ii) imply

$$\frac{m(2m-1)}{3} < \frac{(2m+1)^2}{\pi^2} \sum_{k=1}^{m} \frac{1}{k^2} < m + \frac{m(2m-1)}{3}.$$

Multiplying these inequalities by $\frac{\pi^2}{(2m+1)^2}$ and next letting $m \to \infty$ yield equality (a).

(b) To prove the second identity, note that

$$\sum_{\substack{i,j=1 \\ i \neq j}}^{m} x_i x_j = 2 \frac{\binom{2m+1}{5}}{\binom{2m+1}{1}},$$

where x_k, $k = 1, 2, ..., m$ are the zeros of the polynomial (iv). It then follows by (v) that

$$\sum_{k=1}^{m} \cot^4 \frac{k\pi}{2m+1}$$
$$= \left(\frac{m(2m-1)}{3}\right)^2 - 2\frac{2m(2m-1)(2m-2)(2m-3)}{5!}$$
$$= \frac{m(2m-1)(4m^2+10m-9)}{45}.$$

The inequality $\cot^2 x < \frac{1}{x^2} < 1 + \cot^2 x$ (see (a)) implies that $\cot^4 x < \frac{1}{x^4} < 1 + 2\cot^2 x + \cot^4 x$ for $0 < x < \frac{\pi}{2}$. Consequently,

$$\frac{m(2m-1)(4m^2+10m-9)}{45} < \frac{(2m+1)^4}{\pi^4} \sum_{k=1}^{m} \frac{1}{k^4}$$
$$< m + 2m\frac{2m-1}{3} + \frac{m(2m-1)(4m^2+10m-9)}{45}.$$

Thus (b) is proved.

Remark. It is worth noting here that the above procedure can be applied to calculate the sum of the series $\sum_{n=1}^{\infty} \frac{1}{n^{2k}}$, where $k \in \mathbb{N}$.

3.1. Summation of Series

(c) It follows from DeMoivre's formula, that for $m = 4n$, $n \in \mathbb{N}$,

$$\cos mt = \cos^m t - \binom{m}{2}\cos^{m-2} t \sin^2 t + \ldots + \sin^m t,$$

$$\sin mt = \binom{m}{1}\cos^{m-1} t \sin t + \ldots - \binom{m}{m-1}\cos t \sin^{m-1} t,$$

and consequently,

$$\cot mt = \frac{\cot^m t - \binom{m}{2}\cot^{m-2} t + \ldots - \binom{m}{m-2}\cot^2 t + 1}{\binom{m}{1}\cot^{m-1} t - \binom{m}{3}\cot^{m-3} t + \ldots - \binom{m}{m-1}\cot t}.$$

It follows from the last equality that

$$x_k = \cot\frac{4k\pi + \pi}{4m}, \quad k = 0, \ldots, m-1,$$

are the roots of the equation

$$x^m - \binom{m}{1}x^{m-1} - \binom{m}{2}x^{m-2} + \ldots + \binom{m}{m-1}x + 1 = 0,$$

which implies that

(1) $$\sum_{k=0}^{m-1}\cot\frac{4k\pi + \pi}{4m} = m.$$

Since $m = 4n$,

$$\sum_{k=0}^{m-1}\cot\frac{4k\pi + \pi}{4m} = \sum_{k=0}^{2n-1}\cot\frac{4k\pi + \pi}{4m} + \sum_{k=2n}^{m-1}\cot\frac{4k\pi + \pi}{4m}$$

$$= \sum_{k=0}^{2n-1}\cot\frac{4k\pi + \pi}{4m} - \sum_{k=1}^{2n}\cot\frac{4k\pi - \pi}{4m}.$$

This and (1) give

(2) $$\cot\frac{\pi}{4m} - \cot\frac{3\pi}{4m} + \cot\frac{5\pi}{4m} - \cot\frac{7\pi}{4m}$$
$$+ \ldots + \cot\frac{(2m-3)\pi}{4m} - \cot\frac{(2m-1)\pi}{4m} = m.$$

Since

$$\cot\alpha - \cot\beta = \tan(\beta - \alpha)\left(1 + \frac{1}{\tan\alpha\tan\beta}\right),$$

we get from (2) that m is equal to

$$\tan\frac{\pi}{2m}\left(\frac{m}{2}+\frac{1}{\tan\frac{\pi}{4m}\tan\frac{3\pi}{4m}}+\ldots+\frac{1}{\tan\frac{(2m-3)\pi}{4m}\tan\frac{(2m-1)\pi}{4m}}\right).$$

Hence, by the inequality $\frac{1}{x}>\frac{1}{\tan x}$ for $0<x<\pi/2$,

(3) $\quad m<\tan\dfrac{\pi}{2m}\left(\dfrac{m}{2}+\dfrac{1}{\frac{\pi}{4m}\frac{3\pi}{4m}}+\ldots+\dfrac{1}{\frac{(2m-3)\pi}{4m}\frac{(2m-1)\pi}{4m}}\right).$

On the other hand, we have

$$\cot\alpha-\cot\beta=\frac{\sin(\beta-\alpha)}{\sin\alpha\sin\beta},$$

and as above, using the inequality $\frac{1}{x}<\frac{1}{\sin x}$, we obtain

$$m=\sin\frac{\pi}{2m}\left(\frac{1}{\sin\frac{\pi}{4m}\sin\frac{3\pi}{4m}}+\ldots+\frac{1}{\sin\frac{(2m-3)\pi}{4m}\sin\frac{(2m-1)\pi}{4m}}\right)$$

$$>\sin\frac{\pi}{2m}\left(\frac{1}{\frac{\pi}{4m}\frac{3\pi}{4m}}+\ldots+\frac{1}{\frac{(2m-3)\pi}{4m}\frac{(2m-1)\pi}{4m}}\right).$$

This and (3) yield

$$\left(\frac{m}{\tan\frac{\pi}{2m}}-\frac{m}{2}\right)\frac{\pi^2}{16m^2}<\frac{1}{2\cdot 3}+\ldots+\frac{1}{(2m-3)(2m-1)}$$

$$<\frac{\pi^2}{16m\sin\frac{\pi}{2m}}.$$

Letting $m\to\infty$ we get

$$\frac{1}{2\cdot 3}+\frac{1}{5\cdot 7}+\ldots=\frac{1}{2}\sum_{m=0}^{\infty}(-1)^m\frac{1}{2m+1}=\frac{\pi}{8}.$$

3.1.29. We have $a_{n+1}-1=a_n(a_n-1)$. Hence $\frac{1}{a_{n+1}-1}=-\frac{1}{a_n}+\frac{1}{a_n-1}$. Summing these equalities from $n=1$ to $n=N$, we get

(∗) $\qquad\dfrac{1}{a_1}+\dfrac{1}{a_2}+\ldots+\dfrac{1}{a_N}=1-\dfrac{1}{a_{N+1}-1}.$

It is easy to check that the sequence $\{a_n\}$ increases and diverges to infinity. Thus (∗) implies $\sum\limits_{n=1}^{\infty}\frac{1}{a_n}=1.$

3.1. Summation of Series

3.1.30. By the definition of the sequence,

$$e^{a_1} - 1 = a_1 e^{a_2},$$
$$e^{a_2} - 1 = a_2 e^{a_3},$$
$$\vdots$$

Hence

$$e^{a_1} - 1 = a_1 + a_1 a_2 e^{a_3}$$
$$= \ldots = a_1 + a_1 a_2 + \ldots + a_1 \cdot \ldots \cdot a_n + a_1 \cdot \ldots \cdot a_{n+1} e^{a_{n+2}}.$$

This implies that $\sum_{n=1}^{\infty} b_n = e^{a_1} - 1$, because

$$\lim_{n \to \infty} (a_1 \cdot \ldots \cdot a_{n+1} e^{a_{n+2}}) = 0.$$

Indeed, $\{a_n\}$ is bounded below by zero, is monotonically decreasing, and converges to zero.

3.1.31. We have $S_{n+1} = S_n + a_{n+1} = S_n + \frac{1}{S_n} - \sqrt{2}$. Consider the function $f(x) = x + \frac{1}{x} - \sqrt{2}$, $x > 0$. If the sequence $\{S_n\}$ were convergent to S, we would get $f(S) = S$. The only solution of this equation is $\frac{1}{\sqrt{2}}$. Moreover, the function $x \to f(f(x)) - x$ is monotonically decreasing on the interval $(\frac{1}{\sqrt{2}}, 1)$. So, if $x \in (\frac{1}{\sqrt{2}}, 1)$, then

$$f(f(x)) - x < f(f(\frac{1}{\sqrt{2}})) - \frac{1}{\sqrt{2}} = 0,$$

and since f is monotonically decreasing on the interval $(0, 1)$, we also have $f(f(x)) > \frac{1}{\sqrt{2}}$ for $x \in (\frac{1}{\sqrt{2}}, 1)$. Finally,

$$\frac{1}{\sqrt{2}} < f(f(x)) < x \quad \text{for} \quad x \in (\frac{1}{\sqrt{2}}, 1).$$

This means that the sequence $\{S_{2n-1}\}$ is monotonically decreasing and bounded below. Thus it converges, and its limit is $\frac{1}{\sqrt{2}}$. Furthermore, $\lim_{n \to \infty} S_{2n} = \lim_{n \to \infty} f(S_{2n-1}) = f\left(\frac{1}{\sqrt{2}}\right) = \frac{1}{\sqrt{2}}$. So, the sum of the given series is equal to $\frac{1}{\sqrt{2}}$.

3.1.32.

(a) Note that

$$S_{2n} = 1 - \frac{1}{2} + \frac{1}{3} - \ldots - \frac{1}{2n}$$

$$= 1 + \frac{1}{2} + \frac{1}{3} + \ldots + \frac{1}{2n} - \left(1 + \frac{1}{2} + \frac{1}{3} + \ldots + \frac{1}{n}\right)$$

$$= \frac{1}{n+1} + \frac{1}{n+2} + \ldots + \frac{1}{2n}.$$

So, by 2.5.8 (a), we have $\lim\limits_{n\to\infty} S_{2n} = \ln 2$. Clearly,

$$\lim_{n\to\infty} S_{2n+1} = \lim_{n\to\infty} \left(S_{2n} + \frac{1}{2n+1}\right) = \ln 2.$$

(b) We have $\frac{2n+1}{n(n+1)} = \frac{1}{n} + \frac{1}{n+1}$. Hence by (a),

$$\sum_{n=1}^{\infty} (-1)^{n-1} \frac{2n+1}{n(n+1)} = \sum_{n=1}^{\infty} (-1)^{n-1} \frac{1}{n} + \sum_{n=1}^{\infty} (-1)^{n-1} \frac{1}{n+1}$$

$$= \ln 2 - (\ln 2 - 1) = 1.$$

(c) Denote by S_n the nth partial sum of the given series. Then

$$S_{2n} = \frac{1}{x+2n+1} + \frac{1}{x+2n+2} + \ldots + \frac{1}{x+4n-1} + \frac{1}{x+4n}.$$

As in the proof of 2.5.8, one can show that $\lim\limits_{n\to\infty} S_{2n} = \ln 2$. Obviously, $\lim\limits_{n\to\infty} S_{2n+1} = \ln 2$.

3.1.33. We have

$$S_{2n} = \ln \frac{2}{1} - \ln \frac{3}{2} + \ln \frac{4}{3} - \ln \frac{5}{4} + \ldots + \ln \frac{2n}{2n-1} - \ln \frac{2n+1}{2n}$$

$$= \ln \frac{2 \cdot 4 \cdot \ldots \cdot 2n}{1 \cdot 3 \cdot \ldots \cdot (2n-1)} - \ln \frac{3 \cdot 5 \cdot \ldots \cdot (2n+1)}{2 \cdot 4 \cdot \ldots \cdot 2n}$$

$$= \ln \left(\frac{1}{2n+1} \left(\frac{(2n)!!}{(2n-1)!!}\right)^2\right).$$

By the Wallis formula, $\frac{\pi}{2} = \lim\limits_{n\to\infty} \frac{1}{2n+1} \left(\frac{(2n)!!}{(2n-1)!!}\right)^2$. Hence $\lim\limits_{n\to\infty} S_{2n} = \ln \frac{\pi}{2}$.

3.1. Summation of Series

3.1.34. We have

$$\sum_{n=1}^{\infty}(-1)^{n-1}\ln\left(1-\frac{1}{(n+1)^2}\right)$$
$$=\sum_{n=1}^{\infty}(-1)^{n-1}\left(\ln\left(1+\frac{1}{n+1}\right)+\ln\left(1-\frac{1}{n+1}\right)\right)$$
$$=-\sum_{n=2}^{\infty}(-1)^{n-1}\ln\left(1+\frac{1}{n}\right)-\sum_{n=1}^{\infty}(-1)^{n-1}\ln\left(1+\frac{1}{n}\right).$$

It follows from the foregoing problem that the sum of the series is $\ln 2 - 2\ln\frac{\pi}{2}$.

3.1.35. The nth partial sum of the given series can be written as

$$S_n = 1 + \frac{1}{2} + \ldots + \frac{1}{n} - (\ln 2 - \ln 1 + \ln 3 - \ln 2$$
$$+ \ldots + \ln(n+1) - \ln n) + 1 + \frac{1}{2} + \ldots + \frac{1}{n} - \ln(n+1).$$

Thus, by Problem 2.1.41, the sum of this series is equal to the Euler constant γ.

3.1.36. [20] Write $F(x) = \int\limits_1^x f(t)dt$. It follows from the Taylor theorem that there exist x_k, y_k such that $k < x_k < k+\frac{1}{2}$, $k+\frac{1}{2} < y_k < k+1$ and

$$F\left(k+\frac{1}{2}\right) - F(k) = \frac{1}{2}f(k) + \frac{1}{8}f'(x_k),$$
$$-F\left(k+\frac{1}{2}\right) + F(k+1) = \frac{1}{2}f(k+1) - \frac{1}{8}f'(y_k).$$

Summing the above equations from $k=1$ to $k=n-1$ yields

$$\frac{1}{2}f(1) + f(2) + f(3) + \ldots + f(n-1) + \frac{1}{2}f(n) - F(n)$$
$$= \frac{1}{8}\left(f'(y_1) - f'(x_1) + f'(y_2) - f'(x_2) + \ldots + f'(y_{n-1}) - f'(x_{n-1})\right).$$

The limit of the expression on the right-hand side of the last equality exists, because the series $-f'(x_1) + f'(y_1) - f'(x_2) + f'(y_2) - \ldots$ is convergent (the terms have alternating signs and their absolute values converge monotonically to 0).

If we take $f(x) = \frac{1}{x}$, then we can prove the existence of the limit
$$\lim_{n\to\infty}\left(1 + \frac{1}{2} + \frac{1}{3} + \ldots + \frac{1}{n} - \ln n\right).$$
(Compare with 2.1.41 (a).) Taking $f(x) = \ln x$, we can show that the sequence $\{\ln n! - \left(n + \frac{1}{2}\right)\ln n + n\}$ converges. (By Stirling's formula its limit is $\ln\sqrt{2\pi}$.)

3.1.37. Applying the foregoing problem to $f(x) = \frac{\ln x}{x}$, $x > 0$, we can show the existence of the limit
$$\lim_{n\to\infty}\left(\frac{\ln 1}{1} + \frac{\ln 2}{2} + \ldots + \frac{\ln n}{n} - \frac{(\ln n)^2}{2}\right) = s.$$

Hence
$$\lim_{n\to\infty} S_{2n} = \lim_{n\to\infty}\left(-\frac{\ln 1}{1} + \frac{\ln 2}{2} - \ldots + \frac{\ln 2n}{2n}\right)$$
$$= \lim_{n\to\infty}\left(-\left(\frac{\ln 1}{1} + \frac{\ln 2}{2} + \ldots + \frac{\ln 2n}{2n} - \frac{(\ln 2n)^2}{2}\right)\right.$$
$$\left. + 2\left(\frac{\ln 2}{2} + \frac{\ln 4}{4} + \ldots + \frac{\ln 2n}{2n}\right) - \frac{(\ln 2n)^2}{2}\right)$$
$$= -s + \lim_{n\to\infty}\left(\frac{\ln 1}{1} + \frac{\ln 2}{2} + \ldots + \frac{\ln n}{n} - \frac{(\ln n)^2}{2}\right)$$
$$+ \lim_{n\to\infty}\left(\frac{\ln 2}{1} + \frac{\ln 2}{2} + \ldots + \frac{\ln 2}{n} - \ln 2 \ln n\right) - \frac{(\ln 2)^2}{2}$$
$$= \ln 2\left(\gamma - \frac{\ln 2}{2}\right),$$
where γ is the Euler constant.

3.1.38. By Stirling's formula, $n! = \alpha_n\sqrt{2\pi n}\left(\frac{n}{e}\right)^n$, where $\lim_{n\to\infty}\alpha_n = 1$. Hence
$$S_n = \frac{1}{2}\ln\frac{(2n+1)^{2n}}{((2n-1)!!)^2 e^{2n}} = \frac{1}{2}\ln\frac{(2n+1)^{2n} 2^{2n}(n!)^2}{((2n)!)^2 e^{2n}}$$
$$= \frac{1}{2}\ln\frac{(2n+1)^{2n} 2^{2n}\alpha_n^2 2\pi n\left(\frac{n}{e}\right)^{2n}}{\alpha_{2n}^2 4\pi n\left(\frac{2n}{e}\right)^{4n} e^{2n}} = \frac{1}{2}\ln\left(\left(\frac{2n+1}{2n}\right)^{2n}\frac{\alpha_n^2}{2\alpha_{2n}^2}\right).$$

Therefore $\lim_{n\to\infty} S_n = \frac{1}{2}(1 - \ln 2)$.

3.1. Summation of Series

3.1.39. Assume that the given series converges for x and y, and write

$$S_N(x) = \sum_{n=1}^{N} \left(\frac{1}{(n-1)k+1} + \frac{1}{(n-1)k+2} + \ldots + \frac{1}{nk-1} - \frac{x}{nk} \right).$$

Then $S_N(x) - S_N(y) = \frac{y-x}{k} \sum_{n=1}^{N} \frac{1}{n}$. Therefore the convergence of the series implies that $x = y$. Now, we find the unique value x for which the series converges. We know, by Problem 2.1.41, that the sequence $a_n = 1 + \frac{1}{2} + \ldots + \frac{1}{nk} - \ln(nk)$ converges to the Euler constant γ. Hence

$$S_N(k-1) = a_N + \ln(Nk) - \sum_{n=1}^{N} \frac{1}{nk} - \sum_{n=1}^{N} \frac{k-1}{nk}$$

$$= a_N + \ln k + \left(\ln N - \sum_{n=1}^{N} \frac{1}{n} \right).$$

This implies that $\lim_{N\to\infty} S_N(k-1) = \gamma + \ln k - \gamma = \ln k$. Thus $x = k-1$, and the sum of the given series is equal to $\ln k$.

3.1.40. One can easily check by induction that

$$a_{2n} = 3n+2 \quad \text{for} \quad n = 0, 1, 2, \ldots,$$
$$a_{2n-1} = 3n+1 \quad \text{for} \quad n = 1, 2, \ldots.$$

Thus

$$(*) \quad \begin{aligned} S_{2N} &= \sum_{n=0}^{2N} (-1)^{\left[\frac{n+1}{2}\right]} \frac{1}{a_n^2 - 1} = \sum_{n=0}^{N} \frac{(-1)^n}{a_{2n}^2 - 1} + \sum_{n=1}^{N} \frac{(-1)^n}{a_{2n-1}^2 - 1} \\ &= \sum_{n=0}^{N} (-1)^n \frac{1}{(3n+1)(3n+3)} + \sum_{n=1}^{N} (-1)^n \frac{1}{3n(3n+2)} \\ &= \frac{1}{3} + \frac{1}{2} \sum_{n=1}^{N} (-1)^n \left(\frac{1}{3n+1} - \frac{1}{3n+3} \right) + \frac{1}{2} \sum_{n=1}^{N} (-1)^n \left(\frac{1}{3n} - \frac{1}{3n+2} \right) \\ &= \frac{1}{3} + \frac{1}{2} \sum_{n=1}^{N} (-1)^n \left(\frac{1}{3n} - \frac{1}{3n+3} \right) + \frac{1}{2} \sum_{n=1}^{N} (-1)^n \left(\frac{1}{3n+1} - \frac{1}{3n+2} \right) \end{aligned}$$

$$= \frac{1}{3} - \frac{1}{6} + \sum_{n=2}^{N}(-1)^n \frac{1}{3n} + \frac{(-1)^{N+1}}{6(N+1)} + \frac{1}{2}\sum_{n=1}^{N}(-1)^n \left(\frac{1}{3n+1} - \frac{1}{3n+2}\right).$$

On the other hand, by 3.1.32 (a), we get

$$-\ln 2 = \lim_{N\to\infty} \left(\sum_{n=1}^{3N}(-1)^n \frac{1}{n}\right)$$

$$= \lim_{N\to\infty} \left(\sum_{n=1}^{N}(-1)^n \frac{1}{3n}\right) - \lim_{N\to\infty} \left(\sum_{n=0}^{N-1}(-1)^n \left(\frac{1}{3n+1} - \frac{1}{3n+2}\right)\right)$$

$$= -\frac{1}{3}\ln 2 - \lim_{N\to\infty} \left(\sum_{n=0}^{N-1}(-1)^n \left(\frac{1}{3n+1} - \frac{1}{3n+2}\right)\right).$$

This implies $\lim_{N\to\infty} \sum_{n=0}^{N-1}(-1)^n \left(\frac{1}{3n+1} - \frac{1}{3n+2}\right) = \frac{2}{3}\ln 2$. Finally, by (*), $\lim_{N\to\infty} S_{2N} = \frac{1}{6} - \frac{1}{3}\ln 2 + \frac{1}{3} + \frac{1}{3}\ln 2 - \frac{1}{4} = \frac{1}{4}$. Moreover, since $\lim_{N\to\infty} S_{2N+1} = \lim_{N\to\infty} S_{2N} + \lim_{N\to\infty} \frac{(-1)^{N+1}}{(3N+4)^2-1} = \lim_{N\to\infty} S_{2N}$, the sum of the given series is $\frac{1}{4}$.

3.1.41.

(a) Suppose, contrary to our claim, that the sum S of the given series is a rational number $\frac{p}{q}$. Then $(q-1)!p = q!S = \sum_{n=1}^{q} \frac{q!}{n!} + \sum_{n=q+1}^{\infty} \frac{q!}{n!}$.

This implies that $\sum_{n=q+1}^{\infty} \frac{q!}{n!}$ is an integer. On the other hand,

$$0 < \sum_{n=q+1}^{\infty} \frac{q!}{n!} \leq \frac{1}{q+1} + \frac{1}{(q+1)(q+2)} + \frac{1}{(q+1)(q+2)^2} + \dots$$

$$= \frac{q+2}{(q+1)^2} \leq \frac{3}{4},$$

a contradiction. Thus S is irrational.

(b) The same method as in (a) can be applied.

3.1. Summation of Series

3.1.42. On the contrary, suppose that the sum S of the series is a rational number $\frac{p}{q}$. Then

$$(q-1)!p = q!S = \sum_{n=1}^{q} \frac{q!\epsilon_n}{n!} + \sum_{n=q+1}^{\infty} \frac{q!\varepsilon_n}{n!}.$$

This implies that $\sum_{n=q+1}^{\infty} \frac{q!\varepsilon_n}{n!}$ is an integer. On the other hand,

$$\left| \sum_{n=q+1}^{\infty} \frac{q!\varepsilon_n}{n!} \right| \leq \sum_{n=q+1}^{\infty} \frac{q!}{n!} < 1.$$

To obtain a contradiction, it is enough to show that $\sum_{n=q+1}^{\infty} \frac{q!\varepsilon_n}{n!}$ is not equal to zero. We have

$$\left| \sum_{n=q+1}^{\infty} \frac{\varepsilon_n q!}{n!} \right| \geq \left| \frac{1}{q+1} - \sum_{n=q+2}^{\infty} \frac{q!}{n!} \right| > \frac{1}{q+1} - \frac{1}{q(q+1)} \geq 0,$$

which proves that S is irrational.

3.1.43. Reasoning similar to the above can be applied.

3.1.44. Suppose, contrary to our claim, that $\sum_{i=1}^{\infty} \frac{1}{n_i} = \frac{p}{q}$, $p, q \in \mathbb{N}$. By assumption, there exists a positive integer k such that if $i \geq k$ then $\frac{n_i}{n_1 n_2 \cdot \ldots \cdot n_{i-1}} > 3q$. Thus

$$\sum_{i=1}^{k-1} \frac{n_1 n_2 \cdot \ldots \cdot n_{k-1} q}{n_i} + \sum_{i=k}^{\infty} \frac{n_1 n_2 \cdot \ldots \cdot n_{k-1} q}{n_i} = p n_1 n_2 \cdot \ldots \cdot n_{k-1}.$$

Moreover,

$$\sum_{i=k}^{\infty} \frac{n_1 n_2 \cdot \ldots \cdot n_{k-1} q}{n_i} < \frac{1}{3}\left(1 + \frac{1}{n_k} + \frac{1}{n_k n_{k+1}} + \ldots\right) < 1,$$

a contradiction.

3.1.45. If the sum were a rational number $\frac{p}{q}$, then for any positive integer k_1, we would have $\sum_{k=k_1}^{\infty} \frac{1}{n_k} = \frac{p}{q} - \sum_{k=1}^{k_1-1} \frac{1}{n_k}$. Thus the sum $\sum_{k=k_1}^{\infty} \frac{q \cdot n_1 \cdot n_2 \cdot \ldots \cdot n_{k_1-1}}{n_k}$ would be a positive integer, and we would get

(*) $$\sum_{k=k_1}^{\infty} \frac{1}{n_k} \geq \frac{1}{q \cdot n_1 \cdot n_2 \cdot \ldots \cdot n_{k_1-1}}.$$

Set $\lim\limits_{k\to\infty} \frac{n_k}{n_{k-1}} = l > 1$ and take $\varepsilon > 0$ so small that $\alpha = l - \varepsilon > 1$. Then there exists an index k_0 such that if $k > k_0$, then

(**) $$\frac{n_k}{n_{k-1}} \geq \alpha > 1.$$

Since $\overline{\lim}\limits_{k\to\infty} \frac{n_k}{n_1 \cdot \ldots \cdot n_{k-1}} = +\infty$, there is $k_1 > k_0$ such that $\frac{n_{k_1}}{n_1 \cdot \ldots \cdot n_{k_1-1}} > \frac{\alpha q}{\alpha - 1}$. Thus by (**)

$$\sum_{k=k_1}^{\infty} \frac{1}{n_k} \leq \sum_{j=0}^{\infty} \frac{1}{\alpha^j n_{k_1}} = \frac{\alpha}{(\alpha-1) n_{k_1}} < \frac{1}{q \cdot n_1 \cdot n_2 \cdot \ldots \cdot n_{k_1-1}},$$

which contradicts (*).

3.1.46. On the contrary, suppose that $\sum_{k=1}^{\infty} \frac{1}{n_k} = \frac{p}{q}$, where p and q are positive integers. Then

$$n_1 n_2 \ldots n_{k-1} \sum_{j=0}^{\infty} \frac{1}{n_{k+j}} \geq \frac{1}{q}$$

for all $k \geq 2$. (See the solutions of the preceding problems.) Set $a_k = \sqrt[2^k]{n_k}$. By assumption, $\lim\limits_{k\to\infty} a_k = +\infty$. We claim that there exists an r_1 such that $a_j \leq a_{r_1}$ for $j = 1, 2, \ldots, r_1 - 1$. Indeed, if $a_1 \leq a_2$, then $r_1 = 2$. If it is not the case, then we take the least integer $r_1 > 2$ such that $a_1 \leq a_{r_1}$. In fact, there exist an infinite sequence of integers r_k with the above property, that is, $a_j \leq a_{r_k}$ for $j = 1, 2, \ldots, r_k - 1$. To find r_2 we apply the above procedure to the sequence $\{a_k\}_{n \geq r_1}$ and so on. Denote by r the least positive integer such that $a_{r+j} > q + 1$ for $j = 0, 1, 2, \ldots$ and $a_j \leq a_r$ for

3.1. Summation of Series

$j = 1, 2, ..., r - 1$. Observe also that since $n_r \leq n_{r+j}$, $a_r \leq a_{r+j}^{2^j}$ for $j = 0, 1, 2, ...$ It follows from the above that

$$\frac{n_1 n_2 ... n_{r-1}}{n_{r+j}} \leq \frac{a_r^{2+2^2+...+2^{r-1}}}{n_{r+j}} \leq \frac{a_{r+j}^{2^j(2^r-2)}}{a_{r+j}^{2^{r+j}}} = a_{r+j}^{-2^{j+1}} < (q+1)^{-(j+1)}.$$

Hence

$$n_1 n_2 ... n_{r-1} \sum_{j=0}^{\infty} \frac{1}{n_{r+j}} < \sum_{j=0}^{\infty} (q+1)^{-(j+1)} = \frac{1}{q},$$

a contradiction.

3.1.47. Since the series $\sum_{n=1}^{\infty} \frac{p_n}{q_n}$ converges, $\lim_{n\to\infty} \frac{p_n}{q_n-1} = 0$. It then follows from the given inequality that $\frac{p_m}{q_m-1} \geq \sum_{n=m}^{\infty} \frac{p_n}{q_n}$. Suppose that the set **A** is finite. Then there is an index m such that

$$S = \sum_{n=1}^{\infty} \frac{p_n}{q_n} = \sum_{n=1}^{m-1} \frac{p_n}{q_n} + \frac{p_m}{q_m - 1}.$$

Therefore S is rational. Assume now that

$$S = \sum_{n=1}^{\infty} \frac{p_n}{q_n} = \frac{p}{q} \in \mathbb{Q}.$$

Then

$$r_n = \frac{p}{q} - \sum_{k=1}^{n} \frac{p_k}{q_k} = \sum_{k=n+1}^{\infty} \frac{p_k}{q_k} \leq \frac{p_{n+1}}{q_{n+1} - 1}.$$

Multiplying the inequality by $b_n = q \cdot q_1 \cdot ... \cdot q_n$, we get

$$b_{n+1} r_{n+1} = b_n r_n q_{n+1} - q \cdot q_1 \cdot ... \cdot q_{n+1} \frac{p_{n+1}}{q_{n+1}}$$
$$\leq b_n r_n q_{n+1} - b_n r_n (q_{n+1} - 1) = b_n r_n.$$

This means that the sequence $\{b_n r_n\}$ of positive integers is monotonically decreasing. Thus, beginning with some value of the index n, all its terms are equal, which implies that the set **A** is finite.

3.1.48. Clearly, we can write $n! = 2^{\alpha(n)}\beta(n)$, where $\beta(n)$ is odd. More precisely, a theorem of Legendre asserts that $\alpha(n) = n - \nu(n)$, where $\nu(n)$ denotes the sum of ones in the binary expansion of n. Moreover, we have $\sum_{k=1}^{\infty} \frac{2^{n_k}}{n_k!} = \sum_{n=1}^{\infty} \delta_n \frac{2^n}{n!}$, where $\delta_n = 1$ if $n = n_k$ and $\delta_n = 0$ otherwise. Suppose, contrary to our claim, that $\sum_{n=1}^{\infty} \delta_n \frac{2^n}{n!} = \frac{p}{q}$, $p, q \in \mathbb{N}$. Write $q = 2^s t$, where t is odd. Take $N = 2^r > \max\{t, 2^{s+2}\}$. It then follows that $\frac{\beta(N)}{t} \in \mathbb{N}$. Therefore $2^s \beta(N) \frac{p}{q} = \frac{\beta(N)}{t} p \in \mathbb{N}$. A simple calculation shows that $N! = 2^{N-1} \beta(N)$ (which also follows directly from the Legendre theorem). Multiplying the equality

$$\frac{p}{q} = \sum_{n=1}^{N} \delta_n \frac{2^n}{n!} + \sum_{n=N+1}^{\infty} \delta_n \frac{2^n}{n!}$$

by $2^s \beta(N)$, we get

$$(*) \qquad 2^s \beta(N) \frac{p}{q} = 2^s \beta(N) \sum_{n=1}^{N} \delta_n \frac{2^n}{n!} + 2^s \beta(N) \sum_{n=N+1}^{\infty} \delta_n \frac{2^n}{n!}.$$

Note that

$$2^s \beta(N) \sum_{n=1}^{N} \delta_n \frac{2^n}{n!} = 2^s \sum_{n=1}^{N} \delta_n \frac{\beta(N) 2^n}{2^{\alpha(n)} \beta(n)}.$$

Since $\beta(n)$ divides $\beta(N)$, we see that the first term on the right-hand side of $(*)$ is an integer. To get a contradiction, we will show that $0 < 2^s \beta(N) \sum_{n=N+1}^{\infty} \delta_n \frac{2^n}{n!} < 1$. Indeed,

$$2^s \beta(N) \sum_{n=N+1}^{\infty} \delta_n \frac{2^n}{n!} = 2^{s-N+1} N! \sum_{n=N+1}^{\infty} \delta_n \frac{2^n}{n!}$$

$$= 2^{s+2} \sum_{n=N+1}^{\infty} \delta_n 2^{n-N-1} \frac{N!}{n!} < \frac{2^{s+2}}{N+1} \sum_{n=N+1}^{\infty} \left(\frac{2}{N+2}\right)^{n-N-1}$$

$$= \frac{2^{s+2}}{N+1} \cdot \frac{N+2}{N} < \frac{2^{s+3}}{N+1} < 1.$$

3.2. Series of Nonnegative Terms

3.2.1.

(a) We have
$$a_n = \sqrt{n^2+1} - \sqrt[3]{n^3+1}$$
$$= \frac{1}{\sqrt{n^2+1}+\sqrt[3]{n^3+1}}$$
$$\times \frac{3n^4 - 2n^3 + 3n^2}{(n^2+1)^2 + (n^2+1)\sqrt[3]{(n^3+1)^2} + (n^3+1)\sqrt[3]{n^3+1}}.$$

Hence
$$\lim_{n\to\infty} \frac{a_n}{\frac{1}{n}} = \frac{1}{2}.$$

By the comparison test the series diverges.

(b) $\lim\limits_{n\to\infty} \sqrt[n]{a_n} = \lim\limits_{n\to\infty}\left(1 - \frac{n}{n^2+n+1}\right)^n = \frac{1}{e}$, and the root test gives the divergence of the series.

(c) One can verify by induction that
$$\frac{(2n-3)!!}{(2n-2)!!} > \frac{1}{2n-1} \quad \text{for} \quad n > 2.$$

So by the comparison test, the given series is divergent.

(d) The series is convergent because $\lim\limits_{n\to\infty} \sqrt[n]{a_n} = \frac{1}{e}$.

(e) $1 - \cos\frac{1}{n} = 2\sin^2\frac{1}{2n} < \frac{1}{2n^2}$, and therefore the series is convergent.

(f) $\lim\limits_{n\to\infty} \sqrt[n]{a_n} = 0$, which proves the convergence of the series.

(g) By Problem 2.5.4 (a),
$$\lim_{n\to\infty} \frac{\sqrt[n]{a}-1}{\frac{1}{n}} = \ln a,$$
and therefore the series is divergent.

3.2.2.

(a) The series converges because $\frac{1}{n}\ln\left(1+\frac{1}{n}\right) < \frac{1}{n^2}$.

(b) The convergence of the series follows from the inequality
$$\frac{1}{\sqrt{n}}\ln\frac{n+1}{n-1} < \frac{2}{\sqrt{n(n-1)}} \quad \text{for} \quad n > 1.$$

(c) Using the inequality $\ln n < n$, we get $\frac{1}{n^2-\ln n} < \frac{1}{n(n-1)}$. Hence the given series converges.

(d) We have
$$\frac{1}{(\ln n)^{\ln n}} = \frac{1}{n^{\ln \ln n}}.$$
Thus the series is convergent.

(e) Applying methods of differential calculus, one can prove that for sufficiently large x the inequality $(\ln \ln x)^2 < \ln x$ holds. Therefore
$$\frac{1}{(\ln n)^{\ln \ln n}} = \frac{1}{e^{(\ln \ln n)^2}} > \frac{1}{n}$$
for sufficiently large n. This proves the divergence of the series.

3.2.3. Put $c_n = \frac{a_n}{b_n}$. By our assumption,
$$c_{n+1} = \frac{a_{n+1}}{b_{n+1}} \leq \frac{a_n}{b_n} = c_n, \quad n \geq n_0.$$

Thus the sequence $\{c_n\}$ is monotonically decreasing for $n \geq n_0$. This implies that the sequence is bounded, i.e. there exists $C > 0$ such that $0 < c_n < C$, $n \in \mathbb{N}$. Hence $\sum_{n=1}^{\infty} a_n = \sum_{n=1}^{\infty} c_n b_n < C \sum_{n=1}^{\infty} b_n$, which completes the proof of our statement.

3.2.4.

(a) By Problem 2.1.38, we get
$$\frac{a_{n+1}}{a_n} = \frac{\left(1+\frac{1}{n}\right)^{n-2}}{e} < \left(\frac{n}{n+1}\right)^2 = \frac{\frac{1}{(n+1)^2}}{\frac{1}{n^2}}.$$
Now the convergence of the series follows from the convergence test given in the foregoing problem.

(b) Similarly, by Problem 2.1.38, we obtain
$$\frac{a_{n+1}}{a_n} = \frac{\left(1+\frac{1}{n}\right)^n}{e} > \frac{\left(1+\frac{1}{n}\right)^n}{\left(1+\frac{1}{n}\right)^{n+1}} = \frac{\frac{1}{n+1}}{\frac{1}{n}}.$$

3.2. Series of Nonnegative Terms

If the given series were convergent then, by Problem 3.2.3, the series $\sum_{n=1}^{\infty} \frac{1}{n}$ would also be convergent. Thus the given series diverges.

3.2.5.

(a) By Problem 2.5.4 (a), the series $\sum_{n=1}^{\infty} (\sqrt[n]{a} - 1)^{\alpha}$ and $\sum_{n=1}^{\infty} \frac{1}{n^{\alpha}}$ either both converge or both diverge. Therefore the given series is convergent for $\alpha > 1$ and is divergent for $\alpha \leq 1$.

(b) It follows from the solution of Problem 2.5.4 (b) that $\ln n < n(\sqrt[n]{n} - 1)$. Thus for $n > 3$ and $\alpha > 0$,

$$\frac{1}{n^{\alpha}} < \left(\frac{\ln n}{n}\right)^{\alpha} < (\sqrt[n]{n} - 1)^{\alpha}.$$

This implies that the given series is divergent for $0 < \alpha \leq 1$. Let us also observe that for $\alpha \leq 0$ the necessary condition $a_n \to 0$ for a series to be convergent is not satisfied. For $\alpha > 1$, by Problem 2.5.5, the given series converges if and only if the series $\sum_{n=1}^{\infty} \left(\frac{\ln n}{n}\right)^{\alpha}$ does. The convergence of the latter series follows from Problem 3.2.3, because for sufficiently large n we have

$$\frac{a_{n+1}}{a_n} = \left(\frac{n \ln(n+1)}{(n+1) \ln n}\right)^{\alpha} \leq 2^{\alpha} \left(\frac{n}{n+1}\right)^{\alpha}.$$

(c) By 2.5.5, the given series converges if and only if the series

$$\sum_{n=1}^{\infty} \left(\ln \frac{\left(1 + \frac{1}{n}\right)^{n+1}}{e}\right)^{\alpha}$$

converges. By the inequality $\frac{2x}{2+x} < \ln(1+x) < x$, which is valid if $x > 0$, see Problem 2.5.3, we get

$$\left(\ln\left(1 + \frac{1}{n}\right)^{n+1} - 1\right)^{\alpha} < \frac{1}{n^{\alpha}} \quad \text{for} \quad \alpha > 1$$

and

$$\left(\ln\left(1 + \frac{1}{n}\right)^{n+1} - 1\right)^{\alpha} > \frac{1}{(2n+1)^{\alpha}} \quad \text{for} \quad 0 < \alpha \leq 1.$$

Therefore the given series is convergent for $\alpha > 1$ and divergent for $0 < \alpha \leq 1$. Furthermore, let us observe that for $\alpha \leq 0$ the necessary condition for a series to be convergent is not satisfied.

(d) It is easily verifiable that
$$\lim_{n\to\infty} \frac{1 - n\sin\frac{1}{n}}{\frac{1}{n^2}} = \frac{1}{6}.$$

Hence the given series is convergent if and only if the series $\sum_{n=1}^{\infty} \frac{1}{n^{2\alpha}}$ is convergent. Therefore our series is convergent for $\alpha > \frac{1}{2}$ and divergent for $\alpha \leq \frac{1}{2}$.

3.2.6. By Problem 2.5.5, we get $\lim_{n\to\infty} \frac{a_n \ln a}{a^{a_n} - 1} = 1$, and so our series converges if and only if the series $\sum_{n=1}^{\infty} a_n$ does.

3.2.7.

(a) The convergence of the series $\sum_{n=1}^{\infty} -\ln\left(\cos\frac{1}{n}\right)$ follows from the fact that
$$\lim_{n\to\infty} \frac{-\ln\left(\cos\frac{1}{n}\right)}{\frac{1}{2n^2}} = 1.$$

(b) If $c \neq 0$, then
$$\lim_{n\to\infty} e^{\frac{a \ln n + b}{c \ln n + d}} = e^{\frac{a}{c}} \neq 0.$$
Thus the given series diverges.
If $c = 0$ and $\frac{a}{d} \geq 0$, then the necessary condition $a_n \to 0$ for convergence does not hold. If $c = 0$ and $\frac{a}{d} < 0$, then
$$e^{\frac{a \ln n + b}{d}} = e^{\frac{b}{d}} e^{\frac{a}{d} \ln n} = e^{\frac{b}{d}} n^{\frac{a}{d}}.$$
Therefore, in this case, our series converges if $\frac{a}{d} < -1$ and diverges if $\frac{a}{d} \geq -1$.

(c) We have
$$\frac{n^{2n}}{(n+a)^{n+b}(n+b)^{n+a}} = \frac{1}{(n+a)^b \left(1 + \frac{a}{n}\right)^n (n+b)^a \left(1 + \frac{b}{n}\right)^n}.$$
Thus the series in question converges if and only if $\sum_{n=1}^{\infty} \frac{1}{n^{a+b}}$ does.

3.2. Series of Nonnegative Terms 273

3.2.8. The convergence of the series $\sum_{n=1}^{\infty} \sqrt{a_n a_{n+1}}$ is an immediate consequence of the inequality $\sqrt{a_n a_{n+1}} \leq \frac{1}{2}(a_n + a_{n+1})$. Moreover, if the sequence $\{a_n\}$ is monotonically decreasing then $\sqrt{a_n a_{n+1}} \geq a_{n+1}$. Consequently, the convergence of $\sum_{n=1}^{\infty} a_n$ follows from the convergence of the series $\sum_{n=1}^{\infty} \sqrt{a_n a_{n+1}}$.

Now, let us consider the sequence $\{a_n\}$ defined by setting

$$a_n = \begin{cases} 1 & \text{if } n \text{ is odd,} \\ \dfrac{1}{n^4} & \text{if } n \text{ is even.} \end{cases}$$

Then

$$\sum_{k=1}^{n} \sqrt{a_k a_{k+1}} < \sum_{k=1}^{2n} \sqrt{a_k a_{k+1}} = \frac{1}{4} \sum_{k=1}^{n} \frac{1}{k^2}.$$

Therefore the series $\sum_{n=1}^{\infty} \sqrt{a_n a_{n+1}}$ converges, whereas the series $\sum_{n=1}^{\infty} a_n$ diverges.

3.2.9.

(a) Let us first notice that if the sequence $\{a_n\}$ is bounded above, say by $M > 0$, then

$$\frac{a_n}{1+a_n} \geq \frac{a_n}{1+M}.$$

Therefore the series $\sum_{n=1}^{\infty} \frac{a_n}{1+a_n}$ is divergent. On the other hand, if the sequence $\{a_n\}$ is unbounded above, then there exists a subsequence $\{a_{n_k}\}$ divergent to infinity. Thus

$$\lim_{k \to \infty} \frac{a_{n_k}}{1 + a_{n_k}} = 1,$$

and therefore the necessary condition for convergence fails to hold.

(b) The series $\sum_{n=1}^{\infty} \frac{a_n}{1+na_n}$ can converge or diverge. To show this let us consider the following example:

$$a_n = \begin{cases} 1 & \text{if } n = m^2, \ m = 1, 2, \ldots, \\ \dfrac{1}{n^2} & \text{otherwise.} \end{cases}$$

The series $\sum_{n=1}^{\infty} a_n$ diverges. We have

$$\sum_{k=1}^{n} \frac{a_k}{1+ka_k} < \sum_{k=1}^{n} \frac{1}{1+k^2} + \sum_{k=1}^{n} \frac{1}{k+k^2} \leq 2\sum_{k=1}^{n} \frac{1}{1+k^2}.$$

Hence, in this case, the series $\sum_{n=1}^{\infty} \frac{a_n}{1+na_n}$ converges. If we take $a_n = \frac{1}{n}$, we see that both series $\sum_{n=1}^{\infty} a_n$ and $\sum_{n=1}^{\infty} \frac{a_n}{1+na_n}$ can diverge.

(c) The convergence of the series in question follows from the inequality

$$\frac{a_n}{1+n^2 a_n} \leq \frac{a_n}{n^2 a_n} = \frac{1}{n^2}.$$

(d) If the sequence $\{a_n\}$ is bounded above, say by M, then

$$\frac{a_n}{1+a_n^2} \geq \frac{a_n}{1+M^2}.$$

Therefore, in this case, the series in question is divergent. But, if for example $a_n = n^2$, then the series $\sum_{n=1}^{\infty} \frac{a_n}{1+a_n^2}$ converges.

3.2.10. For any positive integers n and p,

$$\frac{a_{n+1}}{S_{n+1}} + \frac{a_{n+2}}{S_{n+2}} + \ldots + \frac{a_{n+p}}{S_{n+p}} \geq \frac{\sum_{k=n+1}^{n+p} a_k}{S_{n+p}} = \frac{S_{n+p} - S_n}{S_{n+p}}.$$

Since $\lim_{p \to \infty} \frac{S_{n+p} - S_n}{S_{n+p}} = 1$, the sequence of partial sums of the series $\sum_{n=1}^{\infty} \frac{a_n}{S_n}$ is not a Cauchy sequence. Thus this series diverges.

3.2. Series of Nonnegative Terms 275

On the other hand,
$$\frac{a_n}{S_n^2} \le \frac{a_n}{S_n S_{n-1}} = \frac{S_n - S_{n-1}}{S_n S_{n-1}} = \frac{1}{S_{n-1}} - \frac{1}{S_n},$$

and therefore
$$\sum_{k=n+1}^{n+p} \frac{a_k}{S_k^2} \le \sum_{k=n+1}^{n+p} \left(\frac{1}{S_{k-1}} - \frac{1}{S_k}\right) = \frac{1}{S_n} - \frac{1}{S_{n+p}} < \frac{1}{S_n}.$$

Hence the series $\sum_{n=1}^{\infty} \frac{a_n}{S_n^2}$ is convergent by the Cauchy criterion.

3.2.11. We have
$$\frac{a_n}{S_n S_{n-1}^{\beta}} = \frac{S_n - S_{n-1}}{S_n S_{n-1}^{\beta}}.$$

Let p be a positive integer such that $\frac{1}{p} < \beta$. Then, for sufficiently large n, the inequality
$$\frac{a_n}{S_n S_{n-1}^{\beta}} < \frac{a_n}{S_n S_{n-1}^{\frac{1}{p}}}$$

holds. Therefore it is enough to establish the convergence of the series with terms $\frac{a_n}{S_n S_{n-1}^{\frac{1}{p}}}$. To this end we will show that the inequality
$$\frac{S_n - S_{n-1}}{S_n S_{n-1}^{\frac{1}{p}}} \le p\left(\frac{1}{S_{n-1}^{\frac{1}{p}}} - \frac{1}{S_n^{\frac{1}{p}}}\right)$$

is satisfied. The last inequality is equivalent to
$$1 - \frac{S_{n-1}}{S_n} \le p\left(1 - \frac{S_{n-1}^{\frac{1}{p}}}{S_n^{\frac{1}{p}}}\right).$$

This follows immediately from the easily verifiable inequality $1-x^p \le p(1-x)$, which is valid if $0 < x \le 1$. Indeed, it is enough to set $x = \left(\frac{S_{n-1}}{S_n}\right)^{\frac{1}{p}}$. Therefore (see the solution of the foregoing problem) the convergence of the series in question is established for $\beta > 0$.

3.2.12. Let us first assume that $\alpha > 1$. Then for $n \geq 2$,
$$\frac{a_n}{S_n^\alpha} \leq \frac{a_n}{S_n S_{n-1}^{\alpha-1}}.$$
Thus the convergence of the given series follows from the last problem. Now, let $\alpha \leq 1$. Then for sufficiently large n, we have $\frac{a_n}{S_n^\alpha} \geq \frac{a_n}{S_n}$. This and Problem 3.2.10 imply the divergence of our series for $\alpha \leq 1$.

3.2.13.

(a) By assumption, the sequence $\{r_n\}$ is monotonically decreasing and tends to zero. Moreover,
$$\frac{a_n}{r_{n-1}} = \frac{r_{n-1} - r_n}{r_{n-1}}.$$
Hence for any positive integers n and p,
$$\frac{a_{n+1}}{r_n} + \ldots + \frac{a_{n+p}}{r_{n+p-1}} = \frac{r_n - r_{n+1}}{r_n} + \ldots + \frac{r_{n+p-1} - r_{n+p}}{r_{n+p-1}}$$
$$> \frac{r_n - r_{n+p}}{r_n} = 1 - \frac{r_{n+p}}{r_n}.$$

Given n, we have $\lim\limits_{p \to \infty} \left(1 - \frac{r_{n+p}}{r_n}\right) = 1$. Therefore the divergence of our series follows from the Cauchy criterion.

(b) We have
$$\frac{a_n}{\sqrt{r_{n-1}}} = \frac{r_{n-1} - r_n}{\sqrt{r_{n-1}}} = \frac{(\sqrt{r_{n-1}} - \sqrt{r_n})(\sqrt{r_{n-1}} + \sqrt{r_n})}{\sqrt{r_{n-1}}}$$
$$< 2(\sqrt{r_{n-1}} - \sqrt{r_n}).$$
Applying this inequality one can show that the sequence of partial sums of the series $\sum\limits_{n=1}^{\infty} \frac{a_n}{\sqrt{r_{n-1}}}$ is a Cauchy sequence. Thus the series converges.

3.2.14. We first assume that $\alpha \geq 1$. Then, for sufficiently large n,
$$\frac{1}{r_{n-1}^\alpha} \geq \frac{1}{r_{n-1}}.$$
The divergence of the series in question is now derived from part (a) of the foregoing problem.

3.2. Series of Nonnegative Terms 277

Let $\alpha < 1$. Then there exists a positive integer p such that $\alpha < 1 - \frac{1}{p}$. Hence

$$\frac{a_n}{r_{n-1}^\alpha} < \frac{a_n}{(r_{n-1})^{1-\frac{1}{p}}} = \frac{r_{n-1} - r_n}{r_{n-1}} r_{n-1}^{\frac{1}{p}}.$$

Applying the inequality $1 - x^p \leq p(1-x)$, which is valid if $0 < x \leq 1$, with $x = \left(\frac{r_n}{r_{n-1}}\right)^{\frac{1}{p}}$, we get

$$\frac{a_n}{r_{n-1}^\alpha} \leq p\left(r_{n-1}^{\frac{1}{p}} - r_n^{\frac{1}{p}}\right).$$

The convergence of the given series follows from the Cauchy criterion.

3.2.15. For $0 < \alpha < 1$,

$$\lim_{n \to \infty} \frac{a_{n+1} \ln^2 r_n}{\frac{a_{n+1}}{r_n^\alpha}} = 0.$$

Therefore the convergence of the series $\sum\limits_{n=1}^{\infty} a_{n+1} \ln^2 r_n$ follows from the foregoing problem.

3.2.16. We know (see, e.g., Problem 2.1.38) that

$$(*) \qquad \left(1 + \frac{1}{n}\right)^n < e < \left(1 + \frac{1}{n}\right)^{n+1}.$$

Assume that $g > 1$ and let $\varepsilon > 0$ be so small that $g - \varepsilon > 1$. Then there exists n_0 such that $n \ln \frac{a_n}{a_{n+1}} > g - \varepsilon$ for $n \geq n_0$. Thus by (*),

$$n \ln \frac{a_n}{a_{n+1}} > (g - \varepsilon) > n \ln\left(1 + \frac{1}{n}\right),$$

and consequently,

$$\frac{a_{n+1}}{a_n} < \frac{\frac{1}{(n+1)^{g-\varepsilon}}}{\frac{1}{n^{g-\varepsilon}}}.$$

Therefore the series $\sum\limits_{n=1}^{\infty} a_n$ is convergent by the test proved in Problem 3.2.3. Similar arguments apply to the case $g = +\infty$. Analogous reasoning can be used to prove the divergence of the series if $g < 1$.

The following examples show that the test is inconclusive if $g = 1$. Taking $a_n = \frac{1}{n}$, we see that $g = 1$ and $\sum\limits_{n=1}^{\infty} \frac{1}{n}$ diverges.

On the other hand, letting $a_n = \frac{1}{n \ln^2 n}$, we get the convergent series $\sum_{n=1}^{\infty} a_n$ (see Problem 3.2.29). To show that in this case $g = 1$, let us first observe that

$$n \ln \frac{\frac{1}{n \ln^2 n}}{\frac{1}{(n+1) \ln^2(n+1)}} = \ln\left(1 + \frac{1}{n}\right)^n + 2n \ln \frac{\ln(n+1)}{\ln n}.$$

As the first term of the sum tends to 1, it is enough to show that $\lim_{n \to \infty} 2n \ln \frac{\ln(n+1)}{\ln n} = 0$. To this end, note that

$$\lim_{n \to \infty} \left(\frac{\ln(n+1)}{\ln n}\right)^n = \lim_{n \to \infty} \left(1 + \frac{\ln \frac{n+1}{n}}{\ln n}\right)^n = e^0 = 1.$$

3.2.17.
(a) We have

$$\lim_{n \to \infty} n \ln \frac{a_n}{a_{n+1}} = \lim_{n \to \infty} n(\sqrt{n+1} - \sqrt{n}) \ln 2 = +\infty.$$

The convergence of the series follows from the preceding problem.

(b) Similarly,

$$\lim_{n \to \infty} n \ln \frac{a_n}{a_{n+1}} = \lim_{n \to \infty} \ln\left(1 + \frac{1}{n}\right)^n \cdot \ln 2 = \ln 2 < 1.$$

Hence the series diverges.

(c) Likewise,

$$\lim_{n \to \infty} n \ln \frac{a_n}{a_{n+1}} = \ln 3 > 1,$$

which proves the convergence of the series.

(d)
$$\lim_{n \to \infty} n \ln \frac{a_n}{a_{n+1}} = \ln a.$$

Therefore the series is convergent for $a > e$ and divergent for $a < e$. For $a = e$ the series in question is the harmonic divergent series.

(e) We have (see the solution of 3.2.16)

$$\lim_{n \to \infty} n \ln \frac{a_n}{a_{n+1}} = \lim_{n \to \infty} n \ln \frac{\ln(n+1)}{\ln n} \cdot \ln a = 0.$$

3.2. Series of Nonnegative Terms

Therefore, by the convergence test given in the foregoing problem, the series is divergent for any $a > 0$.

3.2.18. Since

$$\lim_{n\to\infty} n \ln \frac{a_n}{a_{n+1}} = \lim_{n\to\infty} n \ln a^{-\frac{1}{n+1}} = \ln \frac{1}{a},$$

the series is convergent for $0 < a < \frac{1}{e}$ and divergent for $a > \frac{1}{e}$ (compare with 3.2.16). If $a = \frac{1}{e}$ then (see, e.g., Problem 2.1.41)

$$\lim_{n\to\infty} \frac{e^{\frac{1}{1+\frac{1}{2}+\ldots+\frac{1}{n}}}}{\frac{1}{n}} = e^{-\gamma},$$

where γ is the Euler constant. The comparison test and the divergence of the harmonic series $\sum\limits_{n=1}^{\infty} \frac{1}{n}$ imply the divergence of the series in question for $a = \frac{1}{e}$.

3.2.19. Applying the inequality $\frac{x}{1+x} \leq \ln(1+x) \leq x$, which is valid if $x > -1$, we get

$$\frac{n\left(\frac{a_n}{a_{n+1}} - 1\right)}{1 + \left(\frac{a_n}{a_{n+1}} - 1\right)} \leq n \ln\left(1 + \frac{a_n}{a_{n+1}} - 1\right) \leq n\left(\frac{a_n}{a_{n+1}} - 1\right).$$

By this inequality, the Raabe test and the test given in Problem 3.2.16 are equivalent for a finite r. It also follows from the above inequality that if $\lim\limits_{n\to\infty} n \ln \frac{a_n}{a_{n+1}} = +\infty$, then $\lim\limits_{n\to\infty} n \left(\frac{a_n}{a_{n+1}} - 1\right) = +\infty$. We will now show that the other implication is also true. Indeed, if $\lim\limits_{n\to\infty} n \left(\frac{a_n}{a_{n+1}} - 1\right) = +\infty$, then for any $A > 0$ there is n_0 such that $\frac{a_n}{a_{n+1}} - 1 > \frac{A}{n}$ for $n > n_0$. Hence

$$n \ln \frac{a_n}{a_{n+1}} = \ln \left(1 + \frac{a_n}{a_{n+1}} - 1\right)^n > \ln\left(1 + \frac{A}{n}\right)^n \xrightarrow[n\to\infty]{} A.$$

Since A can be arbitrarily large, we see that $\lim\limits_{n\to\infty} n \ln \frac{a_n}{a_{n+1}} = +\infty$. Similar arguments apply to the case $r = -\infty$.

3.2.20. Since the sequence $\{a_n\}$ is monotonically increasing,
$$0 < n\left(\frac{\frac{1}{a_n}}{\frac{1}{a_{n+1}}} - 1\right) = \frac{1}{n}\frac{a_{n-1}}{a_n} < \frac{1}{n}.$$
By the Raabe test, the series diverges.

3.2.21. By the definition of the sequence,
$$a_n = a_1 e^{-\sum_{k=1}^{n-1} a_k^\alpha} \quad \text{for} \quad n = 1, 2, \ldots.$$
We will first show that $\sum_{n=1}^{\infty} a_n^\alpha$ is divergent. Indeed, if $\sum_{n=1}^{\infty} a_n^\alpha = S < +\infty$, then $\lim_{n\to\infty} a_n = a_1 e^{-S} > 0$, and so $\lim_{n\to\infty} a_n^\alpha > 0$; this would contradict the necessary condition for the convergence of the series. Therefore the series $\sum_{n=1}^{\infty} a_n^\alpha$ diverges and, by the foregoing, $\lim_{n\to\infty} a_n = 0$.

Now assume that $\beta > \alpha$. We will prove that in this case the series in question is convergent. To this end, we will show that
$$(*) \qquad a_n^{-\alpha} > \alpha(n-1) \quad \text{for} \quad n \geq 1.$$
This inequality is obvious for $n = 1$. Assume that it holds for an arbitrarily fixed n. Then, by the definition of the sequence, we obtain
$$a_{n+1}^{-\alpha} = a_n^{-\alpha} e^{\alpha a_n^\alpha} > a_n^{-\alpha}(1 + \alpha a_n^\alpha) = a_n^{-\alpha} + \alpha > \alpha n.$$
Thus $(*)$ holds for any positive integer n. This inequality is equivalent (for $n \neq 1$) to
$$a_n^\beta < (\alpha(n-1))^{-\frac{\beta}{\alpha}}.$$
Hence, by the comparison test, the given series converges for $\beta > \alpha$. Let $\beta \leq \alpha$. It has already been shown that $\lim_{n\to\infty} a_n = 0$. Therefore, for sufficiently large n, $0 < a_n < 1$. Thus $a_n^\alpha \leq a_n^\beta$. This inequality and the convergence of the series $\sum_{n=1}^{\infty} a_n^\alpha$ imply the convergence of $\sum_{n=1}^{\infty} a_n^\beta$.

3.2. Series of Nonnegative Terms

3.2.22. Note that
$$\lim_{n\to\infty} n\left(\frac{a_n}{a_{n+1}} - 1\right) = \lim_{n\to\infty} \frac{n}{n+1} a = a.$$
By the Raabe test, the series in question is convergent for $a > 1$ and divergent for $0 < a < 1$.
For $a = 1$ the given series reduces to the divergent harmonic series $\sum_{n=1}^{\infty} \frac{1}{n+1}$.

3.2.23. By
$$\lim_{n\to\infty} n\left(\frac{a_n}{a_{n+1}} - 1\right) = \lim_{n\to\infty} \frac{nb_{n+1}}{(n+1)a} = \frac{b}{a}$$
and the Raabe test, the series is convergent for $b > a$ and divergent for $b < a$. In the case where $b = a$, the convergence of the series depends on the sequence $\{b_n\}$. Indeed, if $\{b_n\}$ is a constant sequence, then the given series is the harmonic series $\sum_{n=1}^{\infty} \frac{1}{n+1}$.

We will now show that if $b_n = a + \frac{2a}{\ln(n+1)}$, then the series is convergent. In fact,
$$a_n = \frac{n!}{\left(2 + \frac{2}{\ln 2}\right)\left(3 + \frac{2}{\ln 3}\right)\cdot\ldots\cdot\left(n+1 + \frac{2}{\ln(n+1)}\right)}.$$
Thus
$$a_n(n-1)\ln(n-1) - a_{n+1} n \ln n$$
$$= a_n\left((n-1)\ln(n-1) - \frac{(n+1)n\ln n}{n+2+\frac{2}{\ln(n+2)}}\right).$$
Calculation shows that
$$\lim_{n\to\infty}\left((n-1)\ln(n-1) - \frac{(n+1)n\ln n}{n+2+\frac{2}{\ln(n+2)}}\right) = 1.$$
Therefore, for sufficiently large n,
$$a_n(n-1)\ln(n-1) - a_{n+1} n \ln n \geq (1-\varepsilon)a_n > 0.$$
Hence the positive sequence $a_n(n-1)\ln(n-1)$ is monotonically decreasing and so is convergent. This, in turn, implies the convergence

of the series with the terms $a_n(n-1)\ln(n-1) - a_{n+1}n\ln n$. By the last inequality and by the comparison test, the series $\sum_{n=1}^{\infty} a_n$ converges.

3.2.24. By assumption,
$$a_n((n-1)\ln n - 1) - a_{n+1}n\ln n = (\gamma_n - 1)a_n.$$
Now if $\gamma_n \geq \Gamma > 1$, then

(∗) $\qquad a_n((n-1)\ln n - 1) - a_{n+1}n\ln n \geq (\Gamma - 1)a_n.$

Combining (∗) with the inequality $(n-1)\ln(n-1) > (n-1)\ln n - 1$, we get

(∗∗) $\qquad a_n(n-1)\ln(n-1) - a_{n+1}n\ln n \geq (\Gamma - 1)a_n > 0.$

This means that the sequence $\{a_n(n-1)\ln(n-1)\}$ is monotonically decreasing and so convergent. Therefore the series with the terms $a_n(n-1)\ln(n-1) - a_{n+1}n\ln n$ is convergent. By (∗∗), the series $\sum_{n=1}^{\infty} a_n$ is also convergent.

If $\gamma_n \leq \Gamma < 1$, then $a_n((n-1)\ln n - 1) - a_{n+1}n\ln n \leq (\Gamma - 1)a_n$. Hence
$$a_n(n-1)\ln(n-1) - a_{n+1}n\ln n \leq \left(\Gamma + \ln\left(1 - \frac{1}{n}\right)^{n-1}\right)a_n.$$
Since
$$\lim_{n\to\infty}\left(\Gamma + \ln\left(1 - \frac{1}{n}\right)^{n-1}\right) = \Gamma - 1 < 0,$$
the sequence $\{a_{n+1}n\ln n\}$ is monotonically increasing (except for finitely many n). Therefore there is $M > 0$ such that $a_{n+1}n\ln n > M$. Thus $a_{n+1} > \frac{M}{n\ln n}$, which proves the divergence of the series $\sum_{n=1}^{\infty} a_n$.

3.2.25. We have
$$\lim_{n\to\infty} n\left(\frac{a_n}{a_{n+1}} - 1\right) = \lim_{n\to\infty} \frac{\alpha + \frac{\vartheta_n}{n^{\lambda-1}}}{1 - \frac{\alpha}{n} - \frac{\vartheta_n}{n^\lambda}} = \alpha.$$
Therefore the Raabe test implies the convergence of the series for $\alpha > 1$ and its divergence for $\alpha < 1$. In case $\alpha = 1$, the divergence

3.2. Series of Nonnegative Terms

of the series follows from the test given in the foregoing problem, because
$$\frac{a_{n+1}}{a_n} = 1 - \frac{1}{n} - \frac{\vartheta_n}{n^\lambda} = 1 - \frac{1}{n} - \frac{\gamma_n}{n\ln n},$$
where $\gamma_n = \frac{\vartheta_n \ln n}{n^{\lambda-1}} \leq \Gamma < 1$ for some Γ.

3.2.26. We will apply the criterion of Gauss from the preceding problem. We have
$$\frac{a_{n+1}}{a_n} = \frac{n^2 + (\alpha+\beta)n + \alpha\beta}{n^2 + (1+\gamma)n + \gamma} = 1 - \frac{1+\gamma-\alpha-\beta}{n} - \frac{\vartheta_n}{n^2}.$$
Therefore the series in question converges if $\alpha + \beta < \gamma$ and diverges if $\alpha + \beta \geq \gamma$.

3.2.27. As in the foregoing proof, we will use the Gauss criterion. We have
$$\frac{a_{n+1}}{a_n} = 1 - \frac{\frac{p}{2}}{n} - \frac{\vartheta_n}{n^2}.$$
Hence the series converges if $p > 2$ and diverges if $p \leq 2$.

3.2.28. Let $S_n, \widetilde{S_n}$ denote the nth partial sums of the series $\sum\limits_{n=1}^{\infty} a_n$ and $\sum\limits_{n=1}^{\infty} 2^n a_{2^n}$, respectively. Then for $n \leq 2^k$,
$$S_n \leq a_1 + (a_2 + a_3) + \ldots + (a_{2^k} + \ldots + a_{2^{k+1}-1}) \leq a_1 + 2a_2 + \ldots + 2^k a_{2^k} = \widetilde{S_k}.$$
For $n > 2^k$,
$$S_n \geq a_1 + a_2 + (a_3 + a_4) + \ldots + (a_{2^{k-1}+1} + \ldots + a_{2^k})$$
$$\geq a_2 + 2a_4 + \ldots + 2^{k-1} a_{2^k} = \frac{1}{2}\widetilde{S_k}.$$
Thus the sequences $\{S_n\}$ and $\{\widetilde{S_n}\}$ are either both bounded or both unbounded.

3.2.29.

(a) We will apply Cauchy's condensation test (3.2.28). Since the condensed series is
$$\sum_{n=1}^{\infty} \frac{2^n}{2^n (\ln 2^n)^\alpha} = \sum_{n=1}^{\infty} \frac{1}{(n \ln 2)^\alpha},$$

the given series converges for $\alpha > 1$ and diverges for $0 < \alpha \leq 1$. If $\alpha \leq 0$, the divergence of $\sum\limits_{n=2}^{\infty} \frac{1}{n(\ln n)^\alpha}$ follows immediately from the comparison test.

(b) By the equality

$$\sum_{n=2}^{\infty} \frac{2^n}{2^n \ln 2^n \ln \ln 2^n} = \sum_{n=2}^{\infty} \frac{1}{n \ln 2 \ln(n \ln 2)}$$

and by (a), the given series is convergent.

3.2.30. Reasoning similar to that in the proof of the Cauchy condensation test (3.2.28) can be applied. For $n \leq g_k$,

$$S_n \leq S_{g_k} \leq (a_1 + \ldots a_{g_1-1}) + (a_{g_1} + \ldots + a_{g_2-1}) + \ldots + (a_{g_k} + \ldots + a_{g_{k+1}-1})$$

$$\leq (a_1 + \ldots a_{g_1-1}) + (g_2 - g_1)a_{g_1} + \ldots + (g_{k+1} - g_k)a_{g_k}.$$

For $n > g_k$,

$$cS_n \geq cS_{g_k} \geq c(a_{g_1+1} + \ldots + a_{g_2}) + \ldots + c(a_{g_{k-1}+1} + \ldots + a_{g_k})$$

$$\geq c(g_2-g_1)a_{g_2} + \ldots + c(g_k - g_{k-1})a_{g_k} \geq (g_3-g_2)a_{g_2} + \ldots + (g_{k+1}-g_k)a_{g_k}.$$

These inequalities prove our assertion.

3.2.31.

(a) It is enough to apply the Schlömilch theorem (3.2.30) with $g_n = 3^n$.

(b) Applying the Schlömilch theorem with $g_n = n^2$, we get the equiconvergence of the series $\sum\limits_{n=1}^{\infty} a_n$ and $\sum\limits_{n=1}^{\infty} (2n+1)a_{n^2}$. Since

$$\lim_{n \to \infty} \frac{(2n+1)a_{n^2}}{na_{n^2}} = 2,$$

the series $\sum\limits_{n=1}^{\infty}(2n+1)a_{n^2}$ and $\sum\limits_{n=1}^{\infty} na_{n^2}$ are also equiconvergent.

(c) Compare with (b).

(d) By (b), the series $\sum\limits_{n=1}^{\infty} \frac{1}{2^{\sqrt{n}}}$ and $\sum\limits_{n=1}^{\infty} \frac{n}{2^n}$ are equiconvergent. The latter is convergent, e.g. by the root test. To determine the

3.2. Series of Nonnegative Terms

divergence or convergence of the series $\sum_{n=1}^{\infty} \frac{1}{2^{\ln n}}$, $\sum_{n=1}^{\infty} \frac{1}{3^{\ln n}}$ and $\sum_{n=1}^{\infty} \frac{1}{a^{\ln n}}$, the Cauchy condensation theorem or the test given in (a) can be applied. We will now study the behavior of the series with terms $\frac{1}{a^{\ln \ln n}}$. If $a > 1$, then the convergence of this series is equivalent to the convergence of $\sum_{n=1}^{\infty} \frac{3^n}{a^{\ln n}}$. It is easy to check that the last series diverges, e.g. by the root test. This establishes the divergence of the series $\sum_{n=2}^{\infty} \frac{1}{a^{\ln \ln n}}$ for $a > 1$. Observe that if $0 < a \leq 1$, then the necessary condition for convergence is not satisfied.

3.2.32. By Problem 2.4.13 (a), there exists an $\varepsilon > 0$ such that
$$(a_n)^{\frac{1}{\ln n}} < e^{-1-\varepsilon}, \quad n > k.$$
Hence $\frac{1}{\ln n} \ln a_n < -1 - \varepsilon$, and so $a_n < \frac{1}{n^{1+\varepsilon}}$. The comparison test yields the convergence of $\sum_{n=1}^{\infty} a_n$.

3.2.33. Analysis similar to that in the solution of the preceding problem gives
$$a_n \leq \frac{1}{n(\ln n)^{1+\varepsilon}} \quad \text{for} \quad n > k \quad \text{and for an} \quad \varepsilon > 0.$$
Therefore, by Problem 3.2.29(a), the series $\sum_{n=1}^{\infty} a_n$ is convergent.

3.2.34.
$$S_{2^{n_0+k}-1} - S_{2^{n_0}-1} = (a_{2^{n_0}} + a_{2^{n_0}+1} + \ldots + a_{2^{n_0+1}-1})$$
$$+ (a_{2^{n_0+1}} + \ldots + a_{2^{n_0+2}-1}) + \ldots + (a_{2^{n_0+k-1}} + \ldots + a_{2^{n_0+k}-1})$$
$$\leq 2^{n_0} a_{2^{n_0}} + 2^{n_0+1} a_{2^{n_0+1}} + \ldots + 2^{n_0+k-1} a_{2^{n_0+k-1}}$$
$$\leq g(a_{n_0} + a_{n_0+1} + \ldots + a_{n_0+k-1}).$$
Hence, for sufficiently large k,
$$(1-g) \sum_{n=2^{n_0}}^{2^{n_0+k}-1} a_n \leq g \left(\sum_{n=n_0}^{2^{n_0}-1} a_n - \sum_{n=n_0+k}^{2^{n_0+k}-1} a_n \right) \leq g \sum_{n=n_0}^{2^{n_0}-1} a_n.$$

Thus the sequence of partial sums of $\sum_{n=1}^{\infty} a_n$ is bounded, and so the series converges.

3.2.35. Assume that the series $\sum_{n=1}^{\infty} a_n$ converges. Then

$$\lim_{n\to\infty} \sum_{k=n+1}^{2n} a_k = \lim_{n\to\infty} \sum_{k=n+1}^{2n+1} a_k = 0.$$

Hence, by the monotonicity of $\{a_n\}$,

$$\sum_{k=n+1}^{2n} a_k \geq \sum_{k=n+1}^{2n} a_{2n} = na_{2n} = \frac{1}{2}(2na_{2n})$$

and

$$\sum_{k=n+1}^{2n+1} a_k \geq \sum_{k=n+1}^{2n+1} a_{2n+1} = \frac{n+1}{2n+1}(2n+1)a_{2n+1}.$$

It follows from the above that $\lim_{n\to\infty} na_n = 0$.

Let $a_n = \frac{1}{n\ln(n+1)}$. Then the series with the terms a_n diverges and

$$\lim_{n\to\infty} na_n = \lim_{n\to\infty} \frac{1}{\ln(n+1)} = 0.$$

3.2.36. Let

$$a_n = \begin{cases} \dfrac{1}{n} & \text{for } n = k^2, \ k = 1, 2, \ldots, \\ \dfrac{1}{n^2} & \text{otherwise.} \end{cases}$$

The series $\sum_{n=1}^{\infty} a_n$ is convergent but the limit $\lim_{n\to\infty} na_n$ does not exist.

3.2.37. The condition we are looking for is the convergence of the series $\sum_{n=1}^{\infty} \sqrt{a_n}$. Indeed, if $\sum_{n=1}^{\infty} \sqrt{a_n}$ converges, then we take $b_n = \sqrt{a_n}$.

Now assume that there is a sequence $\{b_n\}$ such that both series $\sum_{n=1}^{\infty} b_n$ and $\sum_{n=1}^{\infty} \frac{a_n}{b_n}$ converge. Then

$$\sqrt{a_n} = \sqrt{b_n \frac{a_n}{b_n}} \leq \frac{1}{2}\left(b_n + \frac{a_n}{b_n}\right),$$

3.2. Series of Nonnegative Terms

and so the series $\sum_{n=1}^{\infty} \sqrt{a_n}$ converges.

3.2.38. Assume that there is a sequence $\{a_n\}$ such that both series $\sum_{n=1}^{\infty} a_n$ and $\sum_{n=1}^{\infty} \frac{1}{n^2 a_n}$ converge. Let

$$\mathbf{A} = \left\{ n_s \in \mathbb{N} : \frac{1}{n_s} \leq \frac{1}{n_s^2 a_{n_s}} \right\} \quad \text{and} \quad \mathbf{A}' = \mathbb{N} \setminus \mathbf{A}.$$

Then $\sum_{n_s \in \mathbf{A}} \frac{1}{n_s} < +\infty$ and so $\sum_{n_s \in \mathbf{A}'} \frac{1}{n_s} = +\infty$ (of course $\mathbf{A}$ can be an empty set).

Now observe that $a_{n_s} > \frac{1}{n_s}$ for $n_s \in \mathbf{A}'$. Therefore the series $\sum_{n=1}^{\infty} a_n$ diverges, contrary to our assumption.

3.2.39. We have

$$\sum_{n=1}^{\infty} \frac{1}{n} \cdot \frac{1 + a_{n+1}}{a_n} = \sum_{n=1}^{\infty} \frac{1}{n a_n} + \sum_{n=1}^{\infty} \frac{a_{n+1}}{n a_n}.$$

We will show that the convergence of the series $\sum_{n=1}^{\infty} \frac{a_{n+1}}{n a_n}$ implies the divergence of $\sum_{n=1}^{\infty} \frac{1}{n a_n}$. By the Cauchy criterion, there is $k \in \mathbb{N}$ such that for any positive integer n, $\sum_{i=k+1}^{k+n} \frac{a_{i+1}}{i a_i} < \frac{1}{4}$. Thus $\frac{n}{n+k} \sum_{i=k+1}^{k+n} \frac{a_{i+1}}{n a_i} < \frac{1}{4}$. Therefore, for $n > k$,

$$\sum_{i=k+1}^{k+n} \frac{a_{i+1}}{n a_i} < \frac{1}{4} \cdot \frac{k+n}{n} < \frac{1}{2}.$$

By the relation between arithmetic and geometric means,

$$\sqrt[n]{\frac{a_{k+n+1}}{a_{k+1}}} < \frac{1}{2}, \quad \text{and so} \quad a_{k+n+1} < \frac{a_{k+1}}{2^n}.$$

Thus

$$\frac{1}{(k+n+1) a_{k+n+1}} > \frac{2^n}{(k+n+1) a_{k+1}}.$$

Therefore the series $\sum_{n=1}^{\infty} \frac{1}{n a_n}$ diverges.

3.2.40. Of course the series $\sum_{n=1}^{\infty} c_n$ can diverge (e.g. if $a_n \leq b_n$ for $n \in \mathbb{N}$). Surprisingly, it can converge. Indeed, take the series with terms
$$1, \frac{1}{2^2}, \frac{1}{2^2}, \frac{1}{2^2}, \frac{1}{2^2}, \frac{1}{2^2}, \frac{1}{7^2}, \frac{1}{8^2}, \frac{1}{9^2}, \ldots$$
and
$$1, \frac{1}{2^2}, \frac{1}{3^2}, \frac{1}{4^2}, \frac{1}{5^2}, \frac{1}{6^2}, \frac{1}{7^2}, \underbrace{\frac{1}{8^2}, \frac{1}{8^2}, \ldots, \frac{1}{8^2}}_{8^2+1 \text{ times}}, \ldots$$

Each of these series contains infinitely many blocks of terms whose sums are greater than 1. Therefore each of them diverges. In this case $c_n = \frac{1}{n^2}$, and so $\sum_{n=1}^{\infty} c_n$ converges.

3.2.41. We will use the Cauchy condensation theorem (3.2.28). The divergence of $\sum_{n=1}^{\infty} \frac{b_n}{n}$ is equivalent to the divergence of the series with the terms
$$b_{2^n} = \min\left\{a_{2^n}, \frac{1}{n \ln 2}\right\}.$$
In turn, the series $\sum_{n=1}^{\infty} b_{2^n}$ diverges if and only if the condensed series with terms
$$2^n b_{2^{2^n}} = \min\left\{2^n a_{2^{2^n}}, \frac{1}{\ln 2}\right\}$$
diverges. We will now show that the latter series diverges. Indeed, if a series $\sum_{n=1}^{\infty} d_n$ is divergent, then $\sum_{n=1}^{\infty} \min\{d_n, c\}$, where $c > 0$, is also divergent. If $\min\{d_n, c\} = c$ for infinitely many n, then the series $\sum_{n=1}^{\infty} \min\{d_n, c\}$ diverges. If $\min\{d_n, c\} = c$ for finitely many n, then the divergence of the series follows from the divergence of $\sum_{n=1}^{\infty} d_n$.

3.2.42. We have
$$1 - \frac{a_n}{a_{n+1}} = \frac{a_{n+1} - a_n}{a_{n+1}} \leq \frac{a_{n+1} - a_n}{a_1}.$$

3.2. Series of Nonnegative Terms

This and the convergence of the telescoping series $\sum_{n=1}^{\infty}(a_{n+1}-a_n)$ imply the convergence of the series in question.

3.2.43. We have
$$1-\frac{a_n}{a_{n+1}}=\frac{a_{n+1}-a_n}{a_{n+1}}.$$
Setting $b_n=a_{n+1}-a_n$ and $S_n=b_1+....+b_n$, we obtain $\frac{b_n}{S_n+a_1}=\frac{a_{n+1}-a_n}{a_{n+1}}$. Thus the divergence of our series follows from 3.2.10.

3.2.44. If the sequence $\{a_n\}$ is unbounded, then the convergence of the series in question follows from Problem 3.2.11. To see this one can apply arguments similar to that given in the solution of the preceding problem. Now assume that the sequence $\{a_n\}$ is bounded. Then
$$\frac{a_{n+1}-a_n}{a_{n+1}a_n^\alpha}\leq\frac{1}{a_2a_1^\alpha}(a_{n+1}-a_n).$$
Hence the convergence of our series follows from the convergence of the telescoping series $\sum_{n=1}^{\infty}(a_{n+1}-a_n)$.

3.2.45. It is enough to take $c_n=\frac{1}{S_n}$, where S_n is the nth partial sum of $\sum_{n=1}^{\infty}a_n$, and apply 3.2.10.

3.2.46. One can set $c_n=\frac{1}{\sqrt{r_{n-1}}}$, where $r_n=a_{n+1}+a_{n+2}+...$, and use 3.2.13 (b).

3.2.47. The sequence $\{r_n\}$ is monotonically decreasing. Hence by 3.2.35, $\lim\limits_{n\to\infty}nr_n=0$. Therefore
$$\lim_{n\to\infty}na_n=\lim_{n\to\infty}n(r_{n-1}-r_n)=\lim_{n\to\infty}((n-1)r_{n-1}-nr_n+r_{n-1})=0.$$

3.2.48.

(a) Since $\lim\limits_{n\to\infty}a_n=+\infty$, $a_n>2$ for n large enough. The convergence of the series in question follows from the inequality $\frac{1}{a_n^n}<\frac{1}{2^n}$, which holds for sufficiently large n.

(b) As in (a), n can be chosen so large that $\frac{1}{a_n^{\ln n}} < \frac{1}{3^{\ln n}}$. Thus by 3.2.17 (c), our series converges.

(c) The series can be either divergent or convergent. Its behavior depends on the sequence $\{a_n\}$. If $a_n = \ln n$, $n \geq 2$, then the series $\sum\limits_{n=1}^{\infty} \frac{1}{a_n^{\ln \ln n}}$ diverges (see 3.2.2 (e)). On the other hand, if $a_n = n$, then for $n > e^e$,

$$\frac{1}{a_n^{\ln \ln n}} = \frac{1}{e^{\ln \ln n \cdot \ln n}} < \frac{1}{n^\alpha}, \quad \text{where} \quad \alpha > 1.$$

In this case the series in question is convergent.

3.2.49. The series diverges because the necessary condition $a_n \to 0$ for convergence is not satisfied (see 2.5.25).

3.2.50. We assume first that $p = 0$. Then by 2.5.22, $\lim\limits_{n \to \infty} \sqrt{n} a_n = \sqrt{3}$ and so our series diverges. Now suppose that $p > 0$. Then $\lim\limits_{n \to \infty} a_n = 0$. Hence $\lim\limits_{n \to \infty} \frac{a_{n+1}}{a_n} = \lim\limits_{n \to \infty} \frac{\sin a_n}{a_n} \cdot \frac{1}{n^p} = 0$. The series converges by the ratio test.

3.2.51. Observe that $a_n \in (n\pi, n\pi + \frac{\pi}{2})$. Hence $\frac{1}{a_n^2} < \frac{1}{n^2 \pi^2}$, and so the series $\sum\limits_{n=1}^{\infty} \frac{1}{a_n^2}$ converges.

3.2.52. Set $b_n = \sqrt{a_n}$; then $b_n \in (n\pi, n\pi + \frac{\pi}{2})$. Hence the series $\sum\limits_{n=1}^{\infty} \frac{1}{a_n} = \sum\limits_{n=1}^{\infty} \frac{1}{b_n^2}$ converges (see the solution of the foregoing problem).

3.2.53. The series diverges because $\lim\limits_{n \to \infty} n a_n = 2$ (see 2.5.29).

3.2.54. For simplicity we introduce the following notation:

$$L_n = a_1 + a_3 + \ldots + a_{2n-1} \quad \text{and} \quad M_n = a_2 + a_4 + \ldots + a_{2n}.$$

By the monotonicity of $\{a_n\}$,

(*) $\qquad\qquad L_n \geq M_n \quad \text{and} \quad L_n - a_1 \leq M_n.$

Hence $2M_n = M_n + M_n \geq M_n + L_n - a_1 = \sum\limits_{k=2}^{n} a_k$. Thus

(**) $\qquad\qquad\qquad \lim\limits_{n \to \infty} M_n = +\infty.$

3.2. Series of Nonnegative Terms

Combining (∗) and (∗∗), we arrive at
$$\frac{L_n}{M_n} - 1 = \frac{L_n - M_n}{M_n} \leq \frac{a_1}{M_n} \xrightarrow[n \to \infty]{} 0.$$

3.2.55. By the definition of k_n, we have $0 \leq S_{k_n} - n < \frac{1}{k_n}$. It is known that $\lim\limits_{n \to \infty} (S_{k_n} - \ln k_n) = \gamma$, where γ is the Euler constant (see 2.1.41). Therefore
$$\lim_{n \to \infty} (n - \ln k_n) = \lim_{n \to \infty} (n + 1 - \ln k_{n+1}) = \gamma.$$
Hence
$$\lim_{n \to \infty} \left(1 - \ln \frac{k_{n+1}}{k_n} \right) = 0,$$
and so
$$\lim_{n \to \infty} \frac{k_{n+1}}{k_n} = e.$$

3.2.56.

(a) [A. J. Kempner, Amer. Math. Monthly 23(1914), 48-50] A k-digit number from **A** can be written in the form
$$10^{k-1} a_1 + 10^{k-2} a_2 + \ldots + a_k, \quad \text{where} \quad 0 < a_i \leq 9, \ i = 1, 2, \ldots, k.$$
For a given k, there are 9^k k-digit numbers in **A**, and each of them exceeds 10^{k-1}. Therefore
$$\sum_{n \in \mathbf{A}} \frac{1}{n} < \sum_{k=1}^{\infty} \frac{9^k}{10^{k-1}} = 90.$$

(b) As in (a) we have
$$\sum_{n \in \mathbf{A}} \frac{1}{n^\alpha} < \sum_{k=1}^{\infty} \frac{9^k}{10^{\alpha(k-1)}}.$$
Therefore if $\alpha > \log_{10} 9$, then the series $\sum\limits_{n \in \mathbf{A}} \frac{1}{n^\alpha}$ converges. Moreover, since
$$\sum_{n \in \mathbf{A}} \frac{1}{n^\alpha} > \sum_{k=1}^{\infty} \frac{9^k}{(10^k - 1)^\alpha} > \sum_{k=1}^{\infty} \frac{9^k}{10^{k\alpha}},$$

the series $\sum_{n\in A} \frac{1}{n^\alpha}$ diverges if $\alpha \leq \log_{10} 9$.

Remark. Let $\mathbf{A_k}$ denote the subset of positive integers that do not contain the digit k in their decimal expansion. In much the same way one can show that the series $\sum_{n\in \mathbf{A_k}} \frac{1}{n^\alpha}$ converges if $\alpha > \log_{10} 9$.

3.2.57. Assume that $-\infty < g < 1$ and take $\varepsilon > 0$ so small that $g + \varepsilon < 1$. Then, for sufficiently large n, $\ln \frac{1}{a_n} < (g+\varepsilon)\ln n$ and $a_n > \frac{1}{n^{g+\varepsilon}}$. Therefore the series diverges. If $g = -\infty$, then (for n large enough), $\ln \frac{1}{a_n} < -1 \cdot \ln n$. Thus $a_n > n$ and the series diverges. The same proof works for $g > 1$. Let us consider two series: $\sum_{n=1}^\infty \frac{1}{n}$ and $\sum_{n=2}^\infty \frac{1}{n \ln^2 n}$. The first one diverges and the second converges, although for both g is equal to 1.

3.2.58. The equivalence of these tests has been shown in the solution of Problem 3.2.19. By 2.5.34, if the Raabe criterion is decisive, then so is the criterion from the foregoing problem. To show that the converse is not true, we consider the series with terms a_n defined by setting $a_{2n-1} = \frac{1}{n^2}$, $a_{2n} = \frac{1}{4n^2}$.

3.2.59. Let $b_n = \underbrace{\sqrt{2 + \sqrt{2 + \ldots + \sqrt{2}}}}_{n-\text{roots}}$. Then $b_n = 2\cos\frac{\pi}{2^{n+1}}$ (compare with 2.5.41). By the definition of $\{a_n\}$, we have $a_n^2 = 2 - b_{n-1}$, and so $a_n = 2\sin\frac{\pi}{2^{n+1}} < \frac{\pi}{2^n}$. Thus the series in question converges.

3.2.60. Assume K is a positive number such that

$$(a_1 - a_n) + (a_2 - a_n) + \ldots + (a_{n-1} - a_n) \leq K \quad \text{for} \quad n \in \mathbb{N}.$$

Hence for every $n \in \mathbb{N}$ we have $a_1 + \ldots + a_n - na_n \leq K$. Let $m \in \mathbb{N}$ be arbitrarily chosen. By the monotonicity of the sequence $\{a_n\}$ and by its convergence to zero, there exists $n_0 \in \mathbb{N}$ such that

(*) $$a_n \leq \frac{1}{2}a_m \quad \text{for} \quad n \geq n_0.$$

We have

$$a_1 + \ldots + a_m - ma_n + a_{m+1} + \ldots + a_n - (n-m)a_n \leq K.$$

3.2. Series of Nonnegative Terms

Again, by the monotonicity of $\{a_n\}$,

$$a_{m+1} + \ldots + a_n \geq (n-m)a_n \quad \text{and} \quad a_1 + \ldots + a_m \geq ma_m.$$

Therefore $m(a_m - a_n) = ma_m - ma_n \leq a_1 + a_2 + \ldots + a_m - ma_n \leq K$. This and (∗) imply $\frac{1}{2}ma_m \leq m(a_m - a_n) \leq K$. Finally,

$$S_m = a_1 + a_2 + \ldots + a_m = S_m - ma_m + ma_m \leq K + ma_m \leq 3K.$$

3.2.61. From the relations

$$a_n = a_{n+1} + a_{n+2} + a_{n+3} + \ldots, \qquad a_{n+1} = a_{n+2} + a_{n+3} + \ldots$$

we gather that $a_{n+1} = \frac{1}{2}a_n$. Now, by induction, $a_n = \frac{1}{2^n}$, $n \in \mathbb{N}$.

3.2.62. [20] Let $r_{n,k} = a_n + a_{n+1} + \ldots + a_{n+k}$, $n = 1, 2, \ldots$, $k = 0, 1, 2, \ldots$, and let $\lim_{k \to \infty} r_{n,k} = r_n$, $n = 1, 2, \ldots$ Assume that $s \in (0, S)$ and that a_{n_1} is the first term of the sequence $\{a_n\}$ for which $a_{n_1} < s$. Either there exists a k_1 such that $r_{n_1,k_1} < s \leq r_{n_1,k_1+1}$, or $r_{n_1} \leq s$. In the second case we have $s \leq a_{n_1-1} \leq r_{n_1} \leq s$, and so $r_{n_1} = s$. In the first case we determine the first term a_{n_2} with $n_2 > n_1 + k_1$, $r_{n_1,k_1} + a_{n_2} < s$. Either there exists a k_2 with

$$r_{n_1,k_1} + r_{n_2,k_2} < s \leq r_{n_1,k_1} + r_{n_2,k_2+1},$$

or $r_{n_1,k_1} + r_{n_2} = s$. This procedure can be repeated, and if the first case occurs at every step, then $s = r_{n_1,k_1} + r_{n_2,k_2} + \ldots$.

3.2.63. [20] Suppose, contrary to our claim, that there is $k \in \mathbb{N}$ such that $a_k = 2p + \sum_{n=k+1}^{\infty} a_n$, where $p > 0$. Then $a_k - p = p + \sum_{n=k+1}^{\infty} a_n = \sum_{n=1}^{\infty} \varepsilon_n a_n$, where ε_n equals either zero or one. By the monotonicity of $\{a_n\}$, $\varepsilon_n = 0$ for $n \leq k$. Hence $a_k - p = \sum_{n=1}^{\infty} \varepsilon_n a_n \leq \sum_{n=k+1}^{\infty} a_n = a_k - 2p$, a contradiction.

3.2.64. By the Stolz theorem (see 2.3.11), we obtain

$$\lim_{n \to \infty} \frac{a_1 S_1^{-1} + a_2 S_2^{-1} + \ldots + a_n S_n^{-1}}{\ln S_n} = \lim_{n \to \infty} \frac{a_n S_n^{-1}}{-\ln(1 - a_n S_n^{-1})} = 1.$$

The last equality follows, e.g., from 2.5.5.

3.2.65. Set $a_n = 1$, $n \in \mathbb{N}$.

3.2.66. Since $\frac{a_1+a_2+\ldots+a_n}{n} > \frac{a_1}{n}$, the series $\sum\limits_{n=1}^{\infty} \frac{a_1+a_2+\ldots+a_n}{n}$ is divergent for any positive sequence $\{a_n\}$. The divergence is independent of the behavior of the series $\sum\limits_{n=1}^{\infty} a_n$.

3.2.67. By assumption,
$$a_2 \leq a_1, \qquad \sum_{k=2^{n-1}+1}^{2^n} a_k \leq \frac{1}{2^{n-1}} \sum_{k=1}^{2^{n-1}} a_k.$$

Hence, by induction,
$$\sum_{k=1}^{2^n} a_k \leq 2\left(1+\frac{1}{2}\right)\left(1+\frac{1}{2^2}\right)\ldots\left(1+\frac{1}{2^{n-1}}\right) a_1.$$

Moreover,
$$\left(1+\frac{1}{2}\right)\left(1+\frac{1}{2^2}\right)\ldots\left(1+\frac{1}{2^{n-1}}\right) = e^{\sum\limits_{k=1}^{n-1} \ln\left(1+\frac{1}{2^k}\right)} \leq e^{\sum\limits_{k=1}^{n-1} \frac{1}{2^k}} < e.$$

3.2.68. Put $c_n = \frac{(n+1)^n}{n^{n-1}} = n\left(\frac{n+1}{n}\right)^n$, $n \in \mathbb{N}$. Then

$(*)$ $\qquad c_1 \cdot \ldots \cdot c_n = (n+1)^n \quad$ and $\quad c_n < ne$.

Using the geometric-arithmetic mean inequality, we arrive at
$$\sqrt[n]{a_1 \cdot \ldots \cdot a_n} = \frac{1}{n+1} \sqrt[n]{a_1 c_1 \cdot \ldots \cdot a_n c_n} \leq \frac{a_1 c_1 + \ldots + a_n c_n}{n(n+1)}.$$

Therefore
$$\sum_{n=1}^{N} \sqrt[n]{a_1 \cdot \ldots \cdot a_n} \leq \sum_{n=1}^{N} \frac{a_1 c_1 + \ldots + a_n c_n}{n(n+1)}$$
$$= a_1 c_1 \left(\frac{1}{1\cdot 2} + \frac{1}{2\cdot 3} + \ldots + \frac{1}{N(N+1)}\right)$$
$$+ a_2 c_2 \left(\frac{1}{2\cdot 3} + \frac{1}{3\cdot 4} + \ldots + \frac{1}{N(N+1)}\right) + \ldots + a_N c_N \frac{1}{N(N+1)}$$
$$< a_1 c_1 + a_2 c_2 \frac{1}{2} + a_3 c_3 \frac{1}{3} + \ldots + a_N c_N \frac{1}{N} \leq 2a_1 + ea_2 + \ldots + ea_N.$$

3.2. Series of Nonnegative Terms

The last inequality follows from $(*)$. Letting $N \to \infty$, we obtain the desired inequality.

3.2.69. Writing

$$c_n = \frac{(n+1)^n \cdot \ldots \cdot (n+k-1)^n (n+k)^n}{n^{n-1} \cdot \ldots \cdot (n+k-2)^{n-1}(n+k-1)^{n-1}}$$

$$= \left(\frac{n+k}{n}\right)^n n(n+1) \cdot \ldots \cdot (n+k-1),$$

we get $c_1 \cdot \ldots \cdot c_n = (n+1)^n \cdot \ldots \cdot (n+k)^n$. Hence, as in the solution of the preceding problem, we obtain

$$\sum_{n=1}^{N} \sqrt[n]{a_1 \cdot \ldots \cdot a_n} \leq \sum_{n=1}^{N} \frac{a_1 c_1 + \ldots + a_n c_n}{n(n+1) \cdot \ldots \cdot (n+k)}$$

$$= a_1 c_1 \left(\frac{1}{1 \cdot 2 \cdot \ldots \cdot (1+k)} + \ldots + \frac{1}{N(N+1) \cdot \ldots \cdot (N+k)}\right)$$

$$+ a_2 c_2 \left(\frac{1}{2 \cdot 3 \cdot \ldots \cdot (2+k)} + \ldots + \frac{1}{N(N+1) \cdot \ldots \cdot (N+k)}\right)$$

$$+ \ldots + a_N c_N \frac{1}{N(N+1) \cdot \ldots \cdot (N+k)}$$

$$< \frac{1}{k}\left(\frac{1}{k!} a_1 c_1 + \frac{1}{2 \cdot 3 \cdot \ldots \cdot (1+k)} a_2 c_2\right.$$

$$\left. + \ldots + \frac{1}{N(N+1) \cdot \ldots \cdot (N+k-1)} a_N c_N\right).$$

The last inequality follows from Problem 3.1.4 (a). Since

$$\frac{1}{l(l+1) \cdot \ldots \cdot (l+k-1)} c_l = \left(\frac{l+k}{l}\right)^l,$$

letting $N \to \infty$, we obtain the desired inequality.

3.2.70. Let $T_n = a_1 + a_2 + \ldots + a_n$ and let S_n denote the nth partial sum of the series in question. Then

$$S_N = \frac{1}{a_1} + \sum_{n=2}^{N} \frac{n^2(T_n - T_{n-1})}{T_n^2} \leq \frac{1}{a_1} + \sum_{n=2}^{N} \frac{n^2(T_n - T_{n-1})}{T_n T_{n-1}}$$

$$= \frac{1}{a_1} + \sum_{n=2}^{N} \frac{n^2}{T_{n-1}} - \sum_{n=2}^{N} \frac{n^2}{T_n} = \frac{1}{a_1} + \sum_{n=1}^{N-1} \frac{(n+1)^2}{T_n} - \sum_{n=2}^{N} \frac{n^2}{T_n}$$

$$\leq \frac{5}{a_1} + \sum_{n=2}^{N-1} \frac{2n}{T_n} + \sum_{n=2}^{N-1} \frac{1}{T_n} \leq \frac{5}{a_1} + \sum_{n=1}^{N} \frac{2n}{T_n} + \sum_{n=1}^{N} \frac{1}{T_n}.$$

Moreover, by the Cauchy inequality (see 1.2.12),

$$\left(\sum_{n=1}^{N} \frac{n}{T_n}\right)^2 \leq \sum_{n=1}^{N} \frac{n^2 a_n}{T_n^2} \sum_{n=1}^{N} \frac{1}{a_n} \leq S_N \cdot M$$

with $M = \sum_{n=1}^{\infty} \frac{1}{a_n}$. Hence

$$\sum_{n=1}^{N} \frac{n}{T_n} \leq \sqrt{S_N} \cdot \sqrt{M}.$$

Consequently, $S_N \leq \frac{5}{a_1} + 2\sqrt{S_N}\sqrt{M} + M$, and so

$$S_N \leq \left(\sqrt{M} + \sqrt{2M + \frac{5}{a_1}}\right)^2.$$

3.2.71. The arithmetic-harmonic mean inequality (see, e.g., 1.2.3) yields

$$\frac{\sum_{n=2^{k-1}+1}^{2^k} \frac{1}{na_n - (n-1)a_{n-1}}}{2^{k-1}} \geq \frac{2^{k-1}}{\sum_{n=2^{k-1}+1}^{2^k} (na_n - (n-1)a_{n-1})}$$

$$= \frac{2^{k-1}}{2^k a_{2^k} - 2^{k-1} a_{2^{k-1}}} \geq \frac{1}{2a_{2^k}}.$$

3.2. Series of Nonnegative Terms

Hence
$$\sum_{n=2^{k-1}+1}^{2^k} \frac{1}{na_n - (n-1)a_{n-1}} \geq \frac{2^k}{4a_{2^k}}.$$

Therefore
$$S_{2^k} \geq \sum_{l=1}^{k} \frac{2^l}{4a_{2^l}}.$$

The divergence of the series follows from the Cauchy condensation theorem (see 3.2.28).

3.2.72. We will show that the series $\sum_{n=1}^{\infty} \frac{1}{p_n}$ is divergent. If it were not, then there would be an n such that $\sum_{m=n+1}^{\infty} \frac{1}{p_m} < \frac{1}{2}$. Let $a = p_1 \cdot p_2 \cdot \ldots \cdot p_n$. Then the number $1 + ka$ with $k \in \mathbb{N}$ can be written as a product of primes. This unique factorization does not contain any of the numbers $p_1, \ldots, p_n$. Hence

$$\sum_{k=1}^{\infty} \frac{1}{1+ka} < \sum_{l=1}^{\infty} \left(\sum_{m=n+1}^{\infty} \frac{1}{p_m} \right)^l < \sum_{l=1}^{\infty} \left(\frac{1}{2} \right)^l = 1,$$

a contradiction.

3.2.73. It is enough to apply the results from the foregoing problem and from Problem 3.2.71.

3.2.74. We get

$$\lim_{n \to \infty} \frac{\sum_{k=2}^{\infty} \frac{1}{k^{n+1}}}{\sum_{k=2}^{\infty} \frac{1}{k^n}} = \lim_{n \to \infty} \frac{\frac{1}{2^{n+1}} \left(1 + \frac{2^{n+1}}{3^{n+1}} + \frac{2^{n+1}}{4^{n+1}} + \ldots \right)}{\frac{1}{2^n} \left(1 + \frac{2^n}{3^n} + \frac{2^n}{4^n} + \ldots \right)} = \frac{1}{2},$$

because the sums in parentheses tend to 1 as n tends to infinity. Indeed,

$$\frac{2^{n+1}}{3^{n+1}} + \frac{2^{n+1}}{4^{n+1}} + \ldots = 2^{n+1} \sum_{k=3}^{\infty} \frac{1}{k^{n+1}}.$$

Moreover,

$$\sum_{k=3}^{\infty} \frac{1}{k^{n+1}} = \frac{1}{3^{n+1}} + \sum_{k=2}^{\infty} \frac{1}{(2k)^{n+1}} + \sum_{k=2}^{\infty} \frac{1}{(2k+1)^{n+1}}$$

$$\leq \frac{1}{3^{n+1}} + 2\sum_{k=2}^{\infty} \frac{1}{(2k)^{n+1}} = \frac{1}{3^{n+1}} + \frac{1}{2^{2n+1}} + \frac{1}{2^n}\sum_{k=3}^{\infty} \frac{1}{k^{n+1}}.$$

Therefore

$$2^{n+1}\sum_{k=3}^{\infty} \frac{1}{k^{n+1}} \leq \frac{\left(\frac{2}{3}\right)^{n+1} + \frac{1}{2^n}}{\left(1 - \frac{1}{2^n}\right)},$$

and so

$$2^{n+1}\sum_{k=3}^{\infty} \frac{1}{k^{n+1}} \xrightarrow[n\to\infty]{} 0.$$

3.2.75. Assume first that the series $\sum_{n=1}^{\infty} a_n$ converges. Then the convergence of the series in question follows from the inequality $\frac{1}{T_n^\alpha} \leq \frac{1}{a_1^\alpha}$. If the series $\sum_{n=1}^{\infty} a_n$ diverges, then there exists a strictly increasing sequence $\{n_m\}$ of positive integers such that $S_{n_m-1} \leq m < S_{n_m}$. Then

$$T_{n_m} = S_1 + \ldots + S_{n_m} \geq S_{n_1} + \ldots + S_{n_m} > \frac{m(m+1)}{2}.$$

Hence

$$\sum_{n=n_2}^{\infty} \frac{a_n}{T_n^\alpha} = \sum_{m=2}^{\infty} \sum_{k=n_m}^{m_{m+1}-1} \frac{a_k}{T_k^\alpha} \leq \sum_{m=2}^{\infty} \frac{S_{n_{m+1}-1} - S_{n_m-1}}{T_{n_m}^\alpha}$$

$$< \sum_{m=2}^{\infty} \frac{1}{T_{n_m}^\alpha} < \sum_{m=2}^{\infty} \frac{1}{\left(\frac{m^2+m}{2}\right)^\alpha}.$$

Therefore the series in question is convergent if $\alpha > \frac{1}{2}$. This series can be divergent if $\alpha \leq \frac{1}{2}$. Indeed, it is enough to take $a_n = 1$, $n \in \mathbb{N}$.

3.2.76. By Problem 3.2.35, $\lim_{n\to\infty} \frac{n}{a_n} = 0$. Take $0 < K < 1$. Then there is n_0 such that $n \leq Ka_n$ for $n \geq n_0$. Hence

$$\frac{\ln^k a_n}{a_n} \geq \ln^k \left(\frac{1}{K}\right) \frac{\ln^k n}{a_n}.$$

3.2. Series of Nonnegative Terms

Thus the convergence of the series $\sum_{n=1}^{\infty} \frac{\ln^k a_n}{a_n}$ implies the convergence of $\sum_{n=1}^{\infty} \frac{\ln^k n}{a_n}$. To prove the other implication put

$$\mathbf{I_1} = \{n \in \mathbb{N}: a_n \leq n^{k+2}\} \quad \text{and} \quad \mathbf{I_2} = \mathbb{N} \setminus \mathbf{I_1}.$$

Then for $n \in \mathbf{I_1}$ we have $\ln a_n \leq (k+2) \ln n$, and so the convergence of $\sum_{n \in \mathbf{I_1}} \frac{\ln^k n}{a_n}$ implies the convergence of $\sum_{n \in \mathbf{I_1}} \frac{\ln^k a_n}{a_n}$. Moreover, for sufficiently large n in $\mathbf{I_2}$,

$$\frac{\ln^k a_n}{a_n} < \frac{a_n^{\frac{k}{k+1}}}{a_n} < \frac{1}{n^{\frac{k+2}{k+1}}}.$$

Hence $\sum_{n \in \mathbf{I_2}} \frac{\ln^k a_n}{a_n} < \infty$, because $\frac{k+2}{k+1} > 1$.

3.2.77. We have

$$\sum_{k=1}^{\varphi(n)-1} f(k) = \sum_{k=1}^{\varphi(1)-1} f(k) + (f(\varphi(1)) + f(\varphi(1)+1) + \ldots + f(\varphi(2)-1))$$
$$+ \ldots + (f(\varphi(n-1)) + f(\varphi(n-1)+1) + \ldots + f(\varphi(n)-1))$$
$$< \sum_{k=1}^{\varphi(1)-1} f(k) + \sum_{k=1}^{n-1} f(\varphi(k))(\varphi(k+1) - \varphi(k)).$$

Inequality (1) is proved. The proof of (2) is analogous.

3.2.78. Assume first that

$$\frac{f(\varphi(n))(\varphi(n+1) - \varphi(n))}{f(n)} \leq q < 1.$$

Then, by (1) in the preceding problem,

$$S_{\varphi(n)-1} < \sum_{k=1}^{\varphi(1)-1} f(k) + qS_{n-1}.$$

Therefore, in view of $\varphi(n) > n$, $(1-q)S_{n-1} < \sum_{k=1}^{\varphi(1)-1} f(k)$. The convergence of $\sum_{n=1}^{\infty} f(n)$ is proved. To prove the second part of the statement we can use inequality (2) from the foregoing problem and proceed analogously.

3.2.79. One can apply the result from the preceding problem with $\varphi(n) = 2n$.

3.2.80. We apply the result from Problem 3.2.78 with $\varphi(n) = 2^n$.

3.2.81. Apply the result from Problem 3.2.77 with

$$\varphi(n) = 3^n, \quad \varphi(n) = n^2 \quad \text{and} \quad \varphi(n) = n^3,$$

respectively.

3.2.82.

(1) We have $a_n b_n - a_{n+1} b_{n+1} \geq c a_{n+1}$. Therefore $\{a_n b_n\}$ is a monotonically decreasing sequence with positive terms, and so it is convergent. Thus the telescoping series $\sum\limits_{n=1}^{\infty} (a_n b_n - a_{n+1} b_{n+1})$ converges. The convergence of $\sum\limits_{n=1}^{\infty} a_n$ follows from the comparison test.

(2) We have

$$\frac{a_{n+1}}{a_n} \geq \frac{\frac{1}{b_{n+1}}}{\frac{1}{b_n}}.$$

Therefore the divergence of $\sum\limits_{n=1}^{\infty} a_n$ follows from the test given in Problem 3.2.3.

3.2.83. To derive the d'Alembert test (the ratio test) we take $b_n = 1$ for $n = 1, 2, \ldots$. Setting $b_n = n$ for $n = 1, 2, \ldots$, we get the Raabe test. Putting $b_n = n \ln n$ for $n = 2, 3, \ldots$, we arrive at the Bertrand test.

3.2.84. [J. Tong, Amer. Math. Monthly, 101(1994), 450-452]

(1) Let $S = \sum\limits_{n=1}^{\infty} a_n$ and let

$$b_n = \frac{S - \sum\limits_{k=1}^{n} a_k}{a_n} = \frac{r_n}{a_n}.$$

Of course, $b_n > 0$ for $n \in \mathbb{N}$. Moreover,

$$b_n \frac{a_n}{a_{n+1}} - b_{n+1} = \frac{r_n}{a_{n+1}} - \frac{r_{n+1}}{a_{n+1}} = \frac{a_{n+1}}{a_{n+1}} = 1.$$

(2) In this case set

$$b_n = \frac{\sum\limits_{k=1}^{n} a_k}{a_n} = \frac{S_n}{a_n}.$$

Then the series $\sum\limits_{n=1}^{\infty} \frac{1}{b_n}$ diverges (see, e.g., Problem 3.2.10). Moreover,

$$b_n \frac{a_n}{a_{n+1}} - b_{n+1} = \frac{S_n}{a_{n+1}} - \frac{S_{n+1}}{a_{n+1}} = \frac{-a_{n+1}}{a_{n+1}} = -1.$$

3.2.85.

(a) It is enough to apply the ratio test to each of the series:

$$\sum_{n=1}^{\infty} a_{kn}, \ \sum_{n=0}^{\infty} a_{1+kn}, \ \ldots, \ \sum_{n=0}^{\infty} a_{(k-1)+kn}.$$

(b) It is enough to apply the Raabe test (see 3.2.19) to each of the series given in the solution of (a).

3.2.86. By assumption, there exists a positive constant K such that

$$\varphi_n \le K \frac{1}{\ln n}, \quad n \ge 2.$$

Let us define the sets of positive integers $\mathbb{N}_1$ and $\mathbb{N}_2$ as follows:

$$\mathbb{N}_1 = \left\{ n : \ a_n \le \frac{1}{n^2} \right\} \quad \text{and} \quad \mathbb{N}_2 = \mathbb{N} \setminus \mathbb{N}_1.$$

For sufficiently large $n \in \mathbb{N}_1$,

$$(1) \qquad a_n^{1-\varphi_n} \le a_n^{1-\frac{K}{\ln n}} = a_n^{\frac{\ln \frac{n}{e^K}}{\ln n}} = \left(\frac{e^K}{n} \right)^{\frac{\ln \frac{1}{a_n}}{\ln n}} \le \frac{e^{2K}}{n^2}.$$

Furthermore, for sufficiently large $n \in \mathbb{N}_2$,

$$(2) \qquad \frac{a_n^{1-\varphi_n}}{a_n} \le a_n^{-\frac{K}{\ln n}} = \left(\frac{1}{a_n} \right)^{\frac{K}{\ln n}} \le n^{\frac{2K}{\ln n}} = e^{2K}.$$

Combining (1) and (2) with the convergence of the series $\sum_{n=2}^{\infty} a_n$, we arrive at

$$\sum_{n\in\mathbb{N}_1} a_n^{1-\varphi_n} < +\infty \quad \text{and} \quad \sum_{n\in\mathbb{N}_2} a_n^{1-\varphi_n} < +\infty.$$

3.3. The Integral Test

3.3.1. For $k-1 \leq x \leq k$, $k \geq 2$, we have $f(x) \geq f(k)$. On the other hand, for $k \leq x \leq k+1$, we have $f(x) \leq f(k)$. Hence

$$\int_k^{k+1} f(x)\,dx \leq f(k) \leq \int_{k-1}^k f(x)\,dx, \quad k = 2, 3, \dots.$$

Summing both sides of the above inequalities from $k=2$ to $k=n$, we get

$$\int_2^{n+1} f(x)\,dx \leq f(2) + f(3) + \dots + f(n) \leq \int_1^n f(x)\,dx,$$

which proves the integral test.

3.3.2. Note that $\frac{f'}{f}$ is a positive and monotonically decreasing function. Therefore, by the integral test, convergence of the given series is equivalent to boundedness of the sequences $\left\{\int_1^n f'(x)\,dx\right\}$ and $\left\{\int_1^n \frac{f'(x)}{f(x)}\,dx\right\}$. Since

$$\int_1^n f'(x)\,dx = f(n) - f(1) \quad \text{and} \quad \int_1^n \frac{f'(x)}{f(x)}\,dx = \ln f(n) - \ln f(1),$$

either both sequences are bounded or both sequences are unbounded.

3.3.3. We have $S_N - I_N - (S_{N+1} - I_{N+1}) = \int_N^{N+1} f(x)\,dx - f(N+1) \geq 0$. Moreover, $f(n) \leq \int_{n-1}^n f(x)\,dx \leq f(n-1)$ for $n = 2, 3, \dots, N$. Summing these inequalities from $n=2$ to $n=N$, we get $S_N - f(1) \leq I_N \leq S_N - f(N)$. Hence $0 < f(N) \leq S_N - I_N \leq f(1)$, which completes the proof.

3.3. The Integral Test

3.3.4. Convergence of the given sequences follows from the foregoing problem. It remains to show that the limits of these sequences belong to $(0,1)$.

(a) Since $f(x) = \frac{1}{x}$ is a strictly decreasing function on the interval $(0,\infty)$, $S_N - I_N < S_2 - I_2 < f(1) = 1$ for $N > 2$ and

$$f(2) + f(3) + \ldots + f(N-1) + f(N)$$
$$> f(2) + f(3) + \ldots + f(N-1) > \int_2^N f(x)dx,$$

or equivalently, $S_N - f(1) > I_N - I_2$. Finally,

$$0 < 1 - I_2 \leq \lim_{N \to \infty} (S_N - I_N) \leq S_2 - I_2 < 1.$$

(See also 2.1.41 and 3.1.36).

(b) The proof is analogous to that in (a).

3.3.5.

(a) Convergence of the series $\sum\limits_{n=2}^{\infty} \frac{1}{n(\ln n)^\alpha}$ is equivalent to boundedness of the sequence $\int\limits_2^n \frac{1}{x(\ln x)^\alpha} dx$. For $\alpha \neq 1$,

$$\int_2^n \frac{1}{x(\ln x)^\alpha} dx = \frac{(\ln n)^{-\alpha+1}}{-\alpha+1} - \frac{(\ln 2)^{-\alpha+1}}{-\alpha+1}.$$

Thus the series converges if $\alpha > 1$ and diverges if $0 < \alpha < 1$. Clearly, if $\alpha \leq 0$, then the series diverges. Finally, if $\alpha = 1$, then $\int\limits_2^n \frac{1}{x \ln x} dx = \ln(\ln n) - \ln(\ln 2)$. Hence the sequence $\int\limits_2^n \frac{1}{x \ln x} dx$ is unbounded and therefore the series diverges.

(b) In this case, we have

$$\int_3^n \frac{1}{x \ln x \ln(\ln x)} dx = \ln(\ln(\ln n)) - \ln(\ln(\ln 3)).$$

Thus by the integral test the series is divergent.

3.3.6.

(a) We have

$$\sum_{n=1}^{N} \frac{a_{n+1}}{S_n \ln S_n} = \sum_{n=1}^{N} \frac{S_{n+1} - S_n}{S_n \ln S_n} \geq \sum_{n=1}^{N} \int_{S_n}^{S_{n+1}} \frac{1}{x \ln x} \, dx$$
$$= \ln \ln S_{N+1} - \ln \ln S_1 \xrightarrow[N \to \infty]{} \infty.$$

(b) As in (a), we have

$$\sum_{n=2}^{N} \frac{a_n}{S_n \ln^2 S_n} = \sum_{n=2}^{N} \frac{S_n - S_{n-1}}{S_n \ln^2 S_n} \leq \sum_{n=2}^{N} \int_{S_{n-1}}^{S_n} \frac{1}{x \ln^2 x} \, dx$$
$$= -\frac{1}{\ln S_N} + \frac{1}{\ln S_1} \xrightarrow[N \to \infty]{} \frac{1}{\ln S_1}.$$

3.3.7. If

$$\frac{\varphi'(x) f(\varphi(x))}{f(x)} \leq q < 1 \quad \text{for} \quad x > x_0,$$

then

$$\int_{\varphi(x_0)}^{\varphi(x)} f(t) \, dt = \int_{x_0}^{x} \varphi'(t) f(\varphi(t)) \, dt \leq q \int_{x_0}^{x} f(t) \, dt.$$

Hence

$$(1-q) \int_{\varphi(x_0)}^{\varphi(x)} f(t) dt \leq q \left(\int_{x_0}^{x} f(t) dt - \int_{\varphi(x_0)}^{\varphi(x)} f(t) dt \right)$$
$$= q \left(\int_{x_0}^{\varphi(x_0)} f(t) dt - \int_{x}^{\varphi(x)} f(t) dt \right) \leq q \int_{x_0}^{\varphi(x_0)} f(t) dt.$$

Thus by the integral test, the series $\sum_{n=1}^{\infty} f(n)$ is convergent.

Now if

$$\frac{\varphi'(x) f(\varphi(x))}{f(x)} \geq 1 \quad \text{for} \quad x > x_0,$$

3.3. The Integral Test

then $\int_{\varphi(x_0)}^{\varphi(x)} f(t)\,dt \geq \int_{x_0}^{x} f(t)\,dt$. As a consequence,

$$\int_{x}^{\varphi(x)} f(t)dt \geq \int_{x_0}^{\varphi(x_0)} f(t)dt.$$

Moreover, since for any n there exists $k_n \in \mathbb{N}$ such that $n < \varphi(n) < n + k_n$, we have

$$I_{n+k_n} - I_n = \int_{n}^{n+k_n} f(t)dt \geq \int_{n}^{\varphi(n)} f(t)dt \geq \int_{x_0}^{\varphi(x_0)} f(t)dt.$$

Therefore $\{I_n\}$ is not a Cauchy sequence, and consequently, it is not bounded. So by the integral test the series diverges.

3.3.8.

(a) If $\lim_{x\to\infty} \left(-g(x)\frac{f'(x)}{f(x)} - g'(x)\right) > 0$, then there exist x_0 and $\delta > 0$ such that

$$-g(x)\frac{f'(x)}{f(x)} - g'(x) \geq \delta \quad \text{for} \quad x \geq x_0.$$

Therefore $-(g(x)f(x))' \geq \delta f(x)$, $x \geq x_0$. Consequently, for sufficiently large n, we get

$$\int_{x_0}^{n} f(x)dx \leq \frac{1}{\delta}\int_{x_0}^{n} -(f(x)g(x))'dx$$
$$= \frac{1}{\delta}(g(x_0)f(x_0) - g(n)f(n)) \leq \frac{1}{\delta}g(x_0)f(x_0).$$

Thus by the integral test, the series converges.

(b) As in (a), we get $-(g(x)f(x))' \leq 0$ for $x \geq x_0$. Thus the function gf is monotonically increasing on $[x_0, \infty)$, and consequently, $g(x)f(x) \geq g(x_0)f(x_0)$ if $x \geq x_0$. This means that $f(x) \geq \frac{f(x_0)g(x_0)}{g(x)}$ for $x > x_0$. Therefore the sequence $\int_{1}^{n} f(x)dx$ is unbounded because, by assumption, the sequence $\int_{1}^{n} \frac{1}{g(x)}dx$ is unbounded.

3.3.9. It is enough to apply the result in the preceding problem to $g(x) = x$.

3.3.10. In 3.3.8, we substitute $g(x) = x \ln x$.

3.3.11.
(a) Set
$$g(x) = \frac{\int\limits_x^\infty f(t)dt}{f(x)}.$$
Then $-g(x)\frac{f'(x)}{f(x)} - g'(x) = 1 > 0$.

(b) Put
$$g(x) = \frac{\int\limits_{1/2}^x f(t)dt}{f(x)}.$$
Then $\int\limits_1^n \frac{1}{g(x)}dx = \ln \int\limits_{1/2}^n f(t)dt - \ln \int\limits_{1/2}^1 f(t)dt$, which means that the sequence $\int\limits_1^n \frac{1}{g(x)}dx$ is unbounded. Moreover,
$$-g(x)\frac{f'(x)}{f(x)} - g'(x) = -1 < 0.$$

3.3.12. We will apply the test proved in 3.3.9. Taking $f(x) = (\ln x)^{-(\ln x)^\gamma}$, $x > 1$, we get
$$-x\frac{f'(x)}{f(x)} = (\ln x)^{\gamma-1}(\gamma \ln \ln x + 1).$$
If $\gamma \geq 1$, then $\lim\limits_{x \to \infty} (\ln x)^{\gamma-1}(\gamma \ln \ln x + 1) = +\infty$, and consequently, the series is convergent. On the other hand, if $0 \leq \gamma < 1$, then we have $\lim\limits_{x \to \infty} (\ln x)^{\gamma-1}(\gamma \ln \ln x + 1) = 0$, which means that the series is divergent.

3.3.13. Set
$$f(x) = \frac{1}{x^{1+\frac{1}{\ln \ln x}} \ln x}, \quad x > e.$$

3.3. The Integral Test

One can show that $\lim\limits_{x\to\infty}\left(-x\frac{f'(x)}{f(x)}\right)=1$. So we cannot apply the test given in 3.3.9. Thus we will apply the test proved in 3.3.10. For sufficiently large x,

$$\left(-\frac{f'(x)}{f(x)}-\frac{1}{x}\right)x\ln x = \frac{\ln x}{\ln\ln x} - \frac{\ln x}{(\ln\ln x)^2} + 1 > 2$$

because $\lim\limits_{x\to\infty}\left(\frac{\ln x}{\ln\ln x} - \frac{\ln x}{(\ln\ln x)^2}\right) = +\infty$.

3.3.14. We have $(\lambda_{n+1}-\lambda_n)\frac{1}{\lambda_{n+1}f(\lambda_{n+1})} \leq \int\limits_{\lambda_n}^{\lambda_{n+1}}\frac{1}{tf(t)}dt$. Hence

$$\sum_{n=1}^{\infty}\left(1-\frac{\lambda_n}{\lambda_{n+1}}\right)\frac{1}{f(\lambda_{n+1})} \leq \int_{\lambda_1}^{\infty}\frac{1}{tf(t)}dt < \infty.$$

Thus we have proved that the series $\sum\limits_{n=1}^{\infty}\left(1-\frac{\lambda_n}{\lambda_{n+1}}\right)\frac{1}{f(\lambda_{n+1})}$ is convergent. Let $\{S_n\}$ and $\{S'_n\}$ denote the sequence of partial sums of the series given in the problem and the sequence of partial sums the above series, respectively. Then

$$S_N - S'_N = \sum_{n=1}^{N}\left(1-\frac{\lambda_n}{\lambda_{n+1}}\right)\left(\frac{1}{f(\lambda_n)}-\frac{1}{f(\lambda_{n+1})}\right)$$

$$\leq \sum_{n=1}^{N}\left(\frac{1}{f(\lambda_n)}-\frac{1}{f(\lambda_{n+1})}\right) < \frac{1}{f(\lambda_1)},$$

which implies convergence of the given series.

3.3.15. By the monotonicity of f,

(∗) $\quad f(\lambda_{n+1})(\lambda_{n+1}-\lambda_n) \leq \int_{\lambda_n}^{\lambda_{n+1}} f(t)dt \leq f(\lambda_n)(\lambda_{n+1}-\lambda_n).$

(a) By the left inequality and by our assumption, we get

$$M\sum_{n=1}^{\infty} f(\lambda_{n+1}) \leq \int_{\lambda_1}^{\infty} f(t)dt < \infty.$$

(b) The right inequality in (∗) implies the divergence of the series $\sum_{n=1}^{\infty} f(\lambda_n)$.

3.3.16. Assume first that the series $\sum_{n=1}^{\infty} \frac{1}{f(n)}$ converges. Then, by the integral test, the improper integral $\int_{1}^{\infty} \frac{1}{f(t)} dt$ also converges. Integration by parts and then integration by substitution give

(∗)
$$\int_{1}^{\infty} \frac{1}{f(t)} dt = \lim_{t \to \infty} \frac{t}{f(t)} - \frac{1}{f(1)} + \int_{1}^{\infty} \frac{t f'(t)}{f^2(t)} dt$$
$$= \lim_{t \to \infty} \frac{t}{f(t)} - \frac{1}{f(1)} + \int_{f(1)}^{\infty} \frac{f^{-1}(t)}{t^2} dt.$$

We will now show that

(∗∗)
$$\lim_{t \to \infty} \frac{t}{f(t)} = 0.$$

The convergence of the improper integral implies $\lim_{t \to \infty} \int_{t}^{2t} \frac{1}{f(x)} dx = 0$. Since $\frac{1}{2} \frac{2t}{f(2t)} = \frac{1}{f(2t)} \int_{t}^{2t} dx < \int_{t}^{2t} \frac{1}{f(x)} dx$, the equality (∗∗) holds. Thus the improper integral $\int_{1}^{\infty} \frac{f^{-1}(t)}{t^2} dt$ converges.

Moreover, we have

$$\sum_{n=1}^{\infty} \frac{f^{-1}(n)}{(n+1)^2} \leq \sum_{n=1}^{\infty} \int_{n}^{n+1} \frac{f^{-1}(t)}{t^2} dt = \int_{1}^{\infty} \frac{f^{-1}(t)}{t^2} dt < \infty,$$

which means that the series $\sum_{n=1}^{\infty} \frac{f^{-1}(n)}{(n+1)^2}$ converges. Obviously, the series $\sum_{n=1}^{\infty} \frac{f^{-1}(n)}{n^2}$ also converges.

To prove the implication in the other direction, assume that the series $\sum_{n=1}^{\infty} \frac{f^{-1}(n)}{n^2}$ converges. In much the same way we can show that the integral $\int_{f(1)}^{\infty} \frac{f^{-1}(t)}{t^2} dt$ converges, and consequently, $\int_{1}^{\infty} \frac{1}{f(t)} dt$ also converges. Thus by the integral test the series $\sum_{n=1}^{\infty} \frac{1}{f(n)}$ is convergent.

3.3.17. First observe that the function φ can be defined, in the same way, on the whole interval $[e, \infty)$. Then $\varphi(x) = 1$ for $x \in [e, e^e)$, $\varphi(x) = 2$ for $x \in [e^e, e^{e^e})$. For simplicity set $\hat{e}^1 = e$ and $\hat{e}^k = e^{\hat{e}^{k-1}}$ for $k > 1$. Thus we have

$$\varphi(x) = k \quad \text{for} \quad x \in [\hat{e}^k, \hat{e}^{k+1}).$$

Let

$$f(x) = \frac{1}{x(\ln_1 x)(\ln_2 x) \cdot \ldots \cdot (\ln_{\varphi(x)} x)};$$

then

$$f(x) = \frac{1}{x(\ln_1 x)(\ln_2 x) \cdot \ldots \cdot (\ln_k x)} \quad \text{for} \quad x \in [\hat{e}^k, \hat{e}^{k+1}).$$

Now, by the integral test, our series diverges because, for $n > \hat{e}^k$,

$$I_n = \int_e^n f(x)dx \geq \int_e^{\hat{e}^k} f(x)dx = \int_e^{\hat{e}^2} \frac{1}{x \ln x} dx + \int_{\hat{e}^2}^{\hat{e}^3} \frac{1}{x(\ln x)(\ln_2 x)} dx$$
$$+ \ldots + \int_{\hat{e}^{k-1}}^{\hat{e}^k} \frac{1}{x(\ln x)(\ln_2 x) \cdot \ldots \cdot (\ln_{k-1} x)} dx = k - 1.$$

3.4. Absolute Convergence. Theorem of Leibniz

3.4.1.
(a) We have

$$\lim_{n \to \infty} \sqrt[n]{\left|\frac{an}{n+1}\right|^n} = |a|.$$

Thus the series converges absolutely if $|a| < 1$ and diverges if $|a| > 1$. If $|a| = 1$, then the series diverges because

$$\lim_{n \to \infty} \left|\frac{an}{n+1}\right|^n = \lim_{n \to \infty} \frac{1}{\left(1 + \frac{1}{n}\right)^n} = \frac{1}{e}.$$

(b) Set $f(x) = \frac{(\ln x)^a}{x}$ for $x > 0$. Then $f'(x) = \frac{(\ln x)^{a-1}(a-\ln x)}{x^2} < 0$ for $x > \max\{1, e^a\}$. Thus by the Leibniz test the series converges for every $a \in \mathbb{R}$. We will now decide whether the series converges absolutely, that is, whether the series $\sum_{n=2}^{\infty} \frac{(\ln n)^a}{n}$ converges. By the theorem of Cauchy (see 3.2.28), the convergence of this series is equivalent to the convergence of $\sum_{n=2}^{\infty} n^a (\ln 2)^a$. Thus our series converges absolutely if $a < -1$.

(c) If $a > 0$, then by the Leibniz test the series converges. If $a < 0$, then
$$\sum_{n=1}^{\infty} (-1)^n \sin \frac{a}{n} = \sum_{n=1}^{\infty} (-1)^{n+1} \sin \frac{|a|}{n}.$$
Applying the Leibniz test again, we see that the series converges for all $a \in \mathbb{R}$. The series does not converge absolutely if $a \neq 0$, because
$$\lim_{n \to \infty} \frac{\sin \frac{|a|}{n}}{\frac{1}{n}} = |a|.$$

(d) The series converges if and only if $-1 \leq \frac{a^2 - 4a - 8}{a^2 + 6a - 16} < 1$, that is, if $a \in [-4, \frac{4}{5}) \cup [3, \infty)$. Clearly, the series converges absolutely if $a \in (-4, \frac{4}{5}) \cup (3, \infty)$.

(e) Since
$$\lim_{n \to \infty} \sqrt[n]{\left| \frac{n^n}{a^{n^2}} \right|} = 0 \quad \text{if} \quad |a| > 1,$$
the series converges absolutely if $|a| > 1$. If $|a| \leq 1$, then the necessary condition for convergence is not satisfied because $\lim_{n \to \infty} \frac{n^n}{|a|^{n^2}} = +\infty$.

(f) Observe that
$$\lim_{n \to \infty} \frac{(\ln n)^{\ln n}}{n^a} = \lim_{n \to \infty} n^{\ln \ln n - a} = +\infty.$$
Thus the necessary condition for convergence is not satisfied.

3.4.2. If $|a| < 1$, then for sufficiently large n,
$$\left| \frac{a^{n-1}}{na^{n-1} + \ln n} \right| < |a|^{n-1}.$$

3.4. Absolute Convergence. Theorem of Leibniz

Thus the series converges absolutely. If $|a| \geq 1$, then

$$\frac{a^{n-1}}{na^{n-1} + \ln n} = \frac{1}{n}\left(\frac{1}{1 + \frac{\ln n}{na^{n-1}}}\right).$$

Therefore for sufficiently large n the terms of the series are positive, and, by the comparison test, its divergence follows from the divergence of $\sum_{n=1}^{\infty} \frac{1}{n}$.

3.4.3. Assume first that $a_n > 0$ for all $n \in \mathbb{N}$. Using differentiation, one can show that $\sin x > x - \frac{x^3}{6}$ for $x > 0$. Hence $1 - \frac{\sin a_n}{a_n} < \frac{1}{6}a_n^2$. Since $a_n^2 < a_n$ for sufficiently large n, the series $\sum_{n=1}^{\infty} a_n^2$ is convergent, which in turn implies the convergence of the given series. If we drop the assumption $a_n > 0$, then the series can diverge or converge. Indeed, take $a_n = (-1)^n \frac{1}{n^\alpha}$ with $\alpha > 0$. Then the series $\sum_{n=1}^{\infty}\left(1 - \frac{\sin a_n}{a_n}\right)$ diverges if $0 < \alpha \leq \frac{1}{2}$ and converges if $\alpha > \frac{1}{2}$.

3.4.4. No, as the following example shows:

$$a_n = \frac{(-1)^n}{n} + \frac{1}{n \ln n}, \quad b_n = \frac{(-1)^n}{n}, \quad n \geq 2.$$

3.4.5. We have $a_n = p_n - q_n$ and $|a_n| = p_n + q_n$. Note also that p_n and q_n are nonnegative. Thus both the series $\sum_{n=1}^{\infty} p_n$ and $\sum_{n=1}^{\infty} q_n$ diverge, because $\sum_{n=1}^{\infty} a_n$ converges and $\sum_{n=1}^{\infty} |a_n|$ diverges.

3.4.6. Set $S_n = a_1 + \ldots + a_n$. By the foregoing problem, we have

$$\lim_{n \to \infty} \frac{P_n}{Q_n} = \lim_{n \to \infty}\left(1 + \frac{S_n}{Q_n}\right) = 1.$$

3.4.7. The series does not converge absolutely. We will show that it converges (conditionally). To this end we group the terms with the same sign and we obtain

$$\sum_{n=1}^{\infty} \frac{(-1)^{\left[\frac{n}{3}\right]}}{n} = \frac{3}{2} + \sum_{n=1}^{\infty}(-1)^n\left(\frac{1}{3n} + \frac{1}{3n+1} + \frac{1}{3n+2}\right).$$

Thus the convergence follows from the Leibniz theorem.

3.4.8. Clearly, the series converges absolutely if $a > 1$ and diverges if $a \leq 0$. We will show that if $0 < a \leq 1$, then the series converges conditionally. Observe that the first three terms of the series are negative, the next five terms are positive, etc. Now grouping the terms of the same sign we get the alternating series $\sum\limits_{n=1}^{\infty} (-1)^n A_n$, where $A_n = \sum\limits_{k=n^2}^{(n+1)^2-1} \frac{1}{k^a}$. Moreover, for $a \neq 1$,

$$A_n < \frac{1}{n^{2a}} + \int_{n^2}^{(n+1)^2} \frac{1}{t^a} dt = \frac{1}{n^{2a}} + \frac{1}{1-a}((n+1)^{2-2a} - n^{2-2a}).$$

Hence (see 2.2.3), $\lim\limits_{n\to\infty} A_n = 0$ if $\frac{1}{2} < a < 1$. For $a = 1$ we have $\frac{2}{n+1} < A_n < \frac{2n+1}{n^2}$, and consequently, $\lim\limits_{n\to\infty} A_n = 0$ for $\frac{1}{2} < a \leq 1$. We will now show that for such a the sequence $\{A_n\}$ is monotonically decreasing. Indeed,

$$A_n - A_{n+1} = \sum_{k=n^2}^{(n+1)^2-1} \frac{1}{k^a} - \sum_{k=(n+1)^2}^{(n+2)^2-1} \frac{1}{k^a}$$

$$= \sum_{k=n^2}^{(n+1)^2-1} \frac{1}{k^a} - \sum_{k'=n^2}^{(n+1)^2+1} \frac{1}{(k'+2n+1)^a}$$

$$= \sum_{k=n^2}^{(n+1)^2-1} \left(\frac{1}{k^a} - \frac{1}{(k+2n+1)^a} \right) - \frac{1}{((n+2)^2-2)^a} - \frac{1}{((n+2)^2-1)^a}$$

$$= \sum_{k=0}^{2n} \left(\frac{1}{(n^2+k)^a} - \frac{1}{((n+1)^2+k)^a} \right) - \frac{1}{((n+2)^2-2)^a}$$

$$- \frac{1}{((n+2)^2-1)^a} > (2n+1) \left(\frac{1}{(n^2+2n)^a} - \frac{1}{((n+1)^2+2n)^a} \right)$$

$$- \frac{1}{((n+2)^2-2)^a} - \frac{1}{((n+2)^2-1)^a},$$

where the last inequality follows from the monotonicity of the function

$$g(x) = \frac{1}{(n^2+x)^a} - \frac{1}{((n+1)^2+x)^a}$$

3.4. Absolute Convergence. Theorem of Leibniz

on the interval $[0, 2n]$. Hence, for sufficiently large n,

$$A_n - A_{n+1} > (2n+1)\left(\frac{1}{(n^2+2n)^a} - \frac{1}{((n+1)^2+2n)^a}\right) - \frac{2}{(n+1)^{2a}}$$

$$= \frac{2}{n^{2a}}\left\{\left(n+\frac{1}{2}\right)\left(\left(1+\frac{2}{n}\right)^{-a} - \left(1+\frac{4}{n}+\frac{1}{n^2}\right)^{-a}\right) - \left(1+\frac{1}{n}\right)^{-2a}\right\}$$

$$\geq n^{-2a}(2a-1) > 0,$$

because $(1+x)^{-a} > 1 - ax$ and $(1+x)^{-a} < 1 - ax + \frac{a(a+1)}{2}x^2$ for $a, x > 0$. (These two inequalities can be proved by differentiation.) Thus by the theorem of Leibniz, the series $\sum_{n=1}^{\infty}(-1)^n A_n$ converges if $\frac{1}{2} < a \leq 1$.

If $0 < a \leq \frac{1}{2}$, then since $A_n > (2n+1)\frac{1}{(n^2+2n)^a}$, the necessary condition for the convergence of $\sum_{n=1}^{\infty}(-1)^n A_n$ is not satisfied.

3.4.9. As in the solutions of 3.4.7 and 3.4.8, we group the terms of the same sign and rewrite the series in the following form:

$$\sum_{n=1}^{\infty}(-1)^{n-1}\left(\frac{1}{[e^{n-1}]+1} + \ldots + \frac{1}{[e^n]}\right).$$

We also observe that

$$\frac{1}{[e^{n-1}]+1} + \ldots + \frac{1}{[e^n]} > \frac{[e^n]-[e^{n-1}]}{[e^n]} = 1 - \frac{[e^{n-1}]}{[e^n]}.$$

Moreover, since

$$\lim_{n \to \infty}\left(1 - \frac{[e^{n-1}]}{[e^n]}\right) = 1 - \frac{1}{e},$$

the necessary condition for the convergence of the series is not satisfied and therefore the series diverges.

3.4.10.

(a) Observe that the series can be written in the form

$$\sum_{n=0}^{\infty}(-1)^n A_n, \quad \text{where} \quad A_n = \sum_{k=2^n}^{2^{n+1}-1}\frac{1}{k}.$$

Since $A_n > 2^n \frac{1}{2^{n+1}-1} \xrightarrow[n\to\infty]{} \frac{1}{2}$, the series diverges.

(b) As in (a), the series can be written in the form

$$\sum_{n=1}^{\infty}(-1)^n A_n, \quad \text{where} \quad A_n = \sum_{k=2^n}^{2^{n+1}-1} \frac{1}{k \ln k}.$$

Moreover,

$$0 < A_n < 2^n \frac{1}{2^n \ln 2^n}.$$

This implies $\lim_{n\to\infty} A_n = 0$. We will now show that $\{A_n\}$ monotonically decreases. Indeed,

$$A_{n+1} = \sum_{k=2^{n+1}}^{2^{n+2}-1} \frac{1}{k \ln k} = \sum_{l=0}^{2^{n+1}-1} \frac{1}{(2^{n+1}+l)\ln(2^{n+1}+l)}$$

$$= \sum_{l=0}^{2^n-1} \left(\frac{1}{(2^{n+1}+2l)\ln(2^{n+1}+2l)} + \frac{1}{(2^{n+1}+2l+1)\ln(2^{n+1}+2l+1)} \right)$$

$$< \sum_{l=0}^{2^n-1} \frac{2}{(2^{n+1}+2l)\ln(2^{n+1}+2l)} < \sum_{l=0}^{2^n-1} \frac{1}{(2^n+l)\ln(2^n+l)} = A_n.$$

3.4.11. We have

$$(-1)^n \frac{\sqrt{n}}{(-1)^n + \sqrt{n}} \sin\frac{1}{\sqrt{n}} = (-1)^n \left(1 - \frac{(-1)^n}{(-1)^n + \sqrt{n}}\right) \sin\frac{1}{\sqrt{n}}$$

$$= (-1)^n \sin\frac{1}{\sqrt{n}} + \frac{(-1)^n}{n-1} \sin\frac{1}{\sqrt{n}} - \frac{\sqrt{n}}{n-1} \sin\frac{1}{\sqrt{n}}.$$

By the Leibniz test, both the series

$$\sum_{n=2}^{\infty}(-1)^n \sin\frac{1}{\sqrt{n}} \quad \text{and} \quad \sum_{n=2}^{\infty} \frac{(-1)^n}{n-1} \sin\frac{1}{\sqrt{n}}$$

converge. But the series

$$\sum_{n=2}^{\infty} \frac{\sqrt{n}}{n-1} \sin\frac{1}{\sqrt{n}}$$

diverges, so the given series also diverges.

3.4. Absolute Convergence. Theorem of Leibniz

3.4.12.

(a) The series converges absolutely (see 3.2.1 (f)).

(b) Convergence of this series follows from the test of Leibniz. The series converges conditionally (see 3.2.1 (g)).

(c) Clearly, the sequence $\{\sqrt[n]{n}\}$, $n \geq 3$, is monotonically decreasing and therefore the series converges. However, it does not converge absolutely (see 3.2.5 (b)).

(d) Convergence follows from the monotonicity of $\{(1+\frac{1}{n})^n\}$ and from the fact that the limit of this sequence is e (see 2.1.38). To prove that the series does not converge absolutely, we use the inequality

$$\ln(1+x) < x - \frac{1}{2}x^2 + \frac{1}{3}x^3, \quad x > 0,$$

with $x = \frac{1}{n}$, and we get $\left(1+\frac{1}{n}\right)^n < e^{1-\frac{1}{2n}+\frac{1}{3n^2}}$. Hence

$$e - \left(1+\frac{1}{n}\right)^n > e\left(1 - e^{-\frac{1}{2n}+\frac{1}{3n^2}}\right) > e\left(1 - e^{-\frac{1}{4n}}\right) \quad \text{for} \quad n > 1.$$

It follows from 2.5.4 (a) that, for sufficiently large n,

$$4n\left(1 - e^{-\frac{1}{4n}}\right) > \frac{1}{2}.$$

Therefore the series $\sum\limits_{n=1}^{\infty} \left(e - \left(1+\frac{1}{n}\right)^n\right)$ diverges.

(e) The convergence of this series follows from the monotonicity of the sequence $\{(1+\frac{1}{n})^{n+1}\}$ and from the fact that e is its limit (see 2.1.38). In view of 3.2.5 (c), the series does not converge absolutely.

3.4.13.

(a) The function

$$f(x) = \frac{(\ln x)^a}{x^b}, \quad x \in (e^{\frac{a}{b}}, +\infty),$$

monotonically decreases to zero as $x \to \infty$. Therefore by the Leibniz test, the series converges. We claim that if $b > 1$, then the series converges absolutely. By the theorem of Cauchy (see 3.2.28), it is enough to show the convergence of

$$\sum_{n=1}^{\infty} 2^n \frac{n^a}{2^{nb}}.$$

Now by the root test this series converges if $b > 1$ and diverges if $0 < b < 1$. Clearly, if $b = 1$, this series diverges.

(b) Note that
$$\frac{(\ln n)^{\ln n}}{n^b} = \frac{e^{(\ln n)(\ln \ln n)}}{n^b} = \frac{n^{\ln \ln n}}{n^b}.$$
Hence the necessary condition for convergence is not satisfied.

3.4.14. By the monotonicity of $\{a_n\}$, we have
$$r_{2n} = (a_{2n+1} - a_{2n+2}) + (a_{2n+3} - a_{2n+4}) + \ldots > 0,$$
$$r_{2n+1} = (-a_{2n+2} + a_{2n+3}) + (-a_{2n+4} + a_{2n+5}) + \ldots < 0$$
and
$$r_{2n} = a_{2n+1} + (-a_{2n+2} + a_{2n+3}) + \ldots < a_{2n+1},$$
$$-r_{2n+1} = a_{2n+2} + (-a_{2n+3} + a_{2n+4}) + \ldots < a_{2n+2}.$$

3.4.15. Note that
$$\sum_{k=1}^{n}(a_k + a_{k+1}) - 2\sum_{k=1}^{n} a_k = a_{n+1} - a_1 \xrightarrow[n\to\infty]{} -a_1.$$

3.4.16. Observe that
$$\sum_{k=1}^{n}(aa_k + ba_{k+1} + ca_{k+2}) - (a+b+c)\sum_{k=1}^{n} a_k$$
$$= b(a_{n+1} - a_1) + c(a_{n+1} + a_{n+2} - a_1 - a_2) \xrightarrow[n\to\infty]{} -ba_1 - c(a_1 + a_2).$$

3.4.17. By assumption, there exist positive constants c and C such that for sufficiently large n, $c < |a_n| \leq C$. Hence
$$\left|\frac{1}{a_{n+1}} - \frac{1}{a_n}\right| \leq \frac{1}{c^2}|a_{n+1} - a_n|,$$
$$|a_{n+1} - a_n| \leq C^2 \left|\frac{1}{a_{n+1}} - \frac{1}{a_n}\right|.$$

3.4. Absolute Convergence. Theorem of Leibniz

Thus our claim follows from the comparison test.

3.4.18. Let S_n and $\widetilde{S}_n$ denote the nth partial sum of $\sum\limits_{n=1}^{\infty} a_n$ and $\sum\limits_{n=1}^{\infty} n(a_n - a_{n+1})$, respectively. Then

$$\widetilde{S}_n = \sum_{k=1}^{n} k(a_k - a_{k+1}) = \sum_{k=1}^{n} ka_k - \sum_{k=1}^{n} (k+1)a_{k+1} + \sum_{k=1}^{n} a_{k+1}$$
$$= -(n+1)a_{n+1} + S_{n+1},$$

which proves our claim.

3.4.19. Convergence follows from the Leibniz test.

3.4.20. If $|a| < 1$, then the series converges absolutely. Indeed, since $|\sin x| \leq |x|$,

$$\left| n! \sin a \sin \frac{a}{2} \cdot \ldots \cdot \sin \frac{a}{n} \right| \leq |a|^n.$$

We now turn to the case $|a| \geq 1$. We claim that in this case the series diverges because the necessary condition for convergence is not satisfied. In fact, for a fixed a there exists n_0 such that $\frac{|a|}{n_0} \leq 1$. Then, setting $C = (n_0 - 1)! \left| \sin a \sin \frac{a}{2} \cdot \ldots \cdot \sin \frac{a}{n_0 - 1} \right|$ and using the inequality $\frac{\sin x}{x} > 1 - \frac{x^2}{6}$, $x > 0$, we get

$$\left| n! \sin a \sin \frac{a}{2} \cdot \ldots \cdot \sin \frac{a}{n} \right| = Cn_0 \cdot \ldots \cdot n \sin \frac{|a|}{n_0} \cdot \ldots \cdot \sin \frac{|a|}{n}$$
$$\geq Cn_0 \cdot \ldots \cdot n \sin \frac{1}{n_0} \cdot \ldots \cdot \sin \frac{1}{n} = C \prod_{k=n_0}^{n} \left(1 - \frac{1}{6k^2}\right)$$
$$\geq C \prod_{k=n_0}^{n} \left(1 - \frac{1}{k^2}\right) = C \frac{(n_0 - 1)(n+1)}{n_0 n} \xrightarrow[n \to \infty]{} C \frac{n_0 - 1}{n_0} > 0.$$

3.4.21. By 2.5.4 (a),

$$\lim_{n \to \infty} \frac{\sqrt[n]{a} - \frac{\sqrt[n]{b} + \sqrt[n]{c}}{2}}{\frac{1}{n}} = \lim_{n \to \infty} \left(\frac{\sqrt[n]{a} - 1}{\frac{1}{n}} - \frac{\sqrt[n]{b} - 1 + \sqrt[n]{c} - 1}{2 \frac{1}{n}} \right)$$
$$= \ln a - \frac{1}{2}(\ln b + \ln c) = \ln \frac{a}{\sqrt{bc}}.$$

Hence if $a > \sqrt{bc}$, then, beginning with some value of the index n, the terms of our series are positive, and by the comparison test it diverges. If $a < \sqrt{bc}$, then the terms of the series are negative and it also diverges. For $a = \sqrt{bc}$, we have

$$\sum_{n=1}^{\infty}\left(\sqrt[n]{a} - \frac{\sqrt[n]{b} + \sqrt[n]{c}}{2}\right) = -\frac{1}{2}\sum_{n=1}^{\infty}\left(\sqrt[2n]{b} - \sqrt[2n]{c}\right)^2.$$

Since

$$\lim_{n\to\infty}\left(\frac{\sqrt[2n]{b} - 1 - \sqrt[2n]{c} + 1}{\frac{1}{2n}}\right)^2 = (\ln b - \ln c)^2,$$

the convergence of our series follows from the convergence of $\sum_{n=1}^{\infty}\frac{1}{n^2}$.

3.4.22.

(a) By 1.1.14, there exist a sequence of integers $\{p_n\}$ and a sequence of positive integers $\{q_n\}$ for which

$$\left|\pi - \frac{p_n}{q_n}\right| < \frac{1}{q_n^2}.$$

Hence $|\cos p_n| = |\cos(\pi q_n - p_n)| > \cos\frac{1}{q_n} = 1 - 2\sin^2\frac{1}{2q_n} > 1 - \frac{1}{2q_n^2}$. Thus

$$(|\cos p_n|)^{p_n} > \left(1 - \frac{1}{2q_n^2}\right)^{p_n} > 1 - \frac{p_n}{q_n}\frac{1}{2q_n}.$$

This means that the subsequence $\{(\cos p_n)^{p_n}\}$ of $\{\cos^n n\}$ does not converge to zero. Therefore the necessary condition for convergence is not satisfied.

(b) By Problem 1.1.22, we know that the sequences $\{p_n\}$ and $\{q_n\}$ mentioned in (a) can be chosen in such a way that all the terms of $\{q_n\}$ are odd. Then by the inequality

$$\left|\frac{\pi}{2} - \frac{p_n}{q_n}\right| < \frac{1}{q_n^2}$$

we get $|\sin p_n| = |\cos\left(\frac{\pi}{2}q_n - p_n\right)| > \cos\frac{1}{q_n} > 1 - \frac{1}{2q_n^2}$. Thus, as in (a), the sequence $(\sin p_n)^{p_n}$ does not converge to zero, and consequently, the series diverges.

3.4. Absolute Convergence. Theorem of Leibniz 319

3.4.23.

(a) By assumption (see also 2.4.13 (b)), there are n_0 and α such that

$$n\left(\frac{a_n}{a_{n+1}} - 1\right) > \alpha > 0 \quad \text{for} \quad n \geq n_0.$$

Hence $\frac{a_{n+1}}{a_n} < \frac{n}{n+\alpha} < 1$, which means that, beginning with the value n_0 of the index n, the sequence $\{a_n\}$ monotonically decreases. We will show that $\lim\limits_{n\to\infty} a_n = 0$. It follows from the above that

$$a_{n+1} = \frac{a_{n+1}}{a_n} \cdot \frac{a_n}{a_{n-1}} \cdot \ldots \cdot \frac{a_{n_0+1}}{a_{n_0}} \cdot a_{n_0} < \frac{n \cdot (n-1) \cdot \ldots \cdot n_0}{(\alpha+n) \cdot \ldots \cdot (\alpha+n_0)} a_{n_0}.$$

Now it is enough to prove that $\lim\limits_{n\to\infty} \frac{n\cdot(n-1)\cdot\ldots\cdot n_0}{(\alpha+n)\cdot\ldots\cdot(\alpha+n_0)} = 0$. Indeed,

$$\lim_{n\to\infty} \frac{n \cdot (n-1) \cdot \ldots \cdot n_0}{(\alpha+n) \cdot \ldots \cdot (\alpha+n_0)} = \lim_{n\to\infty} \frac{1}{\left(1+\frac{\alpha}{n}\right) \cdot \ldots \cdot \left(1+\frac{\alpha}{n_0}\right)} = 0,$$

because (see 1.2.1)

$$\left(1+\frac{\alpha}{n_0}\right) \cdot \ldots \cdot \left(1+\frac{\alpha}{n}\right) > 1 + \frac{\alpha}{n_0} + \ldots + \frac{\alpha}{n} \xrightarrow[n\to\infty]{} \infty.$$

So by the Leibniz test, the series $\sum\limits_{n=1}^{\infty} (-1)^n a_n$ converges.

(b) By the assumption $n\left(\frac{a_n}{a_{n+1}} - 1\right) \leq 0$, the sequence $\{a_n\}$ monotonically increases and consequently the series $\sum\limits_{n=1}^{\infty} (-1)^n a_n$ diverges, because the necessary condition for convergence is not satisfied.

3.4.24. By assumption, $\lim\limits_{n\to\infty} n\left(\frac{a_n}{a_{n+1}} - 1\right) = \alpha$. For $\alpha \neq 0$, the test proved in the foregoing problem can be applied. For $\alpha = 0$, the necessary condition for convergence is not satisfied. Indeed, we have

$$\frac{1}{a_n} = \frac{1}{a_1} \cdot \frac{a_1}{a_2} \cdot \frac{a_2}{a_3} \cdot \ldots \cdot \frac{a_{n-1}}{a_n}$$

$$= \frac{1}{a_1}\left(1+\frac{\beta_1}{1^{1+\varepsilon}}\right)\left(1+\frac{\beta_2}{2^{1+\varepsilon}}\right)\ldots\left(1+\frac{\beta_{n-1}}{(n-1)^{1+\varepsilon}}\right).$$

Moreover, there exists β such that $|\beta_n| \leq \beta$. Hence

$$a_n \geq \frac{a_1}{\left(1 + \frac{\beta}{1^{1+\varepsilon}}\right)\left(1 + \frac{\beta}{2^{1+\varepsilon}}\right) \cdots \left(1 + \frac{\beta}{(n-1)^{1+\varepsilon}}\right)} \geq \frac{a_1}{e^{\beta A}},$$

where $A = \sum_{n=1}^{\infty} \frac{1}{n^{1+\varepsilon}}$.

3.4.25. By 2.5.34, the existence of the limit $\lim_{n\to\infty} n \ln \frac{a_n}{a_{n+1}}$ is equivalent to the existence of $\lim_{n\to\infty} n \left(\frac{a_n}{a_{n+1}} - 1\right)$, and both limits are equal. Set $a_n = \frac{n! e^n}{n^{n+p}}$. Then $\lim_{n\to\infty} n \ln \frac{a_n}{a_{n+1}} = p - \frac{1}{2}$. Hence, by 3.4.23, the series converges if $p > \frac{1}{2}$ and diverges if $p < \frac{1}{2}$. In the case $p = \frac{1}{2}$ the necessary condition for convergence is not satisfied because, by Stirling's formula, $\lim_{n\to\infty} a_n = \sqrt{2\pi}$.

3.4.26. Let $S_n = a_1 + a_2 + \ldots + a_n$. We use the so-called *summation by parts (Abel's transformation)* to obtain

$$a_1 p_1 + a_2 p_2 + \ldots + a_n p_n = \sum_{k=1}^{n-1} S_k (p_k - p_{k+1}) + S_n p_n,$$

and we get

$$\frac{a_1 p_1 + a_2 p_2 + \ldots + a_n p_n}{p_n} = S_n - \sum_{k=1}^{n-1} S_k \frac{p_{k+1} - p_k}{p_n}.$$

Now it is enough to apply the Toeplitz theorem (see 2.3.1).

3.4.27. Apply the result in the last problem to the series $\sum_{n=1}^{\infty} a_n b_n$ and take $p_n = \frac{1}{a_n}$.

3.4.28. The result is contained as a special case in the preceding problem.

3.4.29. If the series were not absolutely convergent, then the subseries of all positive terms and the subseries of all negative terms would diverge (see 3.4.5).

3.4. Absolute Convergence. Theorem of Leibniz

3.4.30. [20] No, as the following example shows.

Take a conditionally convergent series $\sum_{n=1}^{\infty} b_n$ and set

$$a_1 = b_1, \ a_2 = a_3 = \frac{b_2}{2!}, \ a_4 = a_5 = \ldots = a_9 = \frac{b_3}{3!}, \ldots,$$

$$a_{1!+2!+\ldots+(n-1)!+1} = a_{1!+2!+\ldots+(n-1)!+2}$$

$$= \ldots = a_{1!+2!+\ldots+(n-1)!+n!} = \frac{b_n}{n!}, \ldots.$$

Then the series $\sum_{n=1}^{\infty} a_n$ converges conditionally. But for each $k \geq 1$ and $l \geq 2$ the subseries

$$a_k + a_{k+l} + a_{k+2l} + \ldots$$

converges. Indeed, for $n \geq l$ there are $\frac{n!}{l}$ terms of the form $\frac{b_n}{n!}$. Grouping these terms, we get the convergent series

$$constant + \frac{1}{l} \sum_{n=n_0}^{\infty} b_n.$$

3.4.31. Consider the series

$$1 + \frac{1}{2\sqrt[3]{2}} + \frac{1}{2\sqrt[3]{2}} - \frac{1}{\sqrt[3]{2}} + \ldots + \underbrace{\frac{1}{n\sqrt[3]{n}} + \ldots + \frac{1}{n\sqrt[3]{n}}}_{n \text{ times}} - \frac{1}{\sqrt[3]{n}} + \ldots.$$

3.4.32. Yes. Consider the series

$$1 + \frac{1}{2\ln 2} + \frac{1}{2\ln 2} - \frac{1}{\ln 2} + \ldots + \underbrace{\frac{1}{n\ln n} + \ldots + \frac{1}{n\ln n}}_{n \text{ times}} - \frac{1}{\ln n} + \ldots.$$

Then

$$\sum_{n=1}^{\frac{N^2+3N-2}{2}} a_n^k = \begin{cases} 1 + \sum_{n=2}^{N} \left(\frac{1}{n^{k-1}\ln^k n} + \frac{1}{\ln^k n} \right) & \text{if } k \text{ is even,} \\ 1 + \sum_{n=2}^{N} \left(\frac{1}{n^{k-1}\ln^k n} - \frac{1}{\ln^k n} \right) & \text{if } k \text{ is odd.} \end{cases}$$

By the Cauchy theorem (see 3.2.28) the series $\sum_{n=2}^{\infty} \frac{1}{\ln^k n}$ diverges for every $k \in \mathbb{N}$. On the other hand, the series $\sum_{n=2}^{\infty} \frac{1}{n^{k-1} \ln^k n}$ converges if $k \geq 2$.

3.4.33. [20] Suppose, contrary to our claim, that $\lim\limits_{n \to \infty} \frac{\varepsilon_1 + \varepsilon_2 + \ldots + \varepsilon_n}{n} = 2\alpha > 0$. Then, by 2.4.13 (b), there is n_0 such that for $n > n_0$,

(*) $$\varepsilon_1 + \varepsilon_2 + \ldots + \varepsilon_n > \alpha n.$$

Set $E_n = \varepsilon_1 + \varepsilon_2 + \ldots + \varepsilon_n$. Using summation by parts, we get

$$\varepsilon_1 a_1 + \varepsilon_2 a_2 + \ldots + \varepsilon_n a_n = \sum_{k=1}^{n-1} E_k (a_k - a_{k+1}) + E_n a_n.$$

Therefore, by (*),

$$\varepsilon_1 a_1 + \varepsilon_2 a_2 + \ldots + \varepsilon_n a_n$$
$$> \sum_{k=1}^{n_0} E_k (a_k - a_{k+1}) + \alpha \sum_{k=n_0+1}^{n-1} k(a_k - a_{k+1}) + \alpha n a_n$$
$$= \text{constant} + \alpha \sum_{k=n_0+2}^{n} a_k,$$

a contradiction.

3.4.34. [20] Set $E_n = \varepsilon_1 + \varepsilon_2 + \ldots + \varepsilon_n$, $n \in \mathbb{N}$. The sequence $\{E_n\}$ has the property that between two terms with different signs there is a vanishing term. We consider two cases:
  (1) finitely many terms of $\{E_n\}$ vanish,
  (2) infinitely many terms of $\{E_n\}$ vanish.

(1) is contained as a special case in 3.2.35. In case (2), by the Cauchy criterion, for each $\varepsilon > 0$ there is n_0 such that if $n > m > n_0$, then

(*)
$$\varepsilon > \left| \sum_{k=m+1}^{n} \varepsilon_k a_k \right| = \left| \sum_{k=m+1}^{n} ((E_k - E_m) - (E_{k-1} - E_m)) a_k \right|$$
$$= \left| \sum_{k=m+1}^{n} (E_k - E_m)(a_k - a_{k+1}) + (E_n - E_m) a_{n+1} \right|.$$

3.4. Absolute Convergence. Theorem of Leibniz

Assume that $E_m = 0$ and that the terms $E_{m+1}, E_{m+2}, ..., E_n$ have the same sign. Then (∗) and the monotonicity of the sequence $\{a_n\}$ imply that

$$|E_n a_n| < \varepsilon, \qquad n \geq m+1.$$

3.4.35. The proof is analogous to that of 3.4.33. We set $E_n = p_1 b_1 + ... + p_n b_n$ and assume that $\lim\limits_{n \to \infty} \frac{p_1 b_1 + ... + p_n b_n}{n} = 2\alpha > 0$. Then for $n > n_0$ we have $p_1 b_1 + ... + p_n b_n > \alpha n$, and consequently,

$$b_1 + ... + b_n = \frac{1}{p_1}(p_1 b_1) + ... + \frac{1}{p_1}(p_n b_n)$$

$$= \sum_{k=1}^{n-1} E_k \left(\frac{1}{p_k} - \frac{1}{p_{k+1}} \right) + E_n \frac{1}{p_n} > constant + \alpha \sum_{k=n_0+2}^{n} \frac{1}{p_k},$$

a contradiction.

3.4.36. We first show that if $p = q$, then the series converges. We have

$$S_{lp} = \left(1 + \frac{1}{2} + ... + \frac{1}{p}\right) - \left(\frac{1}{p+1} + ... + \frac{1}{2p}\right)$$

$$+ ... + (-1)^{l+1} \left(\frac{1}{(l-1)p+1} + ... + \frac{1}{lp}\right).$$

Therefore S_{lp} is a partial sum of an alternating series. By the Leibniz test the limit $\lim\limits_{l \to \infty} S_{lp}$ exists. Clearly, each partial sum of the form S_{lp+k}, $k = 1, 2, ..., p-1$, tends to the same limit as $l \to \infty$.

Assume now that our series converges. Then, by 3.4.34,

$$\lim_{n \to \infty} \frac{np - nq}{np + nq} = \frac{p-q}{p+q} = 0,$$

which implies that $p = q$.

3.4.37. We note first that if conditions (i)-(iii) are satisfied, then for any convergent sequence $\{a_n\}$, the transformed sequence $\{b_n\}$ is well defined. Now the proof runs in much the same way as in the solutions of Problems 2.3.1 and 2.3.26.

3.5. The Dirichlet and Abel Tests

3.5.1.

(a) Since $\frac{\sin^2 n}{n} = \frac{1}{2n}(1 - \cos(2n))$, it is enough to consider the series

$$\sum_{n=1}^{\infty}(-1)^n \frac{1}{n} \quad \text{and} \quad \sum_{n=1}^{\infty}(-1)^n \frac{1}{n}\cos(2n).$$

By the Leibniz test, the first series converges. Convergence of the second one follows from the Dirichlet test (see, e.g., [12], p.105). Indeed, by the formula (which can be proved by induction)

$$(1) \quad \sum_{k=1}^{n} \cos ka = \frac{\sin \frac{na}{2} \cos \frac{(n+1)a}{2}}{\sin \frac{a}{2}} \quad \text{for} \quad a \neq 2l\pi, \ l \in \mathbb{Z},$$

we obtain

$$\left|\sum_{k=1}^{n}(-1)^k \cos(2k)\right| = \left|\sum_{k=1}^{n}\cos\big((\pi-2)k\big)\right|$$

$$= \left|\frac{\sin\frac{(\pi-2)n}{2} \cos \frac{(n+1)(\pi-2)}{2}}{\cos 1}\right| \leq \frac{1}{\cos 1}.$$

Therefore the sequence of partial sums of $\sum_{n=1}^{\infty}(-1)^n \cos(2n)$ is bounded. Moreover, $\{\frac{1}{n}\}$ tends monotonically to zero. Thus the series $\sum_{n=1}^{\infty}(-1)^n \frac{1}{n}\cos(2n)$ converges.

(b) The sequence

$$a_n = \frac{1 + \frac{1}{2} + \ldots + \frac{1}{n}}{n}$$

of arithmetic means of $\{\frac{1}{n}\}$ converges to zero (see 2.3.2). Moreover, it is easy to check that the sequence $\{a_n\}$ monotonically decreases. By the formula (which can be proved by induction)

$$(2) \quad \sum_{k=1}^{n} \sin ka = \frac{\sin \frac{na}{2} \sin \frac{(n+1)a}{2}}{\sin \frac{a}{2}} \quad \text{for} \quad a \neq 2l\pi, \ l \in \mathbb{Z},$$

3.5. The Dirichlet and Abel Tests

we get
$$\left|\sum_{k=1}^{n}\sin k\right|=\left|\frac{\sin\frac{n}{2}\sin\frac{n+1}{2}}{\sin\frac{1}{2}}\right|\leq\frac{1}{\sin\frac{1}{2}}.$$
Therefore by the Dirichlet test the series converges.

(c) Observe that
$$\cos\left(\pi\frac{n^2}{n+1}\right)=\cos\left(n\pi-\frac{n\pi}{n+1}\right)=(-1)^n\cos\left(\pi-\frac{\pi}{n+1}\right)$$
$$=(-1)^{n+1}\cos\frac{\pi}{n+1}.$$

Thus the given series can be rewritten in the form
$$\sum_{n=2}^{\infty}(-1)^{n+1}\frac{\cos\frac{\pi}{n+1}}{\ln^2 n}.$$

The convergence of the above series follows from the Abel test (see, e.g., [12], p.106), because the series $\sum_{n=2}^{\infty}(-1)^{n+1}\frac{1}{\ln^2 n}$ converges (by the Leibniz test) and $\{\cos\frac{\pi}{1+n}\}$ is a monotonic and bounded sequence.

(d) We have
$$\frac{\sin\frac{n\pi}{4}}{n^a+\sin\frac{n\pi}{4}}=\frac{\sin\frac{n\pi}{4}}{n^a}\left(1-\frac{\frac{\sin\frac{n\pi}{4}}{n^a}}{1+\frac{\sin\frac{n\pi}{4}}{n^a}}\right).$$

The series
$$\sum_{n=1}^{\infty}\frac{\sin\frac{n\pi}{4}}{n^a},\quad a>0,$$
converges (by the Dirichlet test). Now we will study the series (with positive terms)
$$\sum_{n=1}^{\infty}\frac{\frac{\sin^2\frac{n\pi}{4}}{n^{2a}}}{1+\frac{\sin\frac{n\pi}{4}}{n^a}}.$$

There exist positive constants c_a and C_a such that
$$c_a\frac{1}{n^{2a}}<\frac{\frac{\sin^2\frac{n\pi}{4}}{n^{2a}}}{1+\frac{\sin\frac{n\pi}{4}}{n^a}}<C_a\frac{1}{n^{2a}},\quad n\neq 4k,\ k\in\mathbb{N}.$$

Therefore the series converges if $a>\frac{1}{2}$ and diverges if $0<a\leq\frac{1}{2}$.

3.5.2. We have
$$\sum_{n=2}^{N}\frac{\sin\left(n+\frac{1}{n}\right)}{\ln\ln n}=\sum_{n=2}^{N}\frac{\sin n\cos\frac{1}{n}}{\ln\ln n}+\sum_{n=2}^{N}\frac{\cos n\sin\frac{1}{n}}{\ln\ln n}.$$

By formula (2) given in the solution of 3.5.1(b) and by the Dirichlet test we see that the series $\sum_{n=2}^{\infty}\frac{\sin n}{\ln\ln n}$ converges. Since the sequence $\{\cos\frac{1}{n}\}$ is monotonic and bounded, the series $\sum_{n=2}^{\infty}\frac{\sin n\cos\frac{1}{n}}{\ln\ln n}$ converges, by the Abel test. Finally, the convergence of $\sum_{n=2}^{\infty}\frac{\cos n\sin\frac{1}{n}}{\ln\ln n}$ follows from formula (1) given in the solution of 3.5.1(a) and from the Dirichlet test.

3.5.3.

(a) We have
$$2\left|\sum_{k=1}^{n}\sin(k^{2}a)\sin(ka)\right|=\left|\sum_{k=1}^{n}[\cos(k(k-1)a)-\cos(k(k+1)a)]\right|$$
$$=|1-\cos(n(n+1)a)|\leq 2.$$

Thus the convergence of the series follows from the Dirichlet test.

(b) As in (a), the Dirichlet test can be applied.

3.5.4. In view of the formula
$$\frac{\cos n\sin(na)}{n}=\frac{1}{2}\frac{\sin(n(a+1))}{n}+\frac{1}{2}\frac{\sin(n(a-1))}{n},$$
the convergence of our series follows directly from the Dirichlet test (use formula (2) in the solution of 3.5.1(b)).

3.5.5. If $a=k\pi$, $k\in\mathbb{Z}$, then all terms of the series are equal to zero. If $a\neq k\pi$ then, by the inequality $|\sin x|\geq\sin^{2}x=\frac{1}{2}(1-\cos 2x)$, we get
$$\sum_{n=1}^{N}\frac{|\sin(na)|}{n}\geq\frac{1}{2}\sum_{n=1}^{N}\frac{1}{n}-\frac{1}{2}\sum_{n=1}^{N}\frac{\cos(2na)}{n}.$$

Hence in this case the series does not converge absolutely.

3.5. The Dirichlet and Abel Tests

3.5.6. Assume first that $0 < a < \pi$, and set $m = \left[\frac{\sqrt{\pi}}{a}\right]$. Then, for sufficiently large n,

$$\left|\sum_{k=1}^{n}\frac{\sin(ak)}{k}\right| \leq \sum_{k=1}^{m}\left|\frac{\sin(ak)}{k}\right| + \left|\sum_{k=m+1}^{n}\frac{\sin(ak)}{k}\right|.$$

Since $|\sin t| < |t|$ for $t \neq 0$,

(*) $$\sum_{k=1}^{m}\left|\frac{\sin(ak)}{k}\right| < \sum_{k=1}^{m}\frac{ka}{k} = ma \leq \sqrt{\pi}.$$

Moreover, from (2) in the solution of 3.5.1(b) and from the inequality $\sin t > \frac{2}{\pi}t$, $0 < t < \frac{\pi}{2}$, we get

(**) $$\left|\sum_{k=m+1}^{n}\frac{\sin(ak)}{k}\right| \leq \frac{1}{(m+1)|\sin\frac{a}{2}|} < \frac{1}{\frac{a}{\pi}\frac{\sqrt{\pi}}{a}} = \sqrt{\pi}.$$

Combining (*) with (**), we see that the desired inequality holds for $a \in (0, \pi)$. Clearly, since the sine function is odd, it also holds for $a \in (-\pi, 0)$. Moreover, since $\sin k\pi = 0$ and the sine function is periodic, the inequality holds for every $a \in \mathbb{R}$.

3.5.7. The convergence of the series follows from the Abel test, because the series $\sum_{n=1}^{\infty}(-1)^n\frac{1}{\sqrt{n}}$ is convergent and $\{\arctan n\}$ is a monotonically increasing and bounded sequence.

3.5.8. By the Abel test the series converges. Indeed, $\sum_{n=1}^{\infty}(-1)^n\frac{1}{n}$ converges, and the sequence $\{\sqrt[n]{\ln x}\}$ is bounded, strictly decreasing if $x > e$, and strictly increasing if $1 < x < e$.

3.5.9.

(a) Observe first that, by the Abel test, the series $\sum_{n=1}^{\infty}\frac{a_n}{b_n}$ converges.

Moreover, since the series $\sum_{n=1}^{\infty}a_n$ is convergent, the sequence

$\{r_n\}$, $r_n = \sum\limits_{k=n}^{\infty} a_k$, tends to zero. Hence for $p \geq n$,

$$\left| \sum_{k=n}^{p} \frac{a_k}{b_k} \right| = \left| \sum_{k=n}^{p} \frac{r_k - r_{k+1}}{b_k} \right| = \left| \sum_{k=n}^{p} \frac{r_k}{b_k} - \sum_{k=n}^{p} \frac{r_{k+1}}{b_k} \right|$$

$$= \left| \frac{r_n}{b_n} + \sum_{k=n+1}^{p} r_k \left(\frac{1}{b_k} - \frac{1}{b_{k-1}} \right) - \frac{r_{p+1}}{b_p} \right|$$

$$\leq \varepsilon_n \left(\frac{1}{b_n} + \frac{1}{b_n} - \frac{1}{b_p} + \frac{1}{b_p} \right) = \frac{2\varepsilon_n}{b_n},$$

where $\varepsilon_n = \sup\limits_{k \geq n} |r_k|$. Consequently,

$$\left| \sum_{k=n}^{\infty} \frac{a_k}{b_k} \right| \leq 2\varepsilon_n \frac{1}{b_n} = o\left(\frac{1}{b_n} \right).$$

(b) See 3.4.26.

3.5.10. Note that

$$\sum_{k=0}^{\infty} (k+1)c_{n+k} = \sum_{k=1}^{\infty} \frac{k}{n+k-1}(n+k-1)c_{n+k-1}.$$

Thus the Abel test implies the convergence of $\sum\limits_{k=0}^{\infty} (k+1)c_{n+k}$ for each $n \in \mathbb{N}$. Setting $r_n = nc_n + (n+1)c_{n+1} + ...$, we get

$$t_n = \sum_{k=0}^{\infty} (k+1)c_{n+k} = \sum_{k=n}^{\infty} (k-n+1)c_k$$

$$= \sum_{k=n}^{\infty} kc_k - (n-1) \sum_{k=n}^{\infty} \frac{1}{k} kc_k = r_n - (n-1) \sum_{k=n}^{\infty} \frac{1}{k}(r_k - r_{k+1})$$

$$= \frac{1}{n} r_n + (n-1) \sum_{k=n+1}^{\infty} \left(\frac{1}{k-1} - \frac{1}{k} \right) r_k.$$

3.5. The Dirichlet and Abel Tests

Hence

$$|t_n| \le \frac{1}{n}|r_n| + \sup_{k \ge n+1}|r_k|(n-1)\sum_{k=n+1}^{\infty}\left(\frac{1}{k-1}-\frac{1}{k}\right)$$

$$\le \frac{1}{n}|r_n| + \sup_{k \ge n+1}|r_k|\frac{n-1}{n}.$$

This together with $\lim_{n\to\infty} r_n = 0$ yields $\lim_{n\to\infty} t_n = 0$.

3.5.11. By summation by parts,

$$S_n = \sum_{i=1}^{n} a_i b_i^k = \sum_{i=1}^{n-1} A_i(b_i^k - b_{i+1}^k) + A_n b_n^k,$$

where A_n denotes the nth partial sum of the series $\sum_{n=1}^{\infty} a_n$. Given $\varepsilon > 0$, there is n_0 such that $|b_i| < \varepsilon$ for $i \ge n_0$. So if $m > n \ge n_0$ and $|A_n| \le L$, then

$$|S_m - S_n| = \left|\sum_{i=n}^{m-1} A_i(b_i^k - b_{i+1}^k) - A_n b_n^k + A_m b_m^k\right|$$

$$\le \sum_{i=n}^{m-1} |A_i||b_i^k - b_{i+1}^k| + |A_n b_n^k| + |A_m b_m^k|$$

$$\le L\left(\sum_{i=n}^{m-1} |b_i - b_{i+1}||b_i^{k-1} + b_i^{k-2}b_{i+1} + \ldots + b_{i+1}^{k-1}| + |b_n^k| + |b_m^k|\right)$$

$$\le L\left(k\varepsilon^{k-1}\sum_{i=n}^{m-1}|b_i - b_{i+1}| + 2\varepsilon^k\right).$$

Hence the convergence of $\sum_{n=1}^{\infty} a_n b_n$ follows from the Cauchy criterion.

3.5.12. By summation by parts,

$$(*) \qquad S_n = \sum_{i=1}^{n} a_i b_i = \sum_{i=1}^{n-1} A_i(b_i - b_{i+1}) + A_n b_n,$$

where A_n denotes the nth partial sum of $\sum_{n=1}^{\infty} a_n$. Since the series $\sum_{n=1}^{\infty} (b_n - b_{n+1})$ converges absolutely and the sequence $\{A_n\}$ is bounded, the series $\sum_{n=1}^{\infty} A_n(b_n - b_{n+1})$ converges absolutely. The convergence of $\sum_{n=1}^{\infty} (b_n - b_{n+1})$ implies that $\lim_{n\to\infty} b_n$ exists, because $(b_1 - b_2) + (b_2 - b_3) + \ldots + (b_{n-1} - b_n) = b_1 - b_n$. Consequently, $\lim_{n\to\infty} A_n b_n$ also exists and, by (∗), $\sum_{n=1}^{\infty} a_n b_n$ converges.

3.5.13. For $0 \leq x < 1$ the sequence $\{x^n\}$ is monotonically decreasing and bounded, and therefore the Abel test can be applied. For $-1 < x < 0$ both the sequences $\{x^{2n}\}$ and $\{x^{2n-1}\}$ are monotonic and bounded. Consequently, $\sum_{n=1}^{\infty} a_{2n} x^{2n}$ and $\sum_{n=1}^{\infty} a_{2n-1} x^{2n-1}$ are convergent. The convergence of our series follows from the equality

$$\sum_{n=1}^{\infty} a_n x^n = \sum_{n=1}^{\infty} a_{2n} x^{2n} + \sum_{n=1}^{\infty} a_{2n-1} x^{2n-1}.$$

3.5.14. Observe that if $x > x_0$, then

$$\sum_{n=1}^{\infty} \frac{a_n}{n^x} = \sum_{n=1}^{\infty} \frac{a_n}{n^{x_0}} \cdot \frac{1}{n^{x-x_0}}.$$

Now it is enough to apply the Abel test.

3.5.15. We have

$$\sum_{n=1}^{\infty} \frac{n! a_n}{x(x+1)\ldots(x+n)} = \sum_{n=1}^{\infty} \frac{a_n}{n^x} \cdot \frac{n! n^x}{x(x+1)\ldots(x+n)}.$$

Note that for sufficiently large n all the numbers $\frac{n!n^x}{x(x+1)\ldots(x+n)}$ have the same sign. We will show that they form a monotonic sequence. To this end, observe that the ratio of the $(n+1)$th term to the nth term is

$$\frac{(n+1)\left(\frac{n+1}{n}\right)^x}{x+n+1} = \frac{e^{(x+1)\ln\left(1+\frac{1}{n}\right)}}{1+\frac{x+1}{n}}.$$

3.5. The Dirichlet and Abel Tests

Now set $R_n = e^{(x+1)\ln\left(1+\frac{1}{n}\right)} - 1 - \frac{x+1}{n}$. By the result in 2.5.7 we see that

$$R_n = (x+1)\left(\ln\left(1+\frac{1}{n}\right) - \frac{1}{n}\right) + \frac{1}{2!}(x+1)^2 \ln^2\left(1+\frac{1}{n}\right)$$
$$+ \frac{1}{3!}(x+1)^3 \ln^3\left(1+\frac{1}{n}\right) + \ldots$$
$$= \frac{1}{n^2}\left(-\frac{x+1}{2} + \frac{1}{2}(x+1)^2 + O\left(\frac{1}{n}\right)\right)$$
$$+ \frac{1}{3!}(x+1)^3 \ln^3\left(1+\frac{1}{n}\right) + \ldots,$$

where $O(a_n)$ denotes an expression whose value divided by a_n remains bounded as $n \to \infty$. This implies that for sufficiently large n, R_n is positive if $x(x+1) > 0$ and negative if $x(x+1) < 0$. Consequently, for all sufficiently large n the ratio of two consecutive terms of $\left\{\frac{n!n^x}{x(x+1)\ldots(x+n)}\right\}$ is either greater than 1 or less than 1. We will now show that this sequence is convergent for $x \neq 0, -1, -2, \ldots$. To this end, write

$$\frac{n!n^x}{x(x+1)\ldots(x+n)} = \frac{1}{x}\frac{n}{x+n} \prod_{k=1}^{n-1} \frac{\left(1+\frac{1}{k}\right)^x}{1+\frac{x}{k}}.$$

Assume first that $x > 1$. For such x we have $\left(1+\frac{1}{k}\right)^x > 1 + \frac{x}{k}$. Consequently,

$$\ln \prod_{k=1}^{n-1} \frac{\left(1+\frac{1}{k}\right)^x}{1+\frac{x}{k}} = \sum_{k=1}^{n-1} \left(x \ln\left(1+\frac{1}{k}\right) - \ln\left(1+\frac{x}{k}\right)\right),$$

where all terms of the sum are positive. Moreover, note that

$$\lim_{k \to \infty} \frac{x \ln\left(1+\frac{1}{k}\right) - \ln\left(1+\frac{x}{k}\right)}{\frac{1}{k^2}} = \frac{x(x-1)}{2}.$$

As a result, the existence of the limit

$$\lim_{n \to \infty} \ln \prod_{k=1}^{n-1} \frac{\left(1+\frac{1}{k}\right)^x}{1+\frac{x}{k}}$$

follows from the convergence of the series $\sum_{k=1}^{\infty} \frac{1}{k^2}$. Thus the sequence under consideration converges for $x > 1$.

Assume now that $x \in (0,1)$. Then for such x we have $\left(1+\frac{1}{k}\right)^x < 1+\frac{x}{k}$. Therefore one can apply the above reasoning to the sequence with terms

$$-\ln \prod_{k=1}^{n-1} \frac{\left(1+\frac{1}{k}\right)^x}{1+\frac{x}{k}}.$$

Finally, we consider the case $x < 0$, $x \neq -1, -2, -3, \ldots$. Let k_0 be a positive integer such that $1+\frac{x}{k} > 0$ for $k \geq k_0$. To show that the sequence

$$\prod_{k=k_0}^{n-1} \frac{\left(1+\frac{1}{k}\right)^x}{1+\frac{x}{k}}$$

converges, note that

$$\left(1+\frac{1}{k}\right)^x > 1+\frac{x}{k} \quad \text{for} \quad k \geq k_0$$

and proceed as in the case $x > 1$.

3.5.16. It follows immediately from the Abel test that for $|x| < 1$ the convergence of $\sum_{n=1}^{\infty} a_n x^n$ implies the convergence of $\sum_{n=1}^{\infty} a_n x^{2n}$ (see 3.5.13). Since $\{\frac{1}{1-x^{2n}}\}$ is monotonic and bounded, the equality

$$\sum_{n=1}^{\infty} a_n \frac{x^n}{1-x^n} = \sum_{n=1}^{\infty} \left(a_n x^n \frac{1}{1-x^{2n}} + a_n x^{2n} \frac{1}{1-x^{2n}} \right)$$

and the Abel test imply the convergence of $\sum_{n=1}^{\infty} a_n \frac{x^n}{1-x^n}$.

3.5.17. [20] No. Let $\sum_{n=1}^{\infty} b_n$ be a conditionally convergent series. Put $F(x) = 2^{x^2}$ and define the new series $\sum_{n=1}^{\infty} a_n$ by setting

$$a_1 = a_2 = \frac{b_1}{2}, \quad a_k = \frac{b_m}{F(m)-F(m-1)} \quad \text{for} \quad F(m-1) < k \leq F(m).$$

3.6. Cauchy Product of Infinite Series

This series is also conditionally convergent. Now we show that every subseries of the form

(∗) $$a_k + a_{kl} + a_{kl^2} + \ldots$$

converges. Note first that for any positive integer m there exists a unique t_m, $t_m = \left[\log_l \frac{F(m)}{k}\right]$, such that

$$kl^{t_m} \leq F(m) < kl^{t_m+1}.$$

It follows from the definition of t_m that, beginning with some value of the index m, the subseries (∗) has $t_m - t_{m-1}$ terms of the form $\frac{b_m}{F(m)-F(m-1)}$. Grouping these terms, we transform (∗) into the series

$$\text{constant} + \sum_{m=n_1}^{\infty} \frac{t_m - t_{m-1}}{F(m) - F(m-1)} b_m.$$

This series converges by Abel's test, because the sequence with terms

$$c_m = \frac{t_m - t_{m-1}}{F(m) - F(m-1)}$$

is monotonically decreasing. Indeed,

$$c_m > \frac{(2m-1)\log_l 2 - 1}{2m^2 - 2^{(m-1)^2}} \quad \text{and} \quad c_{m+1} < \frac{(2m+1)\log_l 2 + 1}{2^{(m+1)^2} - 2^{m^2}}.$$

Hence for sufficiently large m we have $c_{m+1} < c_m$, because

$$\lim_{m\to\infty} \frac{(2m+1)\log_l 2 + 1}{(2m-1)\log_l 2 - 1} \cdot \frac{2^{m^2} - 2^{(m-1)^2}}{2^{(m+1)^2} - 2^{m^2}} = 0.$$

3.6. Cauchy Product of Infinite Series

3.6.1. Assume that the series $\sum_{n=0}^{\infty} a_n$ converges absolutely. Let A_n, B_n and C_n denote the nth partial sums of $\sum_{n=0}^{\infty} a_n$, $\sum_{n=0}^{\infty} b_n$ and $\sum_{n=0}^{\infty} c_n$, respectively. Then

$$C_n = a_0 b_0 + (a_0 b_1 + a_1 b_0) + \ldots + (a_0 b_n + a_1 b_{n-1} + \ldots + a_n b_0)$$
$$= a_0 B_n + a_1 B_{n-1} + \ldots + a_n B_0.$$

Write
$$B = B_n + r_n, \quad \text{where} \quad \lim_{n\to\infty} r_n = 0.$$
Consequently,
$$C_n = BA_n - (a_0 r_n + a_1 r_{n-1} + \ldots + a_n r_0).$$
We will now show that

(*) $$\lim_{n\to\infty} (a_0 r_n + a_1 r_{n-1} + \ldots + a_n r_0) = 0.$$

To this end, let $\varepsilon > 0$ be chosen arbitrarily and let m, M be such that
$$|r_n| \leq m \text{ for } n \geq 0, \quad M = \sum_{n=0}^{\infty} |a_n|.$$
There exist $k \in \mathbb{N}$ and $l \in \mathbb{N}$ such that if $n \geq k$, then $|r_n| < \frac{\epsilon}{2M}$ and if $n \geq l+1$, then $|a_{l+1}| + \ldots + |a_n| < \frac{\epsilon}{2m}$. Therefore for $n \geq l+k$ we get

$$|a_0 r_n + a_1 r_{n-1} + \ldots + a_n r_0|$$
$$\leq (|a_0||r_n| + \ldots + |a_l||r_{n-l}|) + (|a_{l+1}||r_{n-l-1}| + \ldots + |a_n||r_0|)$$
$$< (|a_0| + |a_1| + \ldots + |a_l|)\frac{\epsilon}{2M} + (|a_{l+1}| + \ldots + |a_n|)m$$
$$< M\frac{\epsilon}{2M} + \frac{\epsilon}{2m}m = \epsilon,$$

which proves (*).

Note that it follows from the above analysis that if both the series converge absolutely, then their Cauchy product also converges absolutely.

3.6.2.

(a) It follows from the Mertens theorem that if $|x| < 1$, then the Cauchy product of the series $\sum_{n=0}^{\infty} x^n$ with itself converges. Moreover,
$$c_n = x^n + x x^{n-1} + \ldots + x^n = (n+1)x^n.$$
Hence
$$\sum_{n=1}^{\infty} n x^{n-1} = \left(\frac{1}{1-x}\right)^2.$$

(b) $\frac{1}{1-x} \cdot \frac{1}{1-y}$.

3.6. Cauchy Product of Infinite Series

(c) The series is the Cauchy product of $\sum_{n=1}^{\infty} \frac{1}{n(n+1)}$ with $\sum_{n=1}^{\infty} \frac{1}{n!}$. The sum of the first series is 1 (see 3.1.4 (b)), and the sum of the second is $e-1$ (see 2.5.6). Therefore, by the Mertens theorem, the sum of our series is $e-1$.

3.6.3.

(a) We have

$$c_n = \sum_{k=0}^{n} \frac{2^k}{k!} \cdot \frac{1}{2^{n-k}(n-k)!} = \frac{1}{n!}\sum_{k=0}^{n} \binom{n}{k} 2^k \frac{1}{2^{n-k}} = \frac{1}{n!}\left(2+\frac{1}{2}\right)^n.$$

By 2.5.7 the sum of the Cauchy product is $e^{\frac{5}{2}}$.

(b) The Cauchy product is the series

$$\sum_{n=1}^{\infty} \frac{1}{3^{n+1}} \sum_{k=1}^{n} \frac{(-3)^k}{k}.$$

By 3.1.32 (a), its sum is $-\frac{1}{2}\ln 2$.

(c) We have

$$c_{2n+1} = x^{2n+1} \sum_{k=0}^{2n+1} (-1)^k (k+1)(2n+1-k+1)$$

$$= x^{2n+1}\Bigg(\sum_{k=0}^{n}(-1)^k(k+1)(2n+1-k+1)$$

$$+ \sum_{k=n+1}^{2n+1}(-1)^k(k+1)(2n+1-k+1)\Bigg)$$

$$= x^{2n+1}\Bigg(\sum_{k=0}^{n}(-1)^k(k+1)(2n+1-k+1)$$

$$-\sum_{k'=0}^{n}(-1)^{k'}(k'+1)(2n+1-k'+1)\Bigg) = 0.$$

Moreover, since $c_{2n+1} = 0$, we get

$$c_{2n} = x^{2n} \sum_{k=0}^{2n} (-1)^{2n-k}(k+1)(2n-k+1)$$

$$= x^{2n}\left(\sum_{k=0}^{2n-1}(-1)^k(k+1)(2n-1-k+1)\right)$$

$$+ \sum_{k=0}^{2n-1}(-1)^k(k+1) + (2n+1)\Big)$$

$$= x^{2n}(0 + (-n) + (2n+1)) = (n+1)x^{2n}.$$

Finally, by 3.6.2 (a),

$$\sum_{n=0}^{\infty}(n+1)x^{2n} = \frac{1}{(1-x^2)^2}.$$

3.6.4. Observe that the series $\sum\limits_{n=0}^{\infty} A_n x^n$ is the Cauchy product of $\sum\limits_{n=0}^{\infty} x^n$ with $\sum\limits_{n=0}^{\infty} a_n x^n$. Therefore it is convergent for $|x|<1$ and its sum is $\frac{1}{1-x}\sum\limits_{n=0}^{\infty} a_n x^n$.

3.6.5. To prove the equality given in the hint it is enough to equate the coefficients of x^n in the formula $(1+x)^n(1+x)^n = (1+x)^{2n}$. Consequently,

$$c_n = (-1)^n x^{2n} \frac{1}{(n!)^2}\sum_{k=0}^{n}\binom{n}{k}^2 = (-1)^n x^{2n}\frac{1}{(n!)^2}\binom{2n}{n}.$$

3.6.6. In view of the relation

$$c_n = \left(\frac{1}{a}\frac{1\cdot 3\cdot\ldots\cdot(2n-1)}{2\cdot 4\cdot\ldots\cdot 2n} + \frac{1}{2\,a+2}\frac{1}{2}\frac{1\cdot 3\cdot\ldots\cdot(2n-3)}{2\cdot 4\cdot\ldots\cdot(2n-2)}\right.$$
$$\left.+\ldots+\frac{1}{a+2n}\frac{1\cdot 3\cdot\ldots\cdot(2n-1)}{2\cdot 4\cdot\ldots\cdot 2n}\right)x^n,$$

it is enough to prove the equality

$$\frac{1}{a}\frac{1\cdot 3\cdot\ldots\cdot(2n-1)}{2\cdot 4\cdot\ldots\cdot 2n} + \frac{1}{2\,a+2}\frac{1}{2}\frac{1\cdot 3\cdot\ldots\cdot(2n-3)}{2\cdot 4\cdot\ldots\cdot(2n-2)}$$
$$+\ldots+\frac{1}{a+2n}\frac{1\cdot 3\cdot\ldots\cdot(2n-1)}{2\cdot 4\cdot\ldots\cdot 2n} = \frac{(a+1)(a+3)\ldots(a+(2n-1))}{a(a+2)(a+4)\ldots(a+2n)}.$$

3.6. Cauchy Product of Infinite Series

To this end, we decompose the right-hand side expression into partial fractions:

$$\frac{(a+1)(a+3)...(a+(2n-1))}{a(a+2)(a+4)...(a+2n)} = \frac{\alpha_0}{a} + \frac{\alpha_1}{a+2} + ... + \frac{\alpha_n}{a+2n}.$$

Multiplying both sides of this equality by $a(a+2)(a+4)...(a+2n)$ and substituting $a=0, a=-2, ..., a=-2k, ..., a=-2n$, we get

$$\alpha_0 = \frac{(2n-1)!!}{(2n)!!},$$

$$\alpha_1 = \frac{-1(2n-3)!!}{-2(2n-2)!!}, ...,$$

$$\alpha_k = \frac{(-2k+1)(-2k+3)...(-1)1 \cdot 3...(2(n-k)-1)}{-2k((-2k+2)...(-2) \cdot 2 \cdot 4...(2(n-k))}$$

$$= \frac{(2k-1)!!(2(n-k)-1)!!}{(2k)!!(2(n-k))!!}, ...,$$

$$\alpha_n = \frac{(2n-1)!!}{(2n)!!},$$

which gives the desired equality.

3.6.7. Let A_n, B_n, C_n denote the nth partial sums of $\sum\limits_{n=0}^{\infty} a_n$, $\sum\limits_{n=0}^{\infty} b_n$ and $\sum\limits_{n=0}^{\infty} c_n$, respectively. It is easy to check that

$$C_n = a_0 B_n + a_1 B_{n-1} + ... + a_n B_0.$$

Therefore

$$C_0 + C_1 + ... + C_n = A_0 B_n + A_1 B_{n-1} + ... + A_n B_0.$$

Dividing both sides of the last equality by $n+1$, using 2.3.2 and 2.3.8, we obtain $C = AB$.

3.6.8. Let $\sum\limits_{n=1}^{\infty} c_n$ be the Cauchy product of $\sum\limits_{n=1}^{\infty}(-1)^{n-1}\frac{1}{n}$ with itself. Then

$$c_n = (-1)^{n-1}\left(\frac{1}{1 \cdot n} + \frac{1}{2(n-1)} + ... + \frac{1}{k(n-k+1)} + ... + \frac{1}{n \cdot 1}\right).$$

Since
$$\frac{1}{k(n-k+1)} = \frac{1}{n+1}\left(\frac{1}{k} + \frac{1}{n-k+1}\right) \quad \text{for} \quad k = 1, 2, ..., n,$$
we can write
$$c_n = (-1)^{n-1}\frac{2}{n+1}\left(1 + \frac{1}{2} + \frac{1}{3} + ... + \frac{1}{n}\right).$$

We know that $\sum_{n=1}^{\infty}(-1)^{n-1}\frac{1}{n} = \ln 2$ (see 3.1.32 (a)) and that the series $\sum_{n=1}^{\infty}(-1)^{n-1}\frac{2}{n+1}\left(1 + \frac{1}{2} + ... + \frac{1}{n}\right)$ is convergent (see 3.4.19). Thus, by the result in the preceding problem,
$$\sum_{n=1}^{\infty}(-1)^{n-1}\frac{2}{n+1}\left(1 + \frac{1}{2} + ... + \frac{1}{n}\right) = (\ln 2)^2.$$

3.6.9. If $\sum_{n=1}^{\infty} c_n$ is the Cauchy product of the series $\sum_{n=1}^{\infty}(-1)^{n-1}\frac{1}{\sqrt{n}}$ with itself, then
$$c_n = (-1)^{n-1}\left(\frac{1}{1 \cdot \sqrt{n}} + ... + \frac{1}{\sqrt{k} \cdot \sqrt{n-k+1}} + ... + \frac{1}{\sqrt{n} \cdot 1}\right).$$
Since each term in the parenthesis is greater than $\frac{1}{n}$, we see that $|c_n| > 1$ for $n > 1$. It then follows that $\sum_{n=1}^{\infty} c_n$ is a divergent series.

3.6.10. We have
$$c_n = a_0 b_n + a_1 b_{n-1} + ... + a_n b_0 > a_0 b_n,$$
and consequently, if the series $\sum_{n=0}^{\infty} b_n$ diverges, then the Cauchy product $\sum_{n=0}^{\infty} c_n$ also diverges.

3.6.11. No. Consider the following two divergent series:
$$1 - \sum_{n=1}^{\infty}\left(\frac{3}{2}\right)^n \quad \text{and} \quad 1 + \sum_{n=1}^{\infty}\left(\frac{3}{2}\right)^{n-1}\left(2^n + \frac{1}{2^{n+1}}\right).$$

3.6. Cauchy Product of Infinite Series

Then
$$c_n = a_0 b_n + b_0 a_n + \sum_{k=1}^{n-1} a_k b_{n-k},$$
where $a_0 = b_0 = 1$, $a_n = -\left(\frac{3}{2}\right)^n$, $b_n = \left(\frac{3}{2}\right)^{n-1}\left(2^n + \frac{1}{2^{n+1}}\right)$. Hence
$$c_n = \left(\frac{3}{2}\right)^{n-1}\left(2^n + \frac{1}{2^{n+1}}\right) - \left(\frac{3}{2}\right)^n$$
$$- \left(\frac{3}{2}\right)^{n-1} \sum_{k=1}^{n-1}\left(2^{n-k} + \frac{1}{2^{n-k+1}}\right) = \left(\frac{3}{4}\right)^n.$$

3.6.12. Let A_n, B_n, C_n denote the nth partial sums of the series $\sum_{n=0}^{\infty} a_n$, $\sum_{n=0}^{\infty} b_n$ and $\sum_{n=0}^{\infty} c_n$, respectively. Then
$$C_n = a_0 B_n + a_1 B_{n-1} + \ldots + a_n B_0.$$
As a result,
$$\sum_{k=1}^{n} a_k (b_n + b_{n-1} + \ldots + b_{n-k+1})$$
$$= a_1(B_n - B_{n-1}) + a_2(B_n - B_{n-2}) + \ldots + a_n(B_n - B_0)$$
$$= B_n(A_n - a_0) - a_1 B_{n-1} - a_2 B_{n-2} - \ldots - a_n B_0 = B_n A_n - C_n.$$

3.6.13. Let $\sum_{n=0}^{\infty} c_n$ be the Cauchy product of the series $\sum_{n=0}^{\infty} (-1)^n a_n$ with $\sum_{n=0}^{\infty} (-1)^n b_n$. Then
$$c_n = (-1)^n (a_0 b_n + a_1 b_{n-1} + \ldots + a_n b_0).$$
Assume first that the series $\sum_{n=0}^{\infty} c_n$ converges. Then $\lim_{n \to \infty} c_n = 0$. By the monotonicity of the sequences $\{a_n\}$ and $\{b_n\}$, we get
$$|c_n| \geq b_n(a_0 + \ldots + a_n) \quad \text{and} \quad |c_n| \geq a_n(b_0 + \ldots + b_n).$$
So
$$\lim_{n \to \infty} a_n(b_0 + b_1 + \ldots + b_n) = 0 \quad \text{and} \quad \lim_{n \to \infty} b_n(a_0 + a_1 + \ldots + a_n) = 0.$$

Assume now that these two equalities hold. Then, by the preceding problem, it is enough to show that

$$\lim_{n\to\infty} \sum_{k=1}^{n} (-1)^k a_k ((-1)^n b_n + (-1)^{n-1} b_{n-1} + \ldots + (-1)^{n-k+1} b_{n-k+1}) = 0.$$

Note that

$$|(-1)^n b_n + (-1)^{n-1} b_{n-1} + \ldots + (-1)^{n-k+1} b_{n-k+1}| \leq b_{n-k+1},$$

and consequently,

$$\left| \sum_{k=1}^{n} (-1)^k a_k \left((-1)^n b_n + (-1)^{n-1} b_{n-1} + \ldots + (-1)^{n-k+1} b_{n-k+1} \right) \right|$$
$$\leq \sum_{k=1}^{n} a_k b_{n-k+1}.$$

Now we show that $\lim_{n\to\infty} \sum_{k=1}^{n} a_k b_{n-k+1} = 0$. Indeed,

$$0 < \sum_{k=1}^{2n} a_k b_{2n-k+1} \leq (a_1 + \ldots + a_n) b_n + (b_1 + \ldots b_n) a_n,$$

which implies that $\lim_{n\to\infty} \sum_{k=1}^{2n} a_k b_{2n-k+1} = 0$. In much the same way we show that $\lim_{n\to\infty} \sum_{k=1}^{2n-1} a_k b_{2n-k} = 0$, which completes the proof.

3.6.14. Observe first that it is enough to consider the case where both α and β do not exceed 1. We will now show that

$$\lim_{n\to\infty} \frac{1}{n^\alpha} \left(1 + \frac{1}{2^\beta} + \ldots + \frac{1}{n^\beta} \right) = 0$$

if and only if $\alpha + \beta > 1$. By the Stolz theorem (see 2.3.11),

$$\lim_{n\to\infty} \frac{1}{n^\alpha} \left(1 + \frac{1}{2^\beta} + \ldots + \frac{1}{n^\beta} \right) = \lim_{n\to\infty} \frac{1}{n^\beta (n^\alpha - (n-1)^\alpha)}$$
$$= \lim_{n\to\infty} \frac{1}{n^{\alpha+\beta} \left(1 - \left(1 - \frac{1}{n}\right)^\alpha\right)}.$$

3.6. Cauchy Product of Infinite Series

By L'Hospital's rule,

$$\lim_{x \to +\infty} \frac{1}{x^{\alpha+\beta}\left(1-\left(1-\frac{1}{x}\right)^\alpha\right)} = \lim_{t \to 0^+} \frac{t^{\alpha+\beta}}{1-(1-t)^\alpha}$$

$$= \lim_{t \to 0^+} \frac{(\alpha+\beta)t^{\alpha+\beta-1}}{\alpha(1-t)^{\alpha-1}}.$$

Hence

$$\lim_{n \to \infty} \frac{1}{n^{\alpha+\beta}\left(1-\left(1-\frac{1}{n}\right)^\alpha\right)} = \begin{cases} 0 & \text{if } \alpha+\beta > 1, \\ \frac{1}{\alpha} & \text{if } \alpha+\beta = 1, \\ +\infty & \text{if } \alpha+\beta < 1. \end{cases}$$

Now the desired result follows from the preceding problem.

3.6.15. Assume that the series $\sum_{n=0}^{\infty} a_n b_n$ converges. By the result in 3.6.13, it is enough to show that

$$\lim_{n \to \infty} a_n(b_0 + b_1 + \ldots + b_n) = 0 \quad \text{and} \quad \lim_{n \to \infty} b_n(a_0 + a_1 + \ldots + a_n) = 0.$$

For an arbitrarily fixed $\varepsilon > 0$ there is $k_0 \in \mathbb{N}$ such that $a_{k_0+1}b_{k_0+1} + a_{k_0+2}b_{k_0+2} + \ldots < \frac{\varepsilon}{2}$. Thus for $n > k_0$,

$$a_n(b_1 + \ldots + b_n) < a_n(b_1 + \ldots + b_{k_0}) + \frac{\varepsilon}{2}.$$

On the other hand, since $\lim_{n \to \infty} a_n = 0$, there is $n_1 > k_0$ such that

$$a_n < \frac{\varepsilon}{2(b_0 + \ldots + b_{k_0})} \quad \text{if } n > n_1,$$

which in turn implies that $a_n(b_0 + \ldots + b_n) < \varepsilon$ for $n > n_1$. Hence we have proved that $\lim_{n \to \infty} a_n(b_0 + \ldots + b_n) = 0$.

Assume now that the Cauchy product is convergent. It then follows from 3.6.13 that $\lim_{n \to \infty} a_n(b_0 + \ldots + b_n) = 0$. As a result, for sufficiently large n,

$$(n+1)a_n b_n < a_n(b_0 + \ldots + b_n) < 1,$$

and consequently,

$$(a_n b_n)^{1+\alpha} < \left(\frac{1}{n+1}\right)^{1+\alpha}.$$

3.7. Rearrangement of Series. Double Series

3.7.1. Let $S_n = a_1 + a_2 + ... + a_n$ be the nth partial sum of $\sum_{n=1}^{\infty} a_n$.
Then
$$b_1 + b_2 + ... + b_n = S_{m_n} \quad \text{for} \quad n \geq 1.$$
Since each subsequence of a convergent sequence converges to the same limit, $\lim_{n\to\infty} S_{m_n} = \lim_{n\to\infty} S_n$.

3.7.2. Denote by $\{T_n\}$ the sequence of partial sums of the rearranged series. Then
$$T_{3n} = \left(1 - \frac{1}{2}\right) - \frac{1}{4} + \left(\frac{1}{3} - \frac{1}{6}\right) - \frac{1}{8} + ...$$
$$+ \left(\frac{1}{2n-1} - \frac{1}{4n-2}\right) - \frac{1}{4n}$$
$$= \frac{1}{2} - \frac{1}{4} + \frac{1}{6} - \frac{1}{8} + ... + \frac{1}{4n-2} - \frac{1}{4n}$$
$$= \frac{1}{2}\left(1 - \frac{1}{2} + \frac{1}{3} - \frac{1}{4} + ... + \frac{1}{2n-1} - \frac{1}{2n}\right).$$

Therefore, by 3.1.32 (a), we get $\lim_{n\to\infty} T_{3n} = \frac{1}{2}\ln 2$. Of course, $\lim_{n\to\infty} T_{3n} = \lim_{n\to\infty} T_{3n+1} = \lim_{n\to\infty} T_{3n+2}$. It then follows that
$$1 - \frac{1}{2} - \frac{1}{4} + \frac{1}{3} - \frac{1}{6} - \frac{1}{8} + \frac{1}{5} - ... = \frac{1}{2}\ln 2.$$

3.7.3. Let $\{T_n\}$ be the sequence of partial sums of the rearranged series. Set $f(n) = 1 + \frac{1}{2} + \frac{1}{3} + \frac{1}{4} + ... + \frac{1}{n-1} + \frac{1}{n}$. Then
$$T_{\alpha+\beta} = 1 + \frac{1}{3} + ... + \frac{1}{2\alpha-1} - \frac{1}{2} - \frac{1}{4} - ... - \frac{1}{2\beta}$$
$$= f(2\alpha-1) - \frac{1}{2}f(\alpha-1) - \frac{1}{2}f(\beta) = f(2\alpha) - \frac{1}{2}f(\alpha) - \frac{1}{2}f(\beta).$$

Now we will prove by induction that
$$T_{n(\alpha+\beta)} = f(2n\alpha) - \frac{1}{2}f(n\alpha) - \frac{1}{2}f(n\beta).$$

3.7. Rearrangement of Series. Double Series

As we have already shown, the equality holds for $n = 1$. If it holds for an $n \in \mathbb{N}$, then

$$T_{(n+1)(\alpha+\beta)} = f(2n\alpha) - \frac{1}{2}f(n\alpha) - \frac{1}{2}f(n\beta) + \frac{1}{2n\alpha+1} + \frac{1}{2n\alpha+3}$$
$$+ \ldots + \frac{1}{2(n+1)\alpha - 1} - \frac{1}{2n\beta + 2} - \frac{1}{2n\beta + 4} - \ldots - \frac{1}{2(n+1)\beta}$$
$$= f(2n\alpha) - \frac{1}{2}f(n\alpha) - \frac{1}{2}f(n\beta) + f(2(n+1)\alpha - 1)$$
$$- \frac{1}{2}f((n+1)\alpha - 1) - f(2n\alpha) + \frac{1}{2}f(n\alpha) - \frac{1}{2}f((n+1)\beta) + \frac{1}{2}f(n\beta)$$
$$= f(2(n+1)\alpha) - \frac{1}{2}f((n+1)\alpha) - \frac{1}{2}f(n+1)\beta.$$

Hence, by 2.1.41,

$$\lim_{n\to\infty} T_{n(\alpha+\beta)} = \lim_{n\to\infty} \left(f(2n\alpha) - \ln(2n\alpha) - \frac{1}{2}f(n\alpha) \right.$$
$$\left. + \frac{1}{2}\ln(n\alpha) - \frac{1}{2}f(n\beta) + \frac{1}{2}\ln(n\beta) \right)$$
$$+ \lim_{n\to\infty} \left(\ln(2n\alpha) - \frac{1}{2}(\ln(n\alpha) + \ln(n\beta)) \right)$$
$$= \lim_{n\to\infty} \ln \frac{2n\alpha}{\sqrt{n^2\alpha\beta}} = \ln 2 + \frac{1}{2}\ln\frac{\alpha}{\beta}.$$

Obviously, for $k = 1, 2, 3, \ldots, (\alpha + \beta) - 1$, we have $\lim_{n\to\infty} T_{n(\alpha+\beta)+k} = \lim_{n\to\infty} T_{n(\alpha+\beta)}$. Consequently, the sum of the series is $\ln 2 + \frac{1}{2}\ln\frac{\alpha}{\beta}$.

3.7.4. Note that this result is contained as a special case ($\alpha = 1$ and $\beta = 4$) in the preceding problem.

3.7.5. It is enough to apply the result in 3.7.3 with $\alpha = 4$ and $\beta = 1$.

3.7.6. Consider the series

(1) $$1 - \frac{1}{2} + \frac{1}{3} + \frac{1}{5} - \frac{1}{4} + \frac{1}{7} + \frac{1}{9} + \frac{1}{11} - \frac{1}{6} + \ldots$$

obtained by rearranging the terms of $\sum_{n=1}^{\infty} \frac{(-1)^{n-1}}{n}$ in such a way that n, $n = 1, 2, 3, \ldots$, positive terms are followed by one negative term.

Collecting the terms of series (1) in the following way:

$$\left(1-\frac{1}{2}\right)+\left(\frac{1}{3}+\frac{1}{5}-\frac{1}{4}\right)+\left(\frac{1}{7}+\frac{1}{9}+\frac{1}{11}-\frac{1}{6}\right)+\dots,$$

we get

(2) $\quad \sum_{n=1}^{\infty}\left(\frac{1}{n^2-n+1}+\frac{1}{n^2-n+3}+\dots+\frac{1}{n^2+n-1}-\frac{1}{2n}\right).$

Let S_n and T_n denote the nth partial sums of the series (1) and (2), respectively. Then

$$T_n = S_{\frac{(n+1)n}{2}+n} > \sum_{k=1}^{n}\left(\frac{k}{k^2+k-1}-\frac{1}{2k}\right) > \frac{1}{4}\sum_{k=1}^{n}\frac{1}{k} \xrightarrow[n\to\infty]{} +\infty.$$

3.7.7. Grouping the terms of our series, we rewrite it in the form

$$\sum_{n=1}^{\infty}\left(\frac{1}{\sqrt{4n-3}}+\frac{1}{\sqrt{4n-1}}-\frac{1}{\sqrt{2n}}\right).$$

Moreover,

$$\frac{1}{\sqrt{4n-3}}+\frac{1}{\sqrt{4n-1}}-\frac{1}{\sqrt{2n}}$$
$$=\frac{\sqrt{(4n-1)2n}+\sqrt{(4n-3)2n}-\sqrt{(4n-3)(4n-1)}}{\sqrt{4n-3}\sqrt{4n-1}\sqrt{2n}}$$
$$>\frac{2\sqrt{2n}-\sqrt{4n-1}}{\sqrt{4n-1}\sqrt{2n}} > \frac{2\sqrt{2n}-\sqrt{4n}}{\sqrt{4n-1}\sqrt{2n}} = \frac{2-\sqrt{2}}{\sqrt{4n-1}}.$$

Thus $\lim\limits_{n\to\infty} S_{3n} = +\infty$, where $\{S_n\}$ denotes the sequence of partial sums of the rearranged series. Consequently, the series diverges.

3.7.8. Assume that the series $\sum\limits_{n=1}^{\infty} a_n$ converges absolutely, let S_n denote its nth partial sum, and set $S = \lim\limits_{n\to\infty} S_n$. Denote by $\{T_n\}$ the sequence of partial sums of a rearranged series. It follows from the absolute convergence of $\sum\limits_{n=1}^{\infty} a_n$ that, given $\varepsilon > 0$, there is $n \in \mathbb{N}$ such that

(1) $\qquad |a_{n+1}|+|a_{n+2}|+\dots < \varepsilon.$

3.7. Rearrangement of Series. Double Series

Let m be so large that all the terms $a_1, a_2, ..., a_n$ appear in T_m. Then, by (1),

$$|S - T_m| \leq |S - S_n| + |S_n - T_m| < 2\varepsilon.$$

3.7.9. [4] Assume first that $l > 0$ and let $n = d + u$, $d > u$; then rearrange the series $\sum\limits_{n=1}^{\infty} (-1)^{n-1} f(n)$ so that the nth partial sum of a new series is

$$T_n = T_{d+u} = (f(1) - f(2) + f(3) - ... - f(2u))$$
$$+ (f(2u+1) + f(2u+3) + ... + f(2d-1)).$$

This sum contains u negative terms, and all remaining terms, d in number, are positive. The sum in the second grouping contains $d - u$ terms, and consequently, this sum is between $(d-u)f(2u)$ and $(d-u)f(2d)$. Since the sum in the first parenthesis converges to S as $u \to \infty$, the change in the sum is equal to the limit of the second parenthesis. Set $\nu(u) = d - u$. Then

(1) $\quad \nu(u)f(2d) < f(2u+1) + f(2u+3) + ... + f(2d-1) < \nu(u)f(2u),$

and the monotonicity of the sequence $\{nf(n)\}$ implies

(2) $\quad \dfrac{u}{u + \nu(u)} < \dfrac{f(2u + 2\nu(u))}{f(2u)} < 1.$

Choose $\nu(u)$ such that

(3) $\quad \lim\limits_{u \to \infty} \nu(u) f(2u) = l.$

(One can take, e.g., $\nu(u) = l \left[\dfrac{1}{f(2u)}\right]$.) Then $\lim\limits_{u \to \infty} \dfrac{\nu(u)}{u} = 0$, because

$$l = \lim_{u \to \infty} \dfrac{1}{2} \dfrac{\nu(u)}{u} 2u f(2u) \quad \text{and} \quad \lim_{u \to \infty} 2u f(2u) = +\infty.$$

Thus (2) implies that $\lim\limits_{u \to \infty} \dfrac{f(2u+2\nu(u))}{f(2u)} = 1$. As a result, (1) and the squeeze principle give

$$\lim_{u \to \infty} (f(2u+1) + f(2u+3) + ... + f(2d-1)) = l.$$

So, we have proved that $\lim\limits_{u \to \infty} T_{2u+\nu(u)} = S + l.$

Now note that if $2u + \nu(u) < k < 2(u+1) + \nu(u+1)$, then
$$0 \leq T_k - T_{2u+\nu(u)} + f(2u+2) \leq T_{2u+2+\nu(u+1)} - T_{2u+\nu(u)} + f(2u+2).$$
Since $f(2u+2) \to 0$ as $u \to \infty$, we see that $\lim\limits_{k\to\infty} T_k = S + l$.

In the case where $l < 0$, we can interchange d and u and proceed analogously.

3.7.10. Given $\varepsilon > 0$, beginning with some value n_0 of the index n, we have

(1) $$\frac{g - \varepsilon}{n} < f(n) < \frac{g + \varepsilon}{n}.$$

Consider the rearranged series whose nth partial sum is (see the solution of 3.7.9)
$$T_n = T_{d+u} = (f(1) - f(2) + f(3) - \ldots - f(2u)) \\ + (f(2u+1) + f(2u+3) + \ldots + f(2d-1)).$$

Moreover, assume that the number d of positive terms is such that $\lim\limits_{u \to \infty} \frac{d}{u} = k$. Then, in the case where $d > u$,

$$\frac{1}{2u+1} + \frac{1}{2u+3} + \ldots + \frac{1}{2d-1}$$
$$= \left(1 + \frac{1}{2} + \ldots + \frac{1}{2u+1} + \ldots + \frac{1}{2d-1} - \ln(2d-1)\right)$$
$$- \left(1 + \frac{1}{2} + \ldots + \frac{1}{2u-1} - \ln(2u-1)\right)$$
$$- \left(\frac{1}{2u} + \frac{1}{2u+2} + \ldots + \frac{1}{2d-2}\right) - \ln \frac{2u-1}{2d-1}.$$

By 2.1.41 each of the first two parentheses tends to the Euler constant γ. As in 2.5.8 (a), we may show that the third parenthesis tends to $\frac{1}{2} \ln k$. Hence

$$\lim_{u \to \infty} \left(\frac{1}{2u+1} + \frac{1}{2u+3} + \ldots + \frac{1}{2d-1}\right) = \frac{1}{2} \ln k.$$

Consequently, (1) implies that

$$\lim_{u \to \infty} (f(2u+1) + f(2u+3) + \ldots + f(2d-1)) = \frac{1}{2} g \ln k.$$

3.7. Rearrangement of Series. Double Series

Thus the change in the sum S of the series is $\frac{1}{2}g\ln k$. Analogous reasoning can be applied to the case $d < u$.

3.7.11. It is enough to apply the rearrangement described in the solution of Problem 3.7.9 with $\nu(u) = l[(2u)^p]$.

3.7.12. Take the rearrangement described in the solution of Problem 3.7.10 with $\lim\limits_{u\to\infty} \frac{d}{u} = \alpha$.

3.7.13. No. Indeed, let $\sum\limits_{k=1}^{\infty} a_{n_k}$ be a rearrangement of a divergent series $\sum\limits_{n=1}^{\infty} a_n$. The monotonicity of the sequence $\{a_n\}$ implies that

$$a_{n_1} + a_{n_2} + \ldots + a_{n_m} \leq a_1 + a_2 + \ldots + a_m.$$

So, it is not possible to accelerate the divergence of this series.

3.7.14. [20] Choose a subsequence $\{a_{r_n}\}$ of $\{a_n\}$ such that $a_{r_n} < \min(2^{-n}, Q_n - Q_{n-1})$, $n = 1, 2, \ldots$, where $Q_0 = 0$. Then

$$a_{r_1} + a_{r_2} + \ldots + a_{r_n} \leq Q_n \quad \text{and} \quad a_{r_1} + a_{r_2} + \ldots + a_{r_n} < 1.$$

Thus since $\lim\limits_{n\to\infty} Q_n = +\infty$, the sequence $\{Q_n - (a_{r_1} + a_{r_2} + \ldots + a_{r_n})\}$ also diverges to infinity. Now, we add the terms of $\sum\limits_{n=1}^{\infty} a_n$ which do not appear in the sequence $\{a_{r_n}\}$ to the sum $a_{r_1} + a_{r_2} + \ldots + a_{r_n}$ in such a way that

$$a_1 + a_2 + \ldots + a_{r_1-1} + a_{r_1} + a_{r_1+1} + \ldots + a_i + a_{r_k} + a_{r_k+1} + \ldots + a_{r_n} \leq Q_n,$$

and a_i is the last term allowed. That is, if we add a term which does not appear in the sequence $\{a_{r_n}\}$ and whose index is greater than i, then the above inequality does not hold.

3.7.15. [W. Sierpiński, Bull. Intern. Acad. Sci. Cracovie, 1911, 149-158] Let $\sum\limits_{n=1}^{\infty} p_n$ and $\sum\limits_{n=1}^{\infty} q_n$ be the complementary subseries of a conditionally convergent series $\sum\limits_{n=1}^{\infty} a_n$ consisting of all successive nonnegative and negative terms, respectively. Let σ be an arbitrarily chosen real number. Since the series $\sum\limits_{n=1}^{\infty} p_n$ diverges to $+\infty$, there exists a least index k_1 for which

$$p_1 + p_2 + \ldots + p_{k_1} > \sigma.$$

Next we choose the least index n_1 for which

$$p_1 + p_2 + \ldots + p_{k_1} + q_1 + q_2 + \ldots + q_{n_1} < \sigma.$$

Then we find the least index k_2 for which

$$p_1 + p_2 + \ldots + p_{k_1} + q_1 + q_2 + \ldots + q_{n_1} + p_{k_1+1} + \ldots + p_{k_2} > \sigma$$

and the least n_2 such that

$$p_1+p_2+\ldots+p_{k_1}+q_1+q_2+\ldots+q_{n_1}+p_{k_1+1}+\ldots+p_{k_2}+q_{n_1+1}+\ldots+q_{n_2} < \sigma.$$

Continuing this process, we define two sequences $k_1, k_2, \ldots$ and $n_1, n_2, \ldots$ and the corresponding rearrangement of our series. Let S_n be the nth partial sum of this rearrangement. Then

$$S_n \leq \sigma \quad \text{for} \quad n < k_1 \quad \text{but} \quad S_n \geq \sigma \quad \text{for} \quad k_1 \leq n < k_1 + n_1.$$

Furthermore,

$$S_n \leq \sigma \quad \text{for} \quad k_m + n_m \leq n < k_{m+1} + n_m,$$
$$S_n \geq \sigma \quad \text{for} \quad k_{m+1} + n_m \leq n < k_{m+1} + n_{m+1},$$

where $m = 1, 2, \ldots$. By the definition of the sequences $\{k_m\}$ and $\{n_m\}$ we also get

$$|S_{k_{m+1}-1+n_m} - \sigma| < p_{k_{m+1}},$$
$$|S_{k_{m+1}+n_m} - \sigma| < p_{k_{m+1}},$$
$$|S_{k_{m+1}+n_m+l} - \sigma| < p_{k_{m+1}} \quad \text{for} \quad l = 1, 2, \ldots, n_{m+1} - n_m - 1,$$
$$|S_{k_{m+1}+n_{m+1}} - \sigma| < |q_{n_{m+1}}|,$$
$$|S_{k_{m+1}+l+n_{m+1}} - \sigma| < |q_{n_{m+1}}| \quad \text{for} \quad l = 1, 2, \ldots, k_{m+2} - k_{m+1} - 1.$$

Since $\lim\limits_{n\to\infty} p_n = \lim\limits_{n\to\infty} q_n = 0$, we conclude that $\lim\limits_{n\to\infty} S_n = \sigma$.

3.7. Rearrangement of Series. Double Series

3.7.16. Denote by $\{S_m\}$ and $\{T_m\}$ the sequences of partial sums of $\sum_{n=1}^{\infty} a_n$ and $\sum_{k=1}^{\infty} a_{n_k}$, respectively. Since $\{n_k - k\}$ is a bounded sequence, there is $l \in \mathbb{N}$ such that $k - l \leq n_k \leq k + l$ for all $k \in \mathbb{N}$. If $m > l$ and $n_k \leq m - l$, then $k - l \leq n_k \leq m - l$. Hence $k \leq m$, and consequently,

(1) $$\{1, 2, ..., m - l\} \subset \{n_1, n_2, ..., n_m\}.$$

Indeed, if s is a positive integer not greater than $m - l$, then there exists a unique $k \in \mathbb{N}$ such that $s = n_k$. It then follows from the above that $k \leq m$, or in other words, $s \in \{n_1, n_2, ..., n_m\}$. By (1), we see that each a_n, with $n = 1, 2, ..., m - l$, appears in T_m. On the other hand, if $k \leq m$, then $n_k \leq k + l \leq m + l$, and consequently, all the terms $a_{n_1}, a_{n_2}, ..., a_{n_m}$ appear in S_{m+l}. Hence

$$|S_m - T_m| \leq |a_{m-l+1}| + ... + |a_{m+l}| \quad \text{for} \quad m > l.$$

Therefore $\lim_{m \to \infty} S_m = \lim_{m \to \infty} T_m$.

If the sequence $\{n_k - k\}$ is unbounded, then the examples given in Problems 3.7.2 - 3.7.6 show that the rearranged series may diverge or may change the sum of the series. Now we give an example of a rearrangement that does not change the sum of the series. To this end, we take a sequence $\{n_k\}$ obtained by the permutation of positive integers that interchanges $\frac{n(n+1)}{2}$ with $\frac{n(n+3)}{2}$ and leaves the other integers unchanged. Since $\frac{n(n+3)}{2} - \frac{n(n+1)}{2} = n$, the sequence $\{n_k - k\}$ is unbounded. Moreover,

$$T_m - S_m = \begin{cases} 0, & \text{if } m = \frac{n(n+3)}{2}, \\ a_{n(n+3)/2} - a_{n(n+1)/2}, & \text{if } \frac{n(n+1)}{2} \leq m < \frac{n(n+3)}{2}. \end{cases}$$

3.7.17. [R. P. Agnew, Proc. Amer. Math. Soc. 6(1955), 563-564] We will apply the Toeplitz theorem (see 3.4.37). Set $S_m = \sum_{k=1}^{m} a_k$ and $T_m = \sum_{k=1}^{m} a_{n_k}$. Assume that m is so large that $1 \in \{n_1, n_2, ..., n_m\}$, and arrange all the members of the set $\{n_1, n_2, ..., n_m\}$ to form an increasing sequence

$$1, 2, 3, ..., \beta_{0,m}, \alpha_{1,m} + 1, \alpha_{1,m} + 2, ..., \beta_{1,m},$$

$$\alpha_{2,m}+1, \alpha_{2,m}+2, ..., \beta_{2,m}, ..., \alpha_{j_m,m}+1, \alpha_{j_m,m}+2, ..., \beta_{j_m,m},$$

where
$$0 < \beta_{0,m} < \alpha_{1,m} < \beta_{1,m} < \alpha_{2,m} < ... < \beta_{j_m,m}.$$

Hence the partial sum T_m of the rearranged series can be written in the following way:

$$T_m = S_{\beta_{0,m}} + (S_{\beta_{1,m}} - S_{\alpha_{1,m}}) + ... + (S_{\beta_{j_m,m}} - S_{\alpha_{j_m,m}}).$$

Consequently, $T_m = \sum_{k=1}^{\infty} c_{m,k} S_k$, where

$$c_{m,k} = \begin{cases} 1, & \text{if } k = \beta_{l,m},\ l = 0, 1, ..., j_m, \\ -1, & \text{if } k = \alpha_{l,m},\ l = 1, 2, ..., j_m, \\ 0 & \text{otherwise.} \end{cases}$$

Since $\lim_{m\to\infty} \beta_{0,m} = +\infty$, $\lim_{m\to\infty} c_{m,k} = 0$ for every $k \in \mathbb{N}$. Moreover, $\sum_{k=1}^{\infty} c_{m,k} = 1$ for $m = 1, 2, ...$, and $\sum_{k=1}^{\infty} |c_{m,k}| = 2B_m - 1$, where B_m denotes the number of disjoint blocks of successive integers in the set $\{n_1, n_2, ..., n_m\}$. Finally, by the Toeplitz theorem, $\lim_{m\to\infty} T_m = \lim_{m\to\infty} S_m$ if and only if there is N such that $B_m \leq N$ for all $m \in \mathbb{N}$.

3.7.18. Assume that the series $\sum_{n=1}^{\infty} c_n$ is absolutely convergent and its sum is S. Then for any $\varepsilon > 0$ there is k_0 such that

$$|c_1 + c_2 + ... + c_{k_0} - S| < \frac{\varepsilon}{2} \quad \text{and} \quad \sum_{l=k_0+1}^{\infty} |c_l| < \frac{\varepsilon}{2}.$$

Let m, n be so large that for each $l \in \{1, 2, ..., k_0\}$ there exist i and k, $i \in \{1, 2, ..., m\}$, $k \in \{1, 2, ..., n\}$, such that $c_l = a_{i,k}$. Then

$$|S_{m,n} - S| < |c_1 + c_2 + ... + c_{k_0} - S| + \sum_{l=k_0+1}^{\infty} |c_l| < \varepsilon.$$

Hence the convergence of the double series to S is proved. Likewise, the absolute convergence of this double series can be established.

3.7. Rearrangement of Series. Double Series

3.7.19. Set

$$S^* = \sum_{i,k=1}^{\infty} |a_{i,k}|, \qquad T^* = \sum_{n=1}^{\infty} |c_n|,$$

$$S^*_{m,n} = \sum_{i=1}^{m}\sum_{k=1}^{n} |a_{i,k}|, \qquad T^*_n = \sum_{k=1}^{n} |c_k|.$$

Arbitrarily fix $\varepsilon > 0$ and $l \in \mathbb{N}$. Take m, n so large that all the terms of T_l^* are in $S^*_{m,n}$ and $|S^* - S^*_{m,n}| < \varepsilon$. Then $T_l^* \leq S^*_{m,n} < S^* + \varepsilon$, which means that the series $\sum_{n=1}^{\infty} c_n$ is absolutely convergent. Denote by T_n and T the nth sum and the sum of $\sum_{n=1}^{\infty} c_n$, respectively. To prove the identity

$$\sum_{i,k=1}^{\infty} a_{i,k} = \sum_{n=1}^{\infty} c_n,$$

fix $\varepsilon > 0$ and take l so large that

$$|T_l^* - T^*| < \frac{\varepsilon}{2} \quad \text{and} \quad |T_l - T| < \frac{\varepsilon}{2}.$$

If $S_{m,n} = \sum_{i=1}^{m}\sum_{k=1}^{n} a_{i,k}$ and if m, n are so large that all terms of T_l are in $S_{m,n}$, then

$$|S_{m,n} - T| \leq |T - T_l| + |T^* - T_l^*| < \varepsilon.$$

3.7.20. This is a corollary of the two preceding problems.

3.7.21. Assume that, e.g., the iterated series $\sum_{i=1}^{\infty}\left(\sum_{k=1}^{\infty} |a_{i,k}|\right)$ converges, and set $\sum_{k=1}^{\infty} |a_{i,k}| = \sigma_i$ and $\sum_{i=1}^{\infty} \sigma_i = \sigma$. Therefore, each of the series $\sum_{k=1}^{\infty} a_{i,k}$, $i = 1, 2, \ldots$, converges, and $\left|\sum_{k=1}^{\infty} a_{i,k}\right| = |S_i| \leq \sigma_i$. This and the convergence of $\sum_{i=1}^{\infty} \sigma_i$ imply the absolute convergence of $\sum_{i=1}^{\infty} S_i$. Consequently, $\sum_{i=1}^{\infty} S_i = \sum_{i=1}^{\infty}\left(\sum_{k=1}^{\infty} a_{i,k}\right)$.

3.7.22. Let $\sum_{i,k=1}^{\infty} a_{i,k} = S$, $\sum_{i,k=1}^{\infty} |a_{i,k}| = S^*$, and set $S_{m,n} = \sum_{i=1}^{m}\sum_{k=1}^{n} a_{i,k}$ and $S^*_{m,n} = \sum_{i=1}^{m}\sum_{k=1}^{n} |a_{i,k}|$. We first show that the iterated series

$$\sum_{i=1}^{\infty}\left(\sum_{k=1}^{\infty} |a_{i,k}|\right)$$

converges to S^*. Indeed, given $\varepsilon > 0$, there is n_0 such that $S^* - \varepsilon < S^*_{m,n} < S^*$ for $m,n > n_0$. Let m be fixed for the moment. Then the sequence $\{S^*_{m,n}\}$ is monotonically increasing and bounded. Thus it is convergent, $\lim_{n\to\infty} S^*_{m,n} = S^*_m$, and consequently, $S^* - \varepsilon \leq S^*_m \leq S^*$ for $m > n_0$. This means that $\lim_{m\to\infty}(\lim_{n\to\infty} S^*_{m,n}) = S^*$. We know from the preceding problem that absolute convergence of the iterated series implies its convergence. Thus $\sum_{k=1}^{\infty} a_{i,k}$ converges for each i, say, to S_i. We will now show that for every $\varepsilon > 0$ there is m_1 such that

$$|(S_1 + S_2 + \ldots + S_m) - S| < \varepsilon \quad \text{for} \quad m > m_1.$$

By the absolute convergence of the double series,

$$|S_{m,n} - S| < \frac{\varepsilon}{2} \quad \text{and} \quad |S^*_{m,n} - S^*| < \frac{\varepsilon}{2} \quad \text{for} \quad m,n > m_1.$$

Therefore, for $m > m_1$,

$$|(S_1 + S_2 + \ldots + S_m) - S| = \left|\sum_{i=1}^{m}\sum_{k=1}^{\infty} a_{i,k} - S\right|$$

$$\leq |S_{m,n} - S| + \left|\sum_{i=1}^{m}\sum_{k=n+1}^{\infty} a_{i,k}\right| \leq |S_{m,n} - S| + |S^* - S^*_{m,n}| < \varepsilon.$$

The proof of the convergence of the other iterated series is analogous.

3.7.23. Note that the series $\sum_{n=1}^{\infty}(a_{n,1} + a_{n-1,2} + a_{n-2,3} + \ldots + a_{1,n})$ is an ordering of the double series. If one of the series

$$\sum_{i,k=1}^{\infty} |a_{i,k}|, \quad \sum_{n=1}^{\infty}(|a_{n,1}| + |a_{n-1,2}| + |a_{n-2,3}| + \ldots + |a_{1,n}|)$$

3.7. Rearrangement of Series. Double Series

converges, then our claim follows directly from 3.7.18, 3.7.19 and 3.7.22. Hence it is enough to show that the absolute convergence of one of the iterated series implies the absolute convergence of any ordering of the double series. To this end, assume that $\sum\limits_{i=1}^{\infty}\left(\sum\limits_{k=1}^{\infty}|a_{i,k}|\right)$ converges, say, to S^*. Let $\{c_n\}$ be a sequence obtained by an enumeration of the infinite matrix $(a_{i,k})_{i,k=1,2,\ldots}$. Then for $l \in \mathbb{N}$ there exist m, n so large that

$$|c_1| + |c_2| + \ldots + |c_l| \leq \sum_{i=1}^{m}\sum_{k=1}^{n}|a_{i,k}| \leq S^*.$$

Thus the series $\sum\limits_{n=1}^{\infty} c_n$ converges absolutely, which in turn implies the absolute convergence of the double series (see 3.7.18).

3.7.24. Since $\sum\limits_{k=0}^{m}\binom{m}{k} = 2^m$, we get $\sum\limits_{\substack{n,k=0 \\ k+n=m}}^{m} \frac{1}{n!k!} = \frac{2^m}{m!}$. Hence

$$\sum_{\substack{n,k=0 \\ k+n=m}}^{m} \frac{1}{n!k!(n+k+1)} = \frac{2^m}{(m+1)!}.$$

Consequently, by 3.7.23,

$$\sum_{n,k=0}^{\infty} \frac{1}{n!k!(n+k+1)} = \sum_{m=0}^{\infty} \sum_{\substack{n,k=0 \\ k+n=m}}^{m} \frac{1}{n!k!(n+k+1)}$$

$$= \sum_{m=0}^{\infty} \frac{2^m}{(m+1)!} = \frac{1}{2}\sum_{m=0}^{\infty} \frac{2^{m+1}}{(m+1)!} = \frac{1}{2}(e^2 - 1),$$

where the last equality follows from 2.5.7.

3.7.25. We have (see 3.7.23)

$$\sum_{n,k=1}^{\infty} \frac{1}{nk(n+k+2)} = \sum_{n=1}^{\infty}\frac{1}{n}\sum_{k=1}^{\infty}\frac{1}{n+2}\left(\frac{1}{k} - \frac{1}{n+k+2}\right)$$

$$= \sum_{n=1}^{\infty} \frac{1}{n(n+2)}\left(1 + \frac{1}{2} + \frac{1}{3} + \ldots + \frac{1}{n+2}\right)$$

$$= \frac{1}{2} \sum_{n=1}^{\infty} \left(\frac{1}{n} - \frac{1}{n+2} \right) \left(1 + \frac{1}{2} + \dots + \frac{1}{n+2} \right)$$

$$= \frac{1}{2} \left\{ 1 + \frac{1}{2} + \frac{1}{3} + \frac{1}{2} \left(1 + \frac{1}{2} + \frac{1}{3} + \frac{1}{4} \right) + \frac{1}{3} \left(\frac{1}{4} + \frac{1}{5} \right) \right.$$

$$\left. + \frac{1}{4} \left(\frac{1}{5} + \frac{1}{6} \right) + \dots \right\} = \frac{7}{4}.$$

3.7.26. It follows from 3.7.23 that

$$\sum_{n,k=0}^{\infty} \frac{n!k!}{(n+k+2)!} = \sum_{k=0}^{\infty} \frac{k!}{k+1} \sum_{n=0}^{\infty} \left(\frac{n!}{(n+k+1)!} - \frac{(n+1)!}{(n+k+2)!} \right)$$

$$= \sum_{k=0}^{\infty} \frac{k!}{k+1} \frac{0!}{(k+1)!} = \sum_{k=0}^{\infty} \frac{1}{(k+1)^2}.$$

Hence, by 3.1.28, the desired equality is proved.

3.7.27. Observe that the sum of each row series of the matrix is finite. Indeed, the sum of the first row is x, of the second is $x(1-x)$, of the third is $x(1-x)^2$, etc. Moreover,

$$x + x(1-x) + x(1-x)^2 + \dots = 1.$$

On the other hand, the sums of the column series are alternately equal to 1 and -1. Therefore the other iterated series diverges. By 3.7.23, we conclude that the iterated series cannot converge absolutely.

3.7.28.

(a) The absolute convergence of $\sum_{i=0}^{\infty} x^i$ and $\sum_{k=0}^{\infty} y^k$ imply the absolute convergence of the iterated series $\sum_{i=0}^{\infty} \left(\sum_{k=0}^{\infty} x^i y^k \right)$, because

$\sum_{i=0}^{\infty} \left(\sum_{k=0}^{\infty} |x^i y^k| \right) = \sum_{i=0}^{\infty} |x^i| \left(\frac{1}{1-|y|} \right) = \frac{1}{(1-|x|)(1-|y|)}$. Consequently, the given double series is absolutely convergent.

(b) Considering the iterated series, we see that our series converges if and only $\alpha > 1$ and $\beta > 1$.

3.7. Rearrangement of Series. Double Series   355

(c) Collecting the terms for which $i+k=n$, we get

$$\sum_{i,k=1}^{\infty} \frac{1}{(i+k)^p} = \sum_{n=2}^{\infty} (n-1)\frac{1}{n^p}.$$

Thus the double series converges if $p>2$ and diverges if $p \leq 2$.

3.7.29.

(a) It is enough to calculate the sum of the iterated series. We have

$$\sum_{i=2}^{\infty}\left(\sum_{k=2}^{\infty} \frac{1}{(p+i)^k}\right) = \sum_{i=2}^{\infty} \left(\frac{1}{p+i} \cdot \frac{1}{p+i-1}\right) = \frac{1}{p+1}.$$

(b) As in (a), we compute the sum of the iterated series:

$$\sum_{k=1}^{\infty}\left(\sum_{i=2}^{\infty} \frac{1}{(2k)^i}\right) = \sum_{k=1}^{\infty} \frac{1}{2k(2k-1)} = \sum_{k=1}^{\infty}\left(\frac{1}{2k-1} - \frac{1}{2k}\right)$$

$$= \sum_{k=1}^{\infty} (-1)^{k-1} \frac{1}{k} = \ln 2.$$

The last equality follows from 3.1.32(a).

(c) As in (b), we have

$$\sum_{i=1}^{\infty}\left(\sum_{k=1}^{\infty} \frac{1}{(4i-1)^{2k}}\right) = \sum_{i=1}^{\infty} \frac{1}{(4i-2)4i} = \frac{1}{4}\ln 2.$$

3.7.30. Since $S_{m,n} = \sum_{i=1}^{m}\sum_{k=1}^{n} a_{i,k} = b_{m,n}$, we see that

$$a_{1,1} = S_{1,1} = b_{1,1},$$
$$a_{1,n} = S_{1,n} - S_{1,n-1} = b_{1,n} - b_{1,n-1}, \quad n > 1,$$
$$a_{m,1} = S_{m,1} - S_{m-1,1} = b_{m,1} - b_{m-1,1}, \quad m > 1.$$

Similarly, for $n, m > 1$, we get

$$a_{m,n} = S_{m,n} - S_{m-1,n} - (S_{m,n-1} - S_{m-1,n-1})$$
$$= b_{m,n} - b_{m-1,n} - (b_{m,n-1} - b_{m-1,n-1}), \quad n, m > 1.$$

3.7.31. We have $S_{m,n} = (-1)^{m+n}\left(\frac{1}{2^m} + \frac{1}{2^n}\right)$. So, for $\varepsilon > 0$, there is n_0 such that if $m, n > n_0$, then

$$|S_{m,n}| < \varepsilon.$$

Therefore the double series converges to zero. However, both the iterated series diverge. Indeed,

$$\sum_{k=1}^{n} a_{i,k} = S_{i,n} - S_{i-1,n} = (-1)^{i+n}\frac{3}{2^i} + (-1)^{i+n}\frac{1}{2^{n-1}},$$

which implies that every series $\sum_{k=1}^{\infty} a_{i,k}$, $i \in \mathbb{N}$, diverges.

3.7.32. We have

$$\sum_{i=1}^{\infty}\left(\sum_{k=1}^{\infty} |x|^{ik}\right) = \sum_{i=1}^{\infty} \frac{|x|^i}{1 - |x|^i}.$$

By the ratio test, the series on the right-hand side of this equality converges. This means that the iterated series converges absolutely. Thus by 3.7.23,

$$\sum_{i,k=1}^{\infty} x^{ik} = \sum_{k=1}^{\infty} \frac{x^k}{1 - x^k}.$$

Collecting the pairs (i, k) with the same value of the product ik, we get

$$\sum_{i,k=1}^{\infty} x^{ik} = \sum_{n=1}^{\infty} \theta(n) x^n,$$

because the number of divisors of n is equal to the number of the pairs (i, k) for which $ik = n$. Moreover, for $n = 2, 3, ...$,

$$S_{n,n} - S_{n-1,n-1}$$
$$= x^n + x^{2n} + ... + x^{(n-1)n} + x^{n^2} + x^{n(n-1)} + ... + x^{n \cdot 2} + x^n$$
$$= 2\frac{x^n - x^{n^2}}{1 - x^n} + x^{n^2}.$$

Obviously, $S_{1,1} = x = 2\frac{x-x}{1-x} + x$. Hence, on account of

$$S_{n,n} = (S_{n,n} - S_{n-1,n-1}) + (S_{n-1,n-1} - S_{n-2,n-2}) + ... + S_{1,1},$$

we see that
$$\sum_{k=1}^{\infty}\frac{x^k}{1-x^k} = 2\sum_{n=1}^{\infty}\frac{x^n-x^{n^2}}{1-x^n}+\sum_{n=1}^{\infty}x^{n^2}.$$

3.7.33. As in the solution of the foregoing problem, we show that the iterated series converges absolutely. Thus the first equality follows directly from 3.7.23. To prove the other equality we consider the ordering of the double series described in the solution of 3.7.32.

3.7.34.

(a) By 3.7.23,
$$\sum_{p=2}^{\infty} S_p = \sum_{n=2}^{\infty}\frac{1}{2^n}+\sum_{n=2}^{\infty}\frac{1}{3^n}+\ldots = \sum_{k=2}^{\infty}\frac{1}{k(k-1)} = 1.$$

(b) As in (a),
$$\sum_{p=2}^{\infty}(-1)^p S_p = \sum_{k=2}^{\infty}\frac{1}{k(k+1)} = \frac{1}{2}.$$

3.7.35. Let **B** denote the set of all integers which are not powers. Then
$$\mathbf{A} = \{k^n : n \in \mathbb{N},\ n\geq 2,\ k\in \mathbf{B}\ \}.$$
Since $\frac{1}{n-1} = \sum\limits_{j=1}^{\infty}\frac{1}{n^j}$, $n\geq 2$, applying 3.7.23 and 3.7.34, we get

$$\sum_{n\in\mathbf{A}}\frac{1}{n-1} = \sum_{n\in\mathbf{A}}\sum_{j=1}^{\infty}\frac{1}{n^j} = \sum_{k\in\mathbf{B}}\sum_{n=2}^{\infty}\sum_{j=1}^{\infty}\frac{1}{k^{nj}}$$
$$= \sum_{k\in\mathbf{B}}\sum_{j=1}^{\infty}\sum_{n=2}^{\infty}\frac{1}{k^{nj}} = \sum_{k=2}^{\infty}\sum_{n=2}^{\infty}\frac{1}{k^n} = 1.$$

3.7.36. [G. T. Williams, Amer. Math. Monthly, 60 (1953), 19-25] The left-hand side of the equality is equal to

(*) $\quad \lim\limits_{N\to\infty}\sum\limits_{j=1}^{N}\sum\limits_{k=1}^{N}\left(\frac{1}{k^2}\frac{1}{j^{2n-2}}+\frac{1}{k^4}\frac{1}{j^{2n-4}}+\ldots+\frac{1}{k^{2n-2}}\frac{1}{j^2}\right).$

Summing the expression in the parentheses, we get

$$(**) \quad \lim_{N\to\infty} \sum_{j=1}^{N} \left(\sum_{\substack{k=1 \\ k\neq j}}^{N} \frac{j^{2-2n} - k^{2-2n}}{k^2 - j^2} + (n-1)\frac{1}{j^{2n}} \right).$$

Note that

$$\sum_{j=1}^{N} \sum_{\substack{k=1 \\ k\neq j}}^{N} \frac{j^{2-2n} - k^{2-2n}}{k^2 - j^2} = \sum_{j=1}^{N} \sum_{\substack{k=1 \\ k\neq j}}^{N} \frac{j^{2-2n}}{k^2 - j^2} - \sum_{j=1}^{N} \sum_{\substack{k=1 \\ k\neq j}}^{N} \frac{k^{2-2n}}{k^2 - j^2}$$

$$= \sum_{j=1}^{N} \sum_{\substack{k=1 \\ k\neq j}}^{N} \frac{1}{j^{2n-2}} \frac{1}{k^2 - j^2} + \sum_{j=1}^{N} \sum_{\substack{k=1 \\ k\neq j}}^{N} \frac{1}{k^{2n-2}} \frac{1}{j^2 - k^2}$$

$$= 2 \sum_{j=1}^{N} \frac{1}{j^{2n-2}} \sum_{\substack{k=1 \\ k\neq j}}^{N} \frac{1}{k^2 - j^2}.$$

Hence

$$(***) \quad \begin{aligned} &\lim_{N\to\infty} \sum_{j=1}^{N} \left(\sum_{\substack{k=1 \\ k\neq j}}^{N} \frac{j^{2-2n} - k^{2-2n}}{k^2 - j^2} + (n-1)\frac{1}{j^{2n}} \right) \\ &= \lim_{N\to\infty} \left(2 \sum_{j=1}^{N} \frac{1}{j^{2n-2}} \sum_{\substack{k=1 \\ k\neq j}}^{N} \frac{1}{k^2 - j^2} + (n-1) \sum_{j=1}^{N} \frac{1}{j^{2n}} \right). \end{aligned}$$

Now, observe that

$$2j \sum_{\substack{k=1 \\ k\neq j}}^{N} \frac{1}{k^2 - j^2} = \sum_{\substack{k=1 \\ k\neq j}}^{N} \frac{1}{k - j} - \sum_{\substack{k=1 \\ k\neq j}}^{N} \frac{1}{k + j}$$

$$= \sum_{k=1}^{j-1} \frac{1}{k-j} + \sum_{k=j+1}^{N} \frac{1}{k-j} - \sum_{k=1}^{N} \frac{1}{k+j} + \frac{1}{2j}$$

$$= -\sum_{k=1}^{j-1} \frac{1}{k} + \sum_{k=1}^{N-j} \frac{1}{k} - \sum_{k=j+1}^{N+j} \frac{1}{k} + \frac{1}{2j} = -\sum_{k=1}^{N+j} \frac{1}{k} + \frac{1}{j} + \sum_{k=1}^{N-j} \frac{1}{k} + \frac{1}{2j}$$

3.7. Rearrangement of Series. Double Series

$$= \frac{3}{2j} - \left(\frac{1}{N-j+1} + \frac{1}{N-j+2} + \ldots + \frac{1}{N+j}\right).$$

Thus by $(***)$,

$$\sum_{j=1}^{N}\left(\sum_{\substack{k=1\\k\neq j}}^{N}\frac{j^{2-2n}-k^{2-2n}}{k^2-j^2} + (n-1)\frac{1}{j^{2n}}\right)$$

$$= \left(n+\frac{1}{2}\right)\sum_{j=1}^{N}\frac{1}{j^{2n}} - \sum_{j=1}^{N}\frac{1}{j^{2n-1}}\left(\frac{1}{N-j+1} + \ldots + \frac{1}{N+j}\right).$$

Moreover, since $0 < \frac{1}{N-j+1} + \ldots + \frac{1}{N+j} < \frac{2j}{N-j+1}$, we see that

$$0 < \sum_{j=1}^{N} \frac{1}{j^{2n-1}}\left(\frac{1}{N-j+1} + \ldots + \frac{1}{N+j}\right)$$

$$< 2\sum_{j=1}^{N}\frac{1}{j^{2n-2}}\frac{1}{N-j+1} \leq 2\sum_{j=1}^{N}\frac{1}{j}\frac{1}{N-j+1}$$

$$= \frac{2}{N+1}\sum_{j=1}^{N}\left(\frac{1}{j} + \frac{1}{N-j+1}\right)$$

$$= \frac{4}{N+1}\sum_{j=1}^{N}\frac{1}{j} \leq \frac{4}{N+1}(\gamma + \ln(N+1)),$$

where γ is the Euler constant (see 2.1.41). Finally, by $(*)$,

$$\lim_{N\to\infty}\sum_{j=1}^{N}\sum_{k=1}^{N}\left(\frac{1}{k^2}\frac{1}{j^{2n-2}} + \frac{1}{k^4}\frac{1}{j^{2n-4}} + \ldots + \frac{1}{k^{2n-2}}\frac{1}{j^2}\right)$$

$$= \lim_{N\to\infty}\left(n+\frac{1}{2}\right)\sum_{j=1}^{N}\frac{1}{j^{2n}} = \left(n+\frac{1}{2}\right)\zeta(2n).$$

3.7.37. Substituting $n = 2$ in the identity given in the foregoing problem, we get

$$\zeta(2)\zeta(2) = \left(2+\frac{1}{2}\right)\zeta(4).$$

Since $\zeta(2) = \frac{\pi^2}{6}$ (see 3.1.28 (a)), we obtain (see also 3.1.28 (b))

$$\zeta(4) = \sum_{n=1}^{\infty} \frac{1}{n^4} = \frac{\pi^4}{90}.$$

Likewise, taking $n = 3$, we find that

$$\zeta(6) = \sum_{n=1}^{\infty} \frac{1}{n^6} = \frac{\pi^6}{945}.$$

Similarly,

$$\zeta(8) = \sum_{n=1}^{\infty} \frac{1}{n^8} = \frac{\pi^8}{9450}.$$

3.8. Infinite Products

3.8.1.

(a) We have

$$P_n = \prod_{k=2}^{n} \left(1 - \frac{1}{k^2}\right) = \prod_{k=2}^{n} \frac{(k-1)(k+1)}{k^2} = \frac{n+1}{2n} \xrightarrow[n \to \infty]{} \frac{1}{2}.$$

(b)

$$\prod_{k=2}^{n} \frac{(k-1)(k^2+k+1)}{(k+1)(k^2-k+1)}$$

$$= \prod_{k=2}^{n} \frac{(k-1)((k+1)^2 - (k+1) + 1)}{(k+1)(k^2-k+1)} = \frac{2(n^2+n+1)}{3n(n+1)} \xrightarrow[n \to \infty]{} \frac{2}{3}.$$

(c) For $x = 0$ the value of the product is 1. If $x \neq 2^m \left(\frac{\pi}{2} + k\pi\right)$, then $\cos \frac{x}{2^m} \neq 0$ and $\sin \frac{x}{2^n} \neq 0$. Hence

$$\prod_{k=1}^{n} \cos \frac{x}{2^k} = \prod_{k=1}^{n} \frac{1}{2} \frac{\sin \frac{x}{2^{k-1}}}{\sin \frac{x}{2^k}} = \frac{\sin x}{2^n \sin \frac{x}{2^n}} \xrightarrow[n \to \infty]{} \frac{\sin x}{x}.$$

(d) On account of the formulas

$$\sinh(2x) = 2\sinh x \cosh x \quad \text{and} \quad \lim_{x \to 0} \frac{\sinh x}{x} = 1,$$

3.8. Infinite Products

as in (c), we get

$$\prod_{n=1}^{\infty} \cosh \frac{x}{2^n} = \begin{cases} \dfrac{\sinh x}{x} & \text{if } x \neq 0, \\ 1 & \text{if } x = 0. \end{cases}$$

(e) We have

$$\prod_{k=0}^{n}\left(1+x^{2^k}\right) = \prod_{k=0}^{n} \frac{1-x^{2^{k+1}}}{1-x^{2^k}} = \frac{1-x^{2^{n+1}}}{1-x} \xrightarrow[n\to\infty]{} \frac{1}{1-x}.$$

(f)

$$\prod_{k=1}^{n}\left(1+\frac{1}{k(k+2)}\right) = \prod_{k=1}^{n} \frac{(k+1)^2}{k(k+2)} = \frac{2(n+1)}{n+2} \xrightarrow[n\to\infty]{} 2.$$

(g) Since

$$\prod_{k=1}^{n} a^{\frac{(-1)^k}{k}} = a^{\sum_{k=1}^{n} \frac{(-1)^k}{k}},$$

the continuity of the exponential function and 3.1.32 (a) imply that $\prod_{n=1}^{\infty} a^{\frac{(-1)^n}{n}} = a^{-\ln 2}$.

(h)

$$\prod_{k=1}^{n} \frac{e^{\frac{1}{k}}}{1+\frac{1}{k}} = \frac{e^{\sum_{k=1}^{n} \frac{1}{k}}}{n+1} = e^{\sum_{k=1}^{n} \frac{1}{k} - \ln n} \cdot \frac{n}{n+1}.$$

Thus by 2.1.41,

$$\prod_{n=1}^{\infty} \frac{e^{\frac{1}{n}}}{1+\frac{1}{n}} = e^{\gamma},$$

where γ is the Euler constant.

(i) We have

$$P_n = \prod_{k=1}^{n} \frac{(3k)^2}{(3k-1)(3k+1)} = \prod_{k=1}^{n} \frac{(3k)^3}{(3k-1)3k(3k+1)} = \frac{3^{3n}(n!)^3}{(3n+1)!}.$$

Using the Stirling formula

$$n! = \alpha_n \sqrt{2\pi n}\left(\frac{n}{e}\right)^n, \quad \text{where} \quad \lim_{n\to\infty} \alpha_n = 1,$$

we get

$$\lim_{n\to\infty} P_n = \lim_{n\to\infty} \frac{3^{3n}(2\pi)^{3/2}n^{3n+3/2}e^{-3n}}{(2\pi)^{1/2}(3n+1)^{3n+1+1/2}e^{-3n-1}}$$

$$= 2\pi e \lim_{n\to\infty} \left(\frac{3n}{3n+1}\right)^{3n}\left(\frac{n}{3n+1}\right)^{3/2} = \frac{2\pi}{3\sqrt{3}}.$$

3.8.2.

(a)

$$P_{2n} = \prod_{k=2}^{2n}\left(1+\frac{(-1)^k}{k}\right) = \frac{3}{2}\cdot\frac{2}{3}\cdot\frac{5}{4}\cdot\frac{4}{5}\cdot\ldots\cdot\left(1+\frac{1}{2n}\right)$$

$$= 1 + \frac{1}{2n} \xrightarrow[n\to\infty]{} 1,$$

$$P_{2n-1} = \frac{3}{2}\cdot\frac{2}{3}\cdot\frac{5}{4}\cdot\frac{4}{5}\cdot\ldots\cdot\frac{2n-1}{2n-2}\cdot\frac{2n-2}{2n-1} = 1.$$

(b) We have

$$P_n = \prod_{k=1}^{n}\left(1+\frac{1}{k}\right) = 2\cdot\frac{3}{2}\cdot\frac{4}{3}\cdot\ldots\cdot\frac{n+1}{n} = n+1 \xrightarrow[n\to\infty]{} +\infty,$$

so that $\prod_{n=1}^{\infty}\left(1+\frac{1}{n}\right)$ diverges.

(c) The product $\prod_{n=1}^{\infty}\left(1-\frac{1}{n}\right)$ diverges, because

$$P_n = \prod_{k=2}^{n}\left(1-\frac{1}{k}\right) = \frac{1}{2}\cdot\frac{2}{3}\cdot\frac{3}{4}\cdot\ldots\cdot\frac{n-1}{n} = \frac{1}{n} \xrightarrow[n\to\infty]{} 0.$$

3.8.3. Note first that for nonnegative a_n,

(1) $a_1 + a_2 + \ldots + a_n \leq (1+a_1)(1+a_2)\ldots(1+a_n).$

Moreover, the inequality $e^x \geq 1+x$, $x \geq 0$, gives

(2) $(1+a_1)(1+a_2)\ldots(1+a_n) \leq e^{a_1+a_2+\ldots+a_n}.$

3.8. Infinite Products

The inequalities (1) and (2) together with the continuity of the exponential function show that the convergence of the product $\prod_{n=1}^{\infty}(1+a_n)$ is equivalent to the convergence of the series $\sum_{n=1}^{\infty} a_n$.

3.8.4. Assume that the series $\sum_{n=1}^{\infty} a_n$ converges. Then for sufficiently large N, $\sum_{n=N}^{\infty} a_n < \frac{1}{2}$. It follows from 1.2.1 that

$$\prod_{k=N}^{n}(1-a_k) \geq 1 - \sum_{k=N}^{n} a_k > \frac{1}{2}.$$

Since $P_n = \prod_{k=1}^{n}(1-a_k) = P_{N-1}\prod_{k=N}^{n}(1-a_k)$, we see that the sequence $\left\{\frac{P_n}{P_{N-1}}\right\}$ is monotonically decreasing and bounded below. Consequently, it converges, say, to P. Then $P \in [\frac{1}{2}, 1]$. Thus $\lim_{n\to\infty} P_n = P_{N-1}P \neq 0$.

To prove the other implication, assume that $\sum_{n=1}^{\infty} a_n$ diverges. If the sequence $\{a_n\}$ does not converge to zero, then the sequence $\{1-a_n\}$ does not converge to 1 and the necessary condition for convergence of $\prod_{n=1}^{\infty}(1-a_n)$ is not satisfied. So we may assume that $\lim_{n\to\infty} a_n = 0$, and consequently, $0 \leq a_n < 1$ beginning with some value N of the index n. In view of the formula (see 2.5.7)

$$e^{-x} = 1 - x + \left(\frac{x^2}{2!} - \frac{x^3}{3!}\right) + \left(\frac{x^4}{4!} - \frac{x^5}{5!}\right) + ...,$$

we get $1-x \leq e^{-x}$ for $0 \leq x < 1$, because all the terms in parentheses are nonnegative. Hence

$$0 \leq \prod_{k=N}^{n}(1-a_k) \leq e^{-\sum_{k=N}^{n} a_k}, \quad n \geq N,$$

and consequently, $\lim_{n\to\infty} \prod_{k=N}^{n}(1-a_k) = 0$. Therefore, $\prod_{n=1}^{\infty}(1-a_n)$ diverges.

3.8.5. Note that

$$\prod_{k=1}^{2n}(1+a_k) = \prod_{k=1}^{n}(1+a_{2k-1})(1+a_{2k})$$

$$= \prod_{k=1}^{n}\left(1+\frac{1}{\sqrt{k}}+\frac{1}{k}\right)\left(1-\frac{1}{\sqrt{k}}\right) = \prod_{k=1}^{n}\left(1-\frac{1}{k\sqrt{k}}\right).$$

Thus, by 3.8.4, the product converges.

3.8.6.

(a) Since $\cos\frac{1}{n} = 1 - (1-\cos\frac{1}{n})$ and $1 \neq 1 - \cos\frac{1}{n} > 0$, $n \in \mathbb{N}$, we can apply the result in 3.8.4. Thus the convergence of the product follows from the convergence of the series $\sum_{n=1}^{\infty}(1-\cos\frac{1}{n})$ (see 3.2.1 (e)).

(b) As in (a), the convergence of the product follows from the convergence of $\sum_{n=1}^{\infty}\left(1-n\sin\frac{1}{n}\right)$ (see 3.2.5 (d)).

(c) We have

$$\tan\left(\frac{\pi}{4}+\frac{1}{n}\right) = \frac{1+\tan\frac{1}{n}}{1-\tan\frac{1}{n}} = 1 + \frac{2\tan\frac{1}{n}}{1-\tan\frac{1}{n}}.$$

Since $\frac{2\tan\frac{1}{n}}{1-\tan\frac{1}{n}} > 0$ for $n \geq 2$ and

$$\lim_{n\to\infty} \frac{\frac{2\tan\frac{1}{n}}{1-\tan\frac{1}{n}}}{\frac{1}{n}} = 2,$$

by 3.8.3 the product diverges.

(d) In view of $\lim_{n\to\infty} \frac{1-n\ln(1+\frac{1}{n})}{\frac{1}{n}} = \frac{1}{2}$, the convergence of the product follows from 3.8.4.

(e) The divergence of the product follows from the divergence of the series $\sum_{n=1}^{\infty}(\sqrt[n]{n}-1)$ (see 3.2.5(a)).

3.8. Infinite Products

(f) Since $\lim\limits_{n\to\infty} \sqrt[n]{n} = 1$, it follows from 2.5.5 that $\lim_{n\to\infty} \frac{\frac{1}{n^2}\ln n}{\sqrt[n]{n}-1} = 1$. Thus the convergence of the product follows from the convergence of the series $\sum\limits_{n=2}^{\infty} \frac{\ln n}{n^2}$.

3.8.7. By assumption, the series $\sum\limits_{n=1}^{\infty} a_n$ converges and, without loss of generality, we can assume that $|a_n| < 1$. Since

$$\text{(1)} \qquad \lim_{n\to\infty} \frac{a_n - \ln(1+a_n)}{a_n^2} = \frac{1}{2}$$

and the series $\sum\limits_{n=1}^{\infty} a_n$ converges, the convergence of $\sum\limits_{n=1}^{\infty} a_n^2$ is equivalent to the convergence of $\sum\limits_{n=1}^{\infty} \ln(1+a_n)$, which in turn is equivalent to the convergence of $\prod\limits_{n=1}^{\infty} (1+a_n)$.

Note that if $\sum\limits_{n=1}^{\infty} a_n^2$ diverges, then by (1),

$$a_n - \ln(1+a_n) > \frac{1}{4}a_n^2 \quad \text{for sufficiently large } n.$$

Thus the series $\sum\limits_{n=1}^{\infty} \ln(1+a_n)$ diverges to $-\infty$, which means that $\prod\limits_{n=1}^{\infty} (1+a_n)$ diverges to zero.

3.8.8. The result follows immediately from 3.8.7.

3.8.9. Apply 3.8.7 or 3.8.8.

3.8.10. We use the equality

$$\lim_{n\to\infty} \frac{|\ln(1+a_n) - a_n + \frac{1}{2}a_n^2|}{|a_n|^3} = \frac{1}{3}$$

and proceed as in the solution of 3.8.7.

3.8.11. No. By the test in the foregoing problem, we see that the product given in the hint converges if $\frac{1}{3} < \alpha$. On the other hand, the series

$$-\frac{1}{2^\alpha} + \left(\frac{1}{2^\alpha} + \frac{1}{2^{2\alpha}}\right) - \frac{1}{3^\alpha} + \left(\frac{1}{3^\alpha} + \frac{1}{3^{2\alpha}}\right) - \frac{1}{4^\alpha} + \ldots$$

and

$$\left(-\frac{1}{2^\alpha}\right)^2 + \left(\frac{1}{2^\alpha} + \frac{1}{2^{2\alpha}}\right)^2 + \left(-\frac{1}{3^\alpha}\right)^2 + \left(\frac{1}{3^\alpha} + \frac{1}{3^{2\alpha}}\right)^2 + \left(-\frac{1}{4^\alpha}\right)^2 + \ldots$$

both diverge if $\alpha \leq \frac{1}{2}$.

3.8.12. Observe that if $\lim\limits_{n\to\infty} a_n = 0$, then

$$\lim_{n\to\infty} \frac{|\ln(1+a_n) - a_n + \frac{1}{2}a_n^2 - \frac{1}{3}a_n^3 + \ldots + \frac{(-1)^k}{k}a_n^k|}{|a_n|^{k+1}} = \frac{1}{k+1}.$$

3.8.13. By the Taylor formula,

$$\ln(1+a_n) = a_n - \frac{1}{2(1+\theta_n)^2}a_n^2 = a_n - \Theta_n a_n^2,$$

where $\frac{2}{9} < \Theta_n < 2$, if $|a_n| < \frac{1}{2}$. Thus if n_1, n_2 are sufficiently large and $n_1 < n_2$, then

$$\sum_{n=n_1}^{n_2} \ln(1+a_n) = \sum_{n=n_1}^{n_2} a_n - \Theta \sum_{n=n_1}^{n_2} a_n^2, \quad \text{where } \Theta \in \left(\frac{2}{9}, 2\right).$$

Hence the convergence of $\sum\limits_{n=1}^{\infty} a_n$ follows from the Cauchy criterion.

3.8.14. If the products $\prod\limits_{n=1}^{\infty}(1+a_n)$ and $\prod\limits_{n=1}^{\infty}(1-a_n)$ both converge, then $\prod\limits_{n=1}^{\infty}(1-a_n^2)$ also converges. Consequently, the series $\sum\limits_{n=1}^{\infty} a_n^2$ converges (see 3.8.4). Now the desired result follows from the preceding problem.

3.8. Infinite Products

3.8.15. Yes. Indeed, since $\{a_n\}$ decreases monotonically to 1, we can write $a_n = 1 + \alpha_n$, where $\{\alpha_n\}$ decreases monotonically to zero. The convergence of the our product is equivalent to the convergence of
$$\sum_{n=1}^{\infty}(-1)^{n-1}\ln(1+\alpha_n).$$
Clearly, by the Leibniz test this series converges.

3.8.16.

(a) Since $\lim\limits_{n\to\infty}(a_n + b_n) = 1 + 1 = 2$, the necessary condition for convergence is not satisfied.

(b) The convergence of $\prod\limits_{n=1}^{\infty} a_n^2$ follows from the convergence of the series $\sum\limits_{n=1}^{\infty} \ln a_n^2$.

(c), (d) The convergence of the products follows from the convergence of the series
$$\sum_{n=1}^{\infty}\ln(a_n b_n) = \sum_{n=1}^{\infty}\ln a_n + \sum_{n=1}^{\infty}\ln b_n$$
and
$$\sum_{n=1}^{\infty}\ln\frac{a_n}{b_n} = \sum_{n=1}^{\infty}\ln a_n - \sum_{n=1}^{\infty}\ln b_n.$$

3.8.17. Suppose that $\sum\limits_{n=1}^{\infty} x_n^2$ converges. Then $\lim\limits_{n\to\infty} x_n = 0$, and the convergence of both the products follows from 3.8.4 and from the equalities
$$\lim_{n\to\infty}\frac{1-\cos x_n}{x_n^2} = \frac{1}{2} \quad \text{and} \quad \lim_{n\to\infty}\frac{1-\frac{\sin x_n}{x_n}}{x_n^2} = \frac{1}{6}.$$
Assume now that one of the products converges. Then $\lim\limits_{n\to\infty} x_n = 0$, and the convergence of $\sum\limits_{n=1}^{\infty} x_n^2$ follows also from the above equalities.

3.8.18. Observe that
$$a_1 \prod_{k=2}^{n}\left(1+\frac{a_k}{S_{k-1}}\right) = a_1 \prod_{k=2}^{n} \frac{S_k}{S_{k-1}} = S_n.$$

3.8.19. See Problem 3.1.9.

3.8.20. See Problem 3.1.9.

3.8.21. Apply the foregoing problem with $a_n = x^n$.

3.8.22. Assume first that the product $\prod_{n=1}^{\infty} a_n$ converges, that is, $\lim_{n\to\infty} P_n = P \neq 0$, where $P_n = \prod_{k=1}^{n} a_k$. This implies that there is $\alpha > 0$ such that $|P_n| \geq \alpha$ for $n \in \mathbb{N}$. The convergent sequence $\{P_n\}$ is a Cauchy sequence. Thus for every $\varepsilon > 0$ there is an integer n_0 such that $|P_{n+k} - P_{n-1}| < \varepsilon\alpha$ if $n \geq n_0$ and $k \in \mathbb{N}$. Therefore
$$\left|\frac{P_{n+k}}{P_{n-1}} - 1\right| < \frac{\varepsilon\alpha}{|P_{n-1}|} \leq \varepsilon \quad \text{for} \quad n \geq n_0.$$

Assume now that for every $\varepsilon > 0$ there is an integer n_0 such that

(∗) $\qquad |a_n a_{n+1} \cdot \ldots \cdot a_{n+k} - 1| < \varepsilon$

for $n \geq n_0$ and $k \in \mathbb{N}$. Taking $\varepsilon = \frac{1}{2}$, we get

(∗∗) $\qquad \dfrac{1}{2} < \dfrac{P_{n-1}}{P_{n_0}} < \dfrac{3}{2} \quad \text{for} \quad n > n_0.$

Next using (∗), with ε replaced by $\frac{2\varepsilon}{3|P_{n_0}|}$, we find an integer n_1 such that
$$\left|\frac{P_{n+k}}{P_{n-1}} - 1\right| < \frac{2\varepsilon}{3|P_{n_0}|} \quad \text{for} \quad n \geq n_1,\ k \in \mathbb{N}.$$
Hence if $n > \max\{n_0, n_1\}$, then
$$|P_{n+k} - P_{n-1}| < \frac{2\varepsilon}{3}\left|\frac{P_{n-1}}{P_{n_0}}\right| < \varepsilon.$$

This means that $\{P_n\}$ is a Cauchy sequence. Moreover, it follows from (∗∗) that its limit is different from zero.

3.8. Infinite Products

3.8.23. We have

$$\prod_{k=1}^{2n}(1+x^k) = \prod_{k=1}^{2n}\frac{1-x^{2k}}{1-x^k} = \frac{\prod_{k=1}^{2n}(1-x^{2k})}{\prod_{k=1}^{2n}(1-x^k)}$$

$$= \frac{\prod_{k=1}^{2n}(1-x^{2k})}{\prod_{k=1}^{n}(1-x^{2k})\prod_{k=1}^{n}(1-x^{2k-1})} = \frac{\prod_{k=n+1}^{2n}(1-x^{2k})}{\prod_{k=1}^{n}(1-x^{2k-1})}.$$

Now, the desired result follows from the Cauchy criterion (3.8.22).

3.8.24. This is a consequence of 3.8.3.

3.8.25. Note that for $a_1, a_2, ..., a_n \in \mathbb{R}$,

$$|(1+a_1)(1+a_2)...(1+a_n) - 1| \leq (1+|a_1|)(1+|a_2|)...(1+|a_n|) - 1$$

and apply the Cauchy criterion (3.8.22).

3.8.26. Set $P_n = (1+a_1)(1+a_2)...(1+a_n)$, $n \in \mathbb{N}$. Then $P_n - P_{n-1} = P_{n-1}a_n$ and

$$P_n = P_1 + (P_2 - P_1) + ... + (P_n - P_{n-1})$$
$$= P_1 + P_1 a_2 + P_2 a_3 + ... + P_{n-1} a_n.$$

Thus

$$P_n = (1+a_1) + a_2(1+a_1) + a_3(1+a_1)(1+a_2)$$
$$+ ... + a_n(1+a_1)(1+a_2)...(1+a_{n-1}),$$

or equivalently,

$$P_n = (1+a_1) + (a_2 + a_1 a_2) + (a_3 + a_1 a_3 + a_2 a_3 + a_1 a_2 a_3)$$
$$+ ... + (a_n + a_1 a_n + ... + a_{n-1} a_n + a_1 a_2 a_n$$
$$+ ... + a_{n-2} a_{n-1} a_n + ... + a_1 a_2 ... a_{n-1} a_n).$$

Note that absolute convergence of $\prod_{n=1}^{\infty}(1+a_n)$ implies the absolute convergence of the series $1 + a_1 + \sum_{n=2}^{\infty} a_n(1+a_1)(1+a_2)...(1+a_{n-1})$.

This series is an ordering of a double series whose terms form the infinite matrix

$$\begin{pmatrix} a_1 & a_2 & a_3 & a_4 & \ldots \\ a_1a_2 & a_1a_3 & a_2a_3 & a_1a_4 & \ldots \\ a_1a_2a_3 & a_1a_2a_4 & a_1a_3a_4 & a_2a_3a_4 & \ldots \\ \ldots & \ldots & \ldots & \ldots & \ldots \end{pmatrix}$$

By 3.7.18 the double series converges absolutely, and by 3.7.22 the iterated series given in the problem converges. Consequently, the desired equality holds.

3.8.27. By the absolute convergence of $\sum_{n=1}^{\infty} a_n$, the series $\sum_{n=1}^{\infty} a_n x$ converges absolutely for every $x \in \mathbb{R}$. Now it is enough to apply the result in the preceding problem.

3.8.28. Obviously, for $|q| < 1$ and $x \in \mathbb{R}$, the product $\prod_{n=1}^{\infty}(1+q^n x)$ converges absolutely. Taking $a_n = q^n$ in the foregoing problem, we get $f(x) = \prod_{n=1}^{\infty}(1+q^n x) = 1 + A_1 x + A_2 x^2 + \ldots$. Now observe that $(1+qx)f(qx) = f(x)$. So equating coefficients of like powers, we obtain

$$A_1 = \frac{q}{1-q} \quad \text{and} \quad A_n = A_{n-1} \frac{q^n}{1-q^n} \quad \text{for} \quad n = 2, 3, \ldots .$$

Finally, by induction, we can show that

$$A_n = \frac{q^{\frac{n(n+1)}{2}}}{(1-q)(1-q^2) \cdot \ldots \cdot (1-q^n)}.$$

3.8.29. Set $f(x) = \prod_{n=1}^{\infty}(1+q^{2n-1}x)$ and note that $(1+qx)f(q^2 x) = f(x)$, and apply reasoning similar to that in 3.8.28.

3.8.30. We have

$$\prod_{n=1}^{\infty}(1+a_n x)\left(1+\frac{a_n}{x}\right) = \left(1+\sum_{k=1}^{\infty} A_k x^k\right)\left(1+\sum_{k=1}^{\infty} \frac{A_k}{x^k}\right)$$

$$= 1 + \sum_{k=1}^{\infty} A_k\left(x^k + \frac{1}{x^k}\right) + \sum_{k=1}^{\infty} A_k x^k \sum_{k=1}^{\infty} \frac{A_k}{x^k}.$$

3.8. Infinite Products

The absolute convergence of $\sum_{k=1}^{\infty} A_k x^k$ and $\sum_{k=1}^{\infty} \frac{A_k}{x^k}$ implies the absolute convergence of their Cauchy product (see the solution of 3.6.1). Observe that this Cauchy product is an ordering of the double series corresponding to the matrix

$$\begin{pmatrix} A_1 A_1 & A_2 A_2 & A_3 A_3 & \dots \\ A_2 A_1 \left(x + \frac{1}{x}\right) & A_3 A_2 \left(x + \frac{1}{x}\right) & A_4 A_3 \left(x + \frac{1}{x}\right) & \dots \\ A_3 A_1 \left(x^2 + \frac{1}{x^2}\right) & A_4 A_2 \left(x^2 + \frac{1}{x^2}\right) & A_5 A_3 \left(x^2 + \frac{1}{x^2}\right) & \dots \\ \dots \end{pmatrix}$$

Therefore, by 3.7.18 and 3.7.22, we get

$$\sum_{k=1}^{\infty} A_k x^k \sum_{k=1}^{\infty} \frac{A_k}{x^k} = (A_1 A_1 + A_2 A_2 + A_3 A_3 + \dots)$$

$$+ (A_2 A_1 + A_3 A_2 + \dots)\left(x + \frac{1}{x}\right) + (A_3 A_1 + A_4 A_2 + \dots)\left(x^2 + \frac{1}{x^2}\right) + \dots.$$

3.8.31. [4] By 3.8.30,

$$\prod_{n=1}^{\infty} \left(1 + q^{2n-1} x\right)\left(1 + \frac{q^{2n-1}}{x}\right) = B_0 + \sum_{n=1}^{\infty} B_n \left(x^n + \frac{1}{x^n}\right).$$

Setting

$$F(x) = \prod_{n=1}^{\infty} \left(1 + q^{2n-1} x\right)\left(1 + \frac{q^{2n-1}}{x}\right)$$

and using the equality $qxF(q^2 x) = F(x)$, we get

$$B_1 = B_0 q, \quad B_n = B_{n-1} q^{2n-1},$$

and inductively,

$$B_n = B_0 q^{n^2}, \quad n = 1, 2, \dots.$$

Thus

$$F(x) = B_0 \left(1 + \sum_{n=1}^{\infty} q^{n^2} \left(x^n + \frac{1}{x^n}\right)\right).$$

To determine B_0 we may use the results in 3.8.29 and 3.8.30. Put $P_n = \prod_{k=1}^{n}(1 - q^{2k})$ and $P = \prod_{n=1}^{\infty}(1 - q^{2n})$. Then

$$B_0 q^{n^2} = B_n = A_n + A_1 A_{n+1} + \dots = \frac{q^{n^2}}{P_n} + \frac{q^{(n+1)^2 + 1}}{P_1 P_{n+1}} + \dots.$$

Hence
$$P_n B_0 - 1 < \frac{q^{2n}}{P^2} + \frac{q^{4n}}{P^2} + \dots .$$
Now, letting $n \to \infty$, we get $B_0 = \frac{1}{P}$.

3.8.32. Apply 3.8.31 with

(a) $x = -1$.

(b) $x = 1$.

(c) $x = q$.

3.8.33. Observe that for $n > 1$,
$$a_n = \frac{1}{2} \left(\prod_{k=1}^{n-1} \frac{x-k}{x+k} - \prod_{k=1}^{n} \frac{x-k}{x+k} \right).$$

Hence
$$S_n = \sum_{k=1}^{n} a_k = \frac{1}{1+x} + \sum_{k=2}^{n} a_k = \frac{1}{2} - \frac{1}{2} \prod_{k=1}^{n} \frac{x-k}{x+k}.$$

If x is a positive integer, then for sufficiently large n, $S_n = \frac{1}{2}$. We now show that for $x \neq 1, 2, \dots$, $\lim_{n \to \infty} S_n = \frac{1}{2}$. Note that for k large enough, $\left|\frac{x-k}{x+k}\right| = 1 - \frac{2x}{x+k}$. Hence, by the result in 3.8.4,
$$\lim_{n \to \infty} \prod_{k=1}^{n} \left|\frac{x-k}{x+k}\right| = 0,$$
which in turn gives $\lim_{n \to \infty} S_n = \frac{1}{2}$, as we have claimed.

3.8.34. Assume that the product $\prod_{n=1}^{\infty}(1+ca_n)$ converges for $c = c_0$ and $c = c_1$, where $c_0 \neq c_1$. Then the products
$$\prod_{n=1}^{\infty}(1+c_1 a_n)^{\frac{c_0}{c_1}} \quad \text{and} \quad \prod_{n=1}^{\infty} \frac{(1+c_1 a_n)^{\frac{c_0}{c_1}}}{1+c_0 a_n}$$
also converge. Moreover,
$$\frac{(1+c_1 a_n)^{\frac{c_0}{c_1}}}{1+c_0 a_n} = 1 + \frac{c_0(c_0-c_1)}{2} a_n^2 (1+\varepsilon_n),$$

3.8. Infinite Products

where $\varepsilon_n \to 0$ as $n \to \infty$. Thus, by 3.8.3 and 3.8.4, the series $\sum_{n=1}^{\infty} a_n^2$ converges. Next, by 3.8.13, $\sum_{n=1}^{\infty} a_n$ also converges. Consequently, for each $c \in \mathbb{R}$, both series $\sum_{n=1}^{\infty} (ca_n)^2$ and $\sum_{n=1}^{\infty} ca_n$ converge. Hence our claim follows from 3.8.7.

3.8.35. Clearly, the series $\sum_{n=1}^{\infty} a_n \prod_{k=0}^{n} (x^2 - k^2)$ converges to zero if x is an integer. Assume now that it converges for a noninteger value x_0. For $x \in \mathbb{R}$, consider the sequence whose terms are given by

$$b_n = \frac{\prod_{k=0}^{n} (x^2 - k^2)}{\prod_{k=0}^{n} (x_0^2 - k^2)}.$$

Then

$$b_n = \prod_{k=0}^{n} \frac{x^2 - k^2}{x_0^2 - k^2} = \prod_{k=0}^{n} \left(1 + \frac{x^2 - x_0^2}{x_0^2 - k^2}\right).$$

From this, we conclude that, beginning with some value of the index n, the sequence $\{b_n\}$ is monotonic. Moreover, since the product $\prod_{k=0}^{\infty} \frac{x^2 - k^2}{x_0^2 - k^2}$ converges, the sequence $\{b_n\}$ is bounded. We have also

$$\sum_{n=1}^{\infty} a_n \prod_{k=0}^{n} (x^2 - k^2) = \sum_{n=1}^{\infty} a_n \prod_{k=0}^{n} (x_0^2 - k^2) b_n.$$

Therefore, by the Abel test, the series under consideration converges for any $x \in \mathbb{R}$.

3.8.36.

(a) We have

$$\left(1 - \frac{1}{p_n^x}\right)^{-1} = 1 + \sum_{k=1}^{\infty} \frac{1}{p_n^{kx}}.$$

Multiplying the first N equalities, we obtain

(i) $\displaystyle\prod_{n=1}^{N} \left(1 - \frac{1}{p_n^x}\right)^{-1} = 1 + \sum_{k=1}^{\infty}{}' \frac{1}{k^x} = \sum_{k=1}^{p_N} \frac{1}{k^x} + \sum_{k=p_N+1}^{\infty}{}' \frac{1}{k^x},$

where $\sum'$ denotes summation over the integers which in their prime factorization contain only prime numbers $p_1, p_2, ..., p_N$. Hence

$$0 < \prod_{n=1}^{N}\left(1-\frac{1}{p_n^x}\right)^{-1} - \sum_{k=1}^{p_N}\frac{1}{k^x} = {\sum_{k=p_N+1}^{\infty}}' \frac{1}{k^x} < \sum_{k=p_N+1}^{\infty}\frac{1}{k^x}.$$

Since $\lim\limits_{N\to\infty} \sum_{k=p_N+1}^{\infty}\frac{1}{k^x} = 0$, we get

$$\prod_{n=1}^{\infty}\left(1-\frac{1}{p_n^x}\right)^{-1} = \sum_{n=1}^{\infty}\frac{1}{n^x}.$$

(b) By (i) in the solution of part (a),

$$\prod_{n=1}^{N}\left(1-\frac{1}{p_n}\right)^{-1} > \sum_{k=1}^{p_N}\frac{1}{k}.$$

Therefore the divergence of $\sum_{n=1}^{\infty}\frac{1}{n}$ implies that $\prod_{n=1}^{\infty}\left(1-\frac{1}{p_n}\right)$ diverges to zero, which in turn is equivalent to the divergence of the series $\sum_{n=1}^{\infty}\frac{1}{p_n}$ (see 3.8.4).

3.8.37. [18]
(a) By DeMoivre's law, $\cos mt + i\sin mt = (\cos t + i\sin t)^m$, with $m = 2n+1$, we get

$$\sin(2n+1)t = (2n+1)\cos^{2n} t \sin t - \binom{2n+1}{3}\cos^{2n-2} t \sin^3 t$$
$$+ ... + (-1)^n \sin^{2n+1} t.$$

So we can write

(1) $\qquad \sin(2n+1)t = \sin t\, W(\sin^2 t),$

where $W(u)$ is a polynomial of degree $\leq n$. Since the function on the left-hand side of the equality vanishes at $t_k = \frac{k\pi}{2n+1}$, $k =$

3.8. Infinite Products

$1, 2, ..., n$, which belong to the interval $(0, \frac{\pi}{2})$, the polynomial $W(u)$ vanishes at $u_k = \sin^2 t_k$, $k = 1, 2, ..., n$. Consequently,

$$W(u) = A \prod_{k=1}^{n} \left(1 - \frac{u}{\sin^2 t_k}\right).$$

Hence, by (1),

(2) $\qquad \sin(2n+1)t = A \sin t \prod_{k=1}^{n} \left(1 - \frac{\sin^2 t}{\sin^2 \frac{k\pi}{2n+1}}\right).$

The task is now to find A. We have $A = \lim\limits_{t \to 0} \frac{\sin(2n+1)t}{\sin t} = 2n+1$. Substituting this value of A into (2) and taking $t = \frac{x}{2n+1}$, we get

(3) $\qquad \sin x = (2n+1) \sin \frac{x}{2n+1} \prod_{k=1}^{n} \left(1 - \frac{\sin^2 \frac{x}{2n+1}}{\sin^2 \frac{k\pi}{2n+1}}\right).$

For fixed $x \in \mathbb{R}$ and $m \in \mathbb{N}$ such that $|x| < (m+1)\pi$, take n greater than m. Then, by (3),

(4) $\qquad \qquad \sin x = P_{m,n} Q_{m,n},$

where

$$P_{m,n} = (2n+1) \sin \frac{x}{2n+1} \prod_{k=1}^{m} \left(1 - \frac{\sin^2 \frac{x}{2n+1}}{\sin^2 \frac{k\pi}{2n+1}}\right),$$

$$Q_{m,n} = \prod_{k=m+1}^{n} \left(1 - \frac{\sin^2 \frac{x}{2n+1}}{\sin^2 \frac{k\pi}{2n+1}}\right).$$

Letting $n \to \infty$, we obtain

(5) $\qquad \lim\limits_{n \to \infty} P_{m,n} = x \prod_{k=1}^{m} \left(1 - \frac{x^2}{k^2 \pi^2}\right).$

It follows from (4) that for $x \neq k\pi$, $\lim\limits_{n \to \infty} Q_{m,n} = Q_m$. To estimate Q_m, we note that by the above assumptions,

$$0 < \frac{|x|}{2n+1} < \frac{k\pi}{2n+1} \leq \frac{n\pi}{2n+1} < \frac{\pi}{2} \quad \text{for } k = m+1, ..., n.$$

Taking into account the inequality $\frac{2}{\pi} u < \sin u < u$, $0 < u < \frac{\pi}{2}$, we see that $\prod_{k=m+1}^{n} \left(1 - \frac{x^2}{4k^2}\right) < Q_{m,n} < 1$. Since the product $\prod_{n=1}^{\infty} \left(1 - \frac{x^2}{4n^2}\right)$ converges, we have

$$\prod_{k=m+1}^{\infty} \left(1 - \frac{x^2}{4k^2}\right) \leq Q_m \leq 1.$$

Consequently,

(6) $$\lim_{m \to \infty} Q_m = 1.$$

Finally, the desired equality follows from (4), (5) and (6).

(b) Apply (a) and the identity $\sin 2x = 2 \sin x \cos x$.

3.8.38. Substitute $x = \frac{\pi}{2}$ in the formula stated in 3.8.37 (a).

3.8.39.

(a) The convergence of the given product is equivalent to the convergence of the series $\sum_{n=1}^{\infty} \left(\ln\left(1 + \frac{x}{n}\right) - \frac{x}{n}\right)$. The absolute convergence of this series follows from the equality

$$\lim_{n \to \infty} \frac{\left|\ln\left(1 + \frac{x}{n}\right) - \frac{x}{n}\right|}{\frac{x^2}{n^2}} = \frac{1}{2}.$$

(b) We have

$$\frac{\left(1 + \frac{1}{n}\right)^x}{1 + \frac{x}{n}} = 1 + \frac{x(x-1)}{2n^2} + o\left(\frac{1}{n^2}\right).$$

So the absolute convergence of the product follows from 3.8.3.

3.8.40. Clearly, the product $\prod_{n=1}^{\infty} (1 + a_n)$, $a_n > -1$, converges if and only if the series $\sum_{n=1}^{\infty} \ln(1 + a_n)$ converges. Moreover, if P is the value of the product and S is the sum of the series, then $P = e^S$.

3.8. Infinite Products

Assume now that the product converges absolutely. Then in view of the equality

(1) $$\lim_{n\to\infty} \frac{|\ln(1+a_n)|}{|a_n|} = 1 \quad \text{(because} \quad \lim_{n\to\infty} a_n = 0\text{)}$$

the series $\sum_{n=1}^{\infty} \ln(1+a_n)$ converges absolutely. Consequently (see 3.7.8), any of its rearrangements converges to the same sum. Finally, by the remark at the beginning of this solution, any rearrangement of the factors of the product does not change its value.

Assume now that the value of the product $\prod_{n=1}^{\infty}(1+a_n)$ does not depend on the order of its factors. This means that the sum of the series $\sum_{n=1}^{\infty} \ln(1+a_n)$ is also independent of the order of its terms. By Riemann's theorem, the series converges absolutely, which in view of (1) implies the convergence of $\sum_{n=1}^{\infty} |a_n|$. Thus the desired result is proved.

3.8.41. [20] Set $R_n = \frac{3}{2} \cdot \frac{5}{4} \cdot \frac{7}{6} \cdot \ldots \cdot \frac{2n+1}{2n} = \frac{(2n+1)!!}{(2n)!!}$. Then

$$\left(1+\frac{1}{2}\right)\left(1+\frac{1}{4}\right)\ldots\left(1+\frac{1}{2\alpha}\right) = R_\alpha,$$

$$\left(1-\frac{1}{3}\right)\left(1-\frac{1}{5}\right)\ldots\left(1-\frac{1}{2\beta+1}\right) = \frac{1}{R_\beta}.$$

Hence the $(\alpha+\beta)$nth partial product is equal to $\frac{R_{n\alpha}}{R_{n\beta}}$. By the Wallis formula (see 3.8.38),

$$\lim_{n\to\infty} \frac{(2n+1)!!}{(2n)!!\sqrt{n}} = \frac{2}{\sqrt{\pi}},$$

and therefore

$$\lim_{n\to\infty} \frac{R_{n\alpha}}{R_{n\beta}} = \sqrt{\frac{\alpha}{\beta}}.$$

3.8.42. If the product $\prod_{n=1}^{\infty}(1+a_n)$ converges, but not absolutely, then the series $\sum_{n=1}^{\infty}\ln(1+a_n)$ converges conditionally (see the solution of 3.8.40). On account of the Riemann theorem, its terms can be rearranged to give either a convergent series whose sum is an arbitrarily preassigned real number S, or a divergent series (to $+\infty$ or to $-\infty$). Thus our claim follows from the relation $P=e^S$ (see the solution of 3.8.40).

Bibliography - Books

References

[1] J. Banaś, S. Wędrychowicz, *Zbiór zadań z analizy matematycznej*, Wydawnictwa Naukowo-Techniczne, Warszawa, 1994.

[2] W. I. Bernuk, I. K. Žuk, O.W. Melnikov, *Sbornik olimpiadnych zadač po matematike*, Narodnaja Asveta, Minsk, 1980.

[3] P. Biler, A. Witkowski, *Problems in Mathematical Analysis*, Marcel Dekker, Inc, New York and Basel, 1990.

[4] T. J. Bromwich, *An Introduction to the Theory of Infinite Series*, Macmillan and Co., Limited, London, 1949.

[5] R. B. Burckel, *An Introduction to Classical Complex Analysis*, Academic Press, New York San Francisco, 1979.

[6] B. P. Demidovič, *Sbornik zadač i upražnenij po matematičeskomu analizu*, Nauka, Moskva, 1969.

[7] A. J. Dorogovcev, *Matematičeskij analiz. Spravočnoe posobe*, Vyščaja Škola, Kiev, 1985.

[8] A. J. Dorogovcev, *Matematičeskij analiz. Sbornik zadač*, Vyščaja Škola, Kiev, 1987.

[9] G. M. Fichtenholz, *Differential- und Integralrechnung, I, II, III*, V.E.B. Deutscher Verlag Wiss., Berlin, 1966-1968.

[10] G. H. Hardy, *A Course of Pure Mathematics*, Cambridge University Press, Cambridge, 1946.

[11] G. H. Hardy, J. E. Littlewood, G. Polya, *Inequalities*, Cambridge University Press, Cambridge, 1967.

[12] G. Klambauer, *Mathematical Analysis*, Marcel Dekker, Inc., New York, 1975.

[13] G. Klambauer, *Problems and Propositions in Analysis*, Marcel Dekker, Inc., New York and Basel, 1979.

[14] K. Knopp, *Theorie und Anwendung der Unendlichen Reihen*, Springer-Verlag, Berlin and Heidelberg, 1947.

[15] L. D. Kudriavtsev, A. D. Kutasov, V. I. Chejlov, M. I. Shabunin, *Problemas de Análisis Matemático. Límite, Continuidad, Derivabilidad*, Mir, Moskva, 1989.

[16] K. Kuratowski, *Introduction to Calculus*, Pergamon Press, Oxford-Edinburgh-New York; Polish Scientific Publishers, Warsaw, 1969.

[17] D. S. Mitrinović, *Elementary Inequalities*, P. Noordhoff Ltd., Groningen, 1964.

[18] D. S. Mitrinović, D. D. Adamović, *Nizovi i Redovi. Definicije, stavovi, zadaci, problemi (Serbo-Croatian)*, Naučna Knjiga, Belgrade, 1971.

[19] A. Ostrowski, *Aufgabensammlung zur Infinitesimalrechnung, Band I: Funktionen einer Variablen*, Birkhäuser Verlag, Basel und Stuttgart, 1964.

[20] G. Pólya, G. Szegö, *Problems and theorems in analysis I*, Spriger-Verlag, Berlin Heidelberg New York, 1978.

[21] Ya. I. Rivkind, *Zadači po matematičeskomu analizu*, Vyšejšaja Škola, Minsk, 1973.

[22] W. I. Rozhkov, G. D. Kurdevanidze, N. G. Panfilov, *Sbornik zadač matematičeskich olimpiad*, Izdat. Univ. Druzhby Narodov, Moskva, 1987.

[23] W. Rudin, *Principles of Mathematical Analysis*, McGraw-Hill Book Company, New York, 1964.

[24] W. A. Sadownicij, A. S. Podkolzin, *Zadači studenčeskich olimpiad po matematike*, Nauka, Moskva, 1978.

[25] W. Sierpiński, *Arytmetyka teoretyczna*, PWN, Warszawa, 1959.

[26] W. Sierpiński, *Działania nieskończone*, Czytelnik, Warszawa, 1948.

[27] H. Silverman, *Complex variables*, Houghton Mifflin Company, Boston, 1975.

[28] G. A. Tonojan, W. N. Sergeev, *Studenčeskije matematičeskije olimpiady*, Izdatelstwo Erevanskogo Universiteta, Erivan, 1985.